NEUROMETHODS ☐ 11

Carbohydrates and Energy Metabolism

NEUROMETHODS

Program Editors: Alan A. Boulton and Glen B. Baker

1. **General Neurochemical Techniques**
   Edited by *Alan A. Boulton and Glen B. Baker,* 1985
2. **Amines and Their Metabolites**
   Edited by *Alan A. Boulton, Glen B. Baker, and Judith M. Baker,* 1985
3. **Amino Acids**
   Edited by *Alan A. Boulton, Glen B. Baker, and James D. Wood,* 1985
4. **Receptor Binding**
   Edited by *Alan A. Boulton, Glen B. Baker, and Pavel D. Hrdina,* 1986
5. **Neurotransmitter Enzymes**
   Edited by *Alan A. Boulton, Glen B. Baker, and Peter H. Yu,* 1986
6. **Peptides**
   Edited by *Alan A. Boulton, Glen B. Baker, and Quentin Pittman,* 1987
7. **Lipids and Related Compounds**
   Edited by *Alan A. Boulton, Glen B. Baker, and Lloyd A. Horrocks,* 1988
8. **Imaging and Correlative Physicochemical Techniques**
   Edited by *Alan A. Boulton, Glen B. Baker, and Donald P. Boisvert,* 1988
9. **The Neuronal Microenvironment**
   Edited by *Alan A. Boulton, Glen B. Baker, and Wolfgang Walz,* 1988
10. **Analysis of Psychiatric Drugs**
    Edited by *Alan A. Boulton, Glen B. Baker, and Ronald T. Coutts,* 1988
11. **Carbohydrates and Energy Metabolism**
    Edited by *Alan A. Boulton, Glen B. Baker, and Roger F. Butterworth,* 1989
12. **Drugs as Tools in Neurotransmitter Research**
    Edited by *Alan A. Boulton, Glen B. Baker, and Augusto V. Juorio,* 1989
13. **Psychopharmacology**
    Edited by *Alan A. Boulton, Glen B. Baker, and Andrew J. Greenshaw,* 1989
14. **Neurophysiology**
    Edited by *Alan A. Boulton, Glen B. Baker, and Case H. Vanderwolf*
15. **Neuropsychology**
    Edited by *Alan A. Boulton, Glen B. Baker, and Merril Hiscock*
16. **Molecular Neurobiological Techniques**
    Edited by *Alan A. Boulton, Glen B. Baker, and Anthony T. Campagnoni*

NEUROMETHODS

Program Editors: *Alan A. Boulton and Glen B. Baker*

NEUROMETHODS ☐ 11

Carbohydrates and Energy Metabolism

Edited by

Alan A. Boulton

University of Saskatchewan, Saskatoon, Canada

Glen B. Baker

University of Alberta, Edmonton, Canada

and

Roger F. Butterworth

University of Montreal, Montreal, Canada

Humana Press • Clifton, New Jersey

Library of Congress Cataloging in Publication Data

Main entry under title:

Carbohydrates and energy metabolism.

(Neuromethods ; 11)
Includes bibliographies and index.
1. Brain—Metabolism—Research—Methodology.
2. Carbohydrates—Metabolism—Research—Methodology.
3. Energy metabolism—Research—Methodology.
I. Boulton, A. A. (Alan A.) II. Baker, Glen B.,
Date . III. Butterworth, Roger F. IV. Series.
[DNLM: 1. Carbohydrates. 2. Energy Metabolism.
W1 NE337G v.11 / QU 75 C2642]
QP376.C35 1988 599'.0188 88-26609
ISBN 0-89603-143-8

Crescent Manor
PO Box 2148
Clifton, NJ 07015

Printed in the United States of America

Preface to the Series

When the President of Humana Press first suggested that a series on methods in the neurosciences might be useful, one of us (A. A. B.) was quite skeptical; only after discussions with G. B. B. and some searching both of memory and library shelves did it seem that perhaps the publisher was right. Although some excellent methods books have recently appeared, notably in neuroanatomy, it is a fact that there is a dearth in this particular field, a fact attested to by the alacrity and enthusiasm with which most of the contributors to this series accepted our invitations and suggested additional topics and areas. After a somewhat hesitant start, essentially in the neurochemistry section, the series has grown and will encompass neurochemistry, neuropsychiatry, neurology, neuropathology, neurogenetics, neuroethology, molecular neurobiology, animal models of nervous disease, and no doubt many more "neuros." Although we have tried to include adequate methodological detail and in many cases detailed protocols, we have also tried to include wherever possible a short introductory review of the methods and/or related substances, comparisons with other methods, and the relationship of the substances being analyzed to neurological and psychiatric disorders. Recognizing our own limitations, we have invited guest editors to join with us on most volumes in order to ensure complete coverage of the field and to add their specialized knowledge and competencies. We anticipate that this series will fill a gap; we can only hope that it will be filled appropriately and with the right amount of expertise with respect to each method, substance or group of substances, and area treated.

Alan A. Boulton
Glen B. Baker

Preface

The mammalian central nervous system depends almost exclusively on glucose as its major energy source. In addition, glucose participates in other cerebral metabolic functions including the biosynthesis of neurotransmitters, such as acetylcholine and the amino acids. This volume of *Neuromethods* assembles currently available methods for the study of cerebral glucose and energy metabolism *in vitro* and *in vivo*.

In the first chapter, Lust et al. describe the various methods available for the appropriate fixation of brain tissue necessary for the study of cerebral energy metabolism. Different fixation methods are compared, and some concerns raised by the USDHHS in their guidelines for the care and use of laboratory animals are addressed. Specific fixation methods pertinent to the various measurements are also covered in other chapters. In vitro preparations have, despite certain limitations, been found to be useful in the study of brain metabolism, since the biochemical environment is amenable to rapid, controlled manipulation. The chapter by Lai and Clark describes methods for the isolation and characterization of metabolically active preparations of synaptic and non-synaptic mitochondria from brain, and studies of enzymes involved in glucose metabolism and glucose-derived neurotransmitter synthesis in these preparations are summarized. The chapter by Whittingham discusses methods of preparations of hippocampal slices for use in the study of energy metabolism. Measurement of glucose and of glycolytic and dicarboxylic acid cycle intermediates in neural tissues are described in the chapter by Bachelard.

Increasing attention has been focused in recent years on the use of quantitative autoradiography for the study of local cerebral glucose utilization (LCGU). Such methods allow the comparison of the relative rates of glucose utilization in various structures of the brain. Two chapters are devoted to the subject of LCGU using either ^{14}C-deoxyglucose (Sokoloff et al.) or ^{14}C-glucose (Hawkins and Mans). In addition to providing information on rates of glucose utilization in different brain structures, the autoradiographs afford pictorial representations of relative cerebral function in these structures. Details of experimental procedures, as well as methods of calculation of rates of LCGU and recent advances in computerized image processing are provided in these chapters.

Three chapters deal with the measurement of enzymes involved in cerebral glucose metabolism. The chapter by Clark and Lai focuses on spectrophotometric methods for the study of glycolytic, tricarboxylic acid cycle, and ketone body-metabolizing enzymes. Some phylogenetic and ontogenic aspects are discussed. In view of the difficulties frequently encountered in their assay, a single chapter is devoted to the measurement of the five enzymic components of the pyruvate dehydrogenase complex (Butterworth); both radiometric and spectrophotometric methods are included. The chapter by Patel describes methods available for the study of CO_2-fixing enzymes in brain and discusses their significance in brain metabolism.

No volume on methods for the study of brain energy metabolism would be complete without a chapter on the measurement of ATP and other high-energy phosphates. In the chapter by Dworsky et al., methods for the fluorimetric measurement of high-energy phosphates in microdissected, microgram, and submicrogram samples of brain tissue are described.

Roger F. Butterworth

Contents

ISOLATION AND CHARACTERIZATION OF SYNAPTIC AND NONSYNAPTIC MITOCHONDRIA FROM MAMMALIAN BRAIN
James C. K. Lai and John B. Clark

THE USE OF HIPPOCAMPAL SLICES FOR THE STUDY OF ENERGY METABOLISM
Tim S. Whittingham

THE [^{14}C]DEOXYGLUCOSE METHOD FOR MEASUREMENT OF LOCAL CEREBRAL GLUCOSE UTILIZATION
Louis Sokoloff, Charles Kennedy, and Carolyn B. Smith

DETERMINATION OF CEREBRAL GLUCOSE USE IN RATS USING [^{14}C]GLUCOSE
Richard A. Hawkins and Anke M. Mans

GLYCOLYTIC, TRICARBOXYLIC ACID CYCLE AND RELATED ENZYMES IN BRAIN

John B. Clark and James C. K. Lai

Contributors

HERMAN S. BACHELARD • *Division of Biochemistry, St. Thomas's Hospital, London, U. K.*

GLEN B. BAKER • *Neurochemical Research Unit, Department of Psychiatry, University of Alberta, Edmonton, Alberta, Canada*

ALAN A. BOULTON • *Neuropsychiatric Research Unit, University of Saskatchewan, Saskatoon, Saskatchewan, Canada*

ROGER F. BUTTERWORTH • *André-Viallet Clinical Research Center, Hôpital Saint-Luc, University of Montreal, Montreal, Quebec, Canada*

JOHN B. CLARK • *Medical College of St. Bartholomew's Hospital, University of London, London, U. K.*

SUZANNE DWORSKY • *University of Health Sciences, The Chicago Medical School, North Chicago, Illinois*

RICHARD A. HAWKINS • *College of Medicine, The Pennsylvania State University, Hershey, Pennsylvania*

CHARLES KENNEDY • *Department of Pediatrics, Georgetown University School of Medicine, Washington, D. C.*

JAMES C. K. LAI • *Laboratory of Cerebral Metabolism, Cornell University Medical College, New York, New York*

W. DAVID LUST • *Case Western Reserve University School of Medicine, Cleveland, Ohio*

ANKE M. MANS • *College of Medicine, The Pennsylvania State University, Hershey, Pennsylvania*

DAVID W. MCCANDLESS • *University of Health Sciences, The Chicago Medical School, North Chicago, Illinois*

MARIAN NAMOVIC • *University of Health Sciences, The Chicago Medical School, North Chicago, Illinois*

MULCHAND S. PATEL • *Case Western Reserve University School of Medicine, Cleveland, Ohio*

ROBERT A. RATCHESON • *Case Western Reserve University School of Medicine, Cleveland, Ohio*

ANTHONY J. RICCI • *Case Western Reserve University School of Medicine, Cleveland, Ohio*

WARREN R. SELMAN • *Case Western Reserve University School of Medicine, Cleveland, Ohio*

CAROLYN B. SMITH • *National Institute of Mental Health, U. S. Department of Health and Human Services, Bethesda, MD*

LOUIS SOKOLOFF • *National Institute of Mental Health, U. S. Department of Health and Human Services, Bethesda, MD*

TIM S. WHITTINGHAM • *Case Western Reserve University School of Medicine, Cleveland, Ohio*

Methods of Fixation of Nervous Tissue for Use in the Study of Cerebral Energy Metabolism

W. David Lust, Anthony J. Ricci, Warren R. Selman, and Robert A. Ratcheson

1. Introduction

The purpose of this chapter is to describe the various methods that are presently available for the proper fixation of the brain in the investigation of cerebral energy metabolism. The advantages, disadvantages, and procedures of each method will be described in detail for those investigators not having experience in these procedures. There have been a number of in-depth reviews on brain fixation during the last ten years (Siesjö, 1978; Jones and Stavinoha, 1979; Passonneau et al., 1979; Lust et al, 1980; Lenox et al., 1982), and the amount of new information on the subject has been modest. However, another issue peripheral to proper fixation has reemerged in recent years (Rhodes, 1892).

The guidelines in the care and use of laboratory animals recently published by the US Department of Health and Human Services have raised some questions concerning several of the accepted methods of euthanasia (NIH Publ. # 85-23, 1985). Since these guidelines rely almost totally on the report of the American Veterinary Medical Association Panel on euthanasia (JAVMA, 1986), relevant sections of this report will be presented in this chapter. There are additional guidelines based on the Health Research Extension Act of 1985 (Public Law 99-158, November 20, 1985), and this report is provided by the NIH Office for Protection from Research Risks. Moreover, a new Animal Welfare Act is under consideration by the Congress of the US Government. Since many of these guidelines are subject to review and modification, it is strongly suggested that investigators keep current through their Animal Care and Use Committee.

1.1. Rationale for Rapid Brain Fixation

Many of the methods described in this book will require the use of rapid fixation to avoid artifacts that arise when the supply of nutrients to the brain has been interrupted. The following characteristics of brain tissue make it particularly susceptible to postmortem changes in the concentrations of cerebral energy-related metabolites:

1. Blood flow and oxygen consumption are disproportionately higher compared to equivalent weights of other tissues
2. Glucose is the predominant substrate under normal circumstances
3. Energy reserves in the form of *P*-creatine and glycogen are relatively low and
4. Brain energy metabolism is tightly coupled to function.

The amount of energy consumed in the brain is a function of synaptic transmission, sodium and potassium resting fluxes, and other unidentified anabolic events (Michenfelder and Milde, 1975; Astrup 1982). Although the amount of energy utilization required for electrolyte leakage and biosynthetic processes is a relative constant, changes in synaptic activity can have a pronounced effect on cerebral metabolic rate (Sokoloff, 1977). Thus, any neurotropic agent that affects excitability could result in an artifact in the profile of energy-related compounds secondary to a change in energy demands. For example, anesthetics generally depress CNS excitability and cause an increase in the levels of energy stores uncharacteristic of the normal in vivo condition (Brunner et al., 1971).

In the absence of blood flow, the energy reserves are rapidly exhausted (Lowry et al., 1964) and can only sustain the ionic gradients necessary for excitability for approximately 1 min (Hansen, 1981), suggesting that brain function is almost totally dependent on a continual supply of exogenous nutrients. To make things even more complex, the brain is encased in bone. This makes quick access to the brain difficult. It also decreases the likelihood of obtaining samples without some external perturbation that may alter the metabolic state of the tissue. Collectively, these characteristics of the brain have made fixation a major obstacle in the study of cerebral energy metabolism. Dealing with this issue has culminated in the development of specialized equipment that seemingly captures the energy-related metabolites as they exist in vivo.

1.2. Criteria for Use of Rapid Fixation

The need for rapid fixation will be determined by the objectives set forth in the experiment. When the goal is to measure the activity of an enzyme involved in energy metabolism, rapid fixation may be unimportant, unless the enzyme is activated or inhibited by postmortem events. For example, glycogen phosphorylase exists in an active form, phosphorylase *a*, and an inactive form, phosphorylase *b*, and the conversion of phosphorylase *a* and *b* occurs quite rapidly in unfixed tissue (Breckenridge and Norman, 1962; Breckenridge and Norman, 1965; Kakiuchi and Rall, 1968; Lust and Passonneau, 1976). In this case, it is important to fix the brain as rapidly as possible to avoid the interconversions of the two forms of the glycogen phosphorylase. Similarly, the measurement of energy-related metabolites requires rapid fixation, although there are exceptions.

Lowry and coworkers have examined the concentrations of the energy-related metabolites at various times following decapitation in the mouse brain. There was a rapid derangement in the substrate and cofactor profile of both the glycolytic pathway and the tricarboxylic acid (TCA) cycle (Lowry et al., 1964; Goldberg et al., 1966). Other metabolites such as aspartate and glutamate have subsequently been shown to remain unchanged following decapitation (Siesjö, 1978). If a question remains about the postmortem stability of either a metabolite or an enzyme, measurements of the metabolite or of enzyme activity at various intervals after decapitation will provide an answer concerning the need for rapid fixation. It should be noted that, in most cases, the artifacts of improper fixation can be mimicked by decapitation.

1.3. Methods of Rapid Fixation in the Brain

The goal of fixation is to trap the metabolites as they exist in vivo by rapidly stopping metabolic processes. Although specific inhibitors and tissue acidification have been used in an attempt to suspend metabolism, generally, the most successful approach to fixation has been by means of changing tissue temperature to attain a condition unfavorable to metabolic activity. Generally, increases in temperature to 80°C or above will denature the proteins and irreversibly prevent secondary metabolic events, even if the tissue is returned to 37°C.

In contrast, decreases in temperature to 0°C or less will essentially stop most metabolic activity, but this suspension of metabolism may be readily reversible upon warming. A study on the fate of energy metabolites in frozen tissue warmed to varying temperatures demonstrates that metabolic reactions quite similar to those observed after decapitation occur (*see* Section 4.2). Thus, extraction of frozen tissue should be considered as part of the fixation procedure, and extreme care must be taken to keep the tissue at –20°C or lower until the enzymes have been irreversibly inactivated by a denaturing agent. For this reason, a section on extraction procedures for the brain is included.

1.4. Evolution of Brain Fixation Methods

Historically, the importance of brain fixation first became evident from studies demonstrating increasing cerebral acidification following decapitation (Langendorf, 1886). The evolution of fixation procedures probably reflects as much on the improved methods to measure brain metabolites as on the need to fix the brain properly (Avery et al., 1935; Lowry and Passonneau, 1972). As more sophisticated assays for brain metabolites become available, the impetus to fix the brain rapidly increased. Once studies on the levels of P-creatine and lactate demonstrated marked postmortem changes, freezing methods using either liquid air, liquid nitrogen, or Freon were tried (Kerr, 1935). For a long time, freezing intact or immediately following decapitation were the most acceptable methods of fixation. However, there were obvious limitations to the freezing procedure that eventually restricted their use. Although it would be expected that the decapitated head would freeze at a faster rate than the entire animal, it was demonstrated that decapitation apparently stimulated the CNS, causing an increase in energy demands at a time when substrate delivery and energy production were severely compromised. Freezing the experimental animal intact was equally flawed because hypoxemia occurred immediately upon plunging the animal into the coolant, which resulted in autolytic artifact.

The *in situ* fixation technique first described by Kerr in 1935 and later perfected by Ponten et al. (1973a) became the method of choice. Blood flow to and oxygenation of the tissue proceeded throughout the entire fixation procedure. This approach minimized the hypoxia elicited by whole animal freezing, but allowed regional studies of the brain. Unfortunately, such a procedure was somewhat confounded by the following: deafferentation phenomenon in lower regions of the brain, periods of hypothermia, and the use of anesthesia. It became evident that the relatively slow rate of freezing in an object the size of the brain, whether it be of a rat or a cat, limited the usefulness of freezing unless a greater area of the brain could be, in some fashion, exposed to the coolant. In the last 20 yr, several methods including freeze blowing, freeze clamping, and brain chopping have been developed (McCandless et al., 1976; Quistorff, 1975; Veech et al., 1973). These techniques combine the freezing process with either smashing, cutting, or crushing of the brain. Although these procedures do achieve a rapid fixation of the brain, all have limitations, not the least of which is that specialized equipment is required.

It was during this period that Stavinoha and coworkers (1970) first examined the possibility of using microwave irridiation for the fixation of the brain. Although their goal was to examine brain levels of acetylcholine, this method has been developed to a state where it can be used for studies on brain energy metabolism. Many of the early pitfalls of microwave irradiation related to deficiencies in the power supply, fine tuning, and focusing of the energy have been resolved, and along with funnel freezing, they are considered the optimal methods for brain fixation.

Since fixation has been such a major obstacle to the investigation of energy metabolism in the brain, many investigators have chosen to eliminate the need for fixation by developing other strategies such as positron emission tomography (PET), magnetic resonance imaging (MRI), and the 2-deoxyglucose technique. Others have resorted to the investigation of energy metabolism in brain slices (*see* Whittingham, this volume). Nevertheless, the measurement of energy metabolites in fixed brain is a worthwhile approach to the investigation of cerebral energy metabolism, and it is hoped that the following descriptions of the methods will help to guide the investigator in selecting the appropriate fixation procedure.

2. Fixation Techniques by Rapid Freezing

The rate of freezing is determined by the coolant, the size of the object to be frozen, the surface area exposed to the coolant, and the extent of the temperature gradient between object and coolant.

Experience with various coolants has shown that the rate of freezing is comparable when the temperature difference between the object and the coolant approaches 200°C. In addition, the rate of freezing is a time-dependent phenomenon that decreases at distances further from the coolant-object interface. Since the possibility of finding a more effective coolant than those already tried is unlikely, the only way to improve the rate of fixation in the brain is to increase the area of coolant exposure.

2.1. Coolants

A list of freezing agents that have been used to rapidly fix the brain is presented in Table 1. There was controversy concerning the advantages and disadvantages of the various coolants, and several papers have been published that address this issue (Swaab, 1971; Ferrendelli et al., 1972). Presently, liquid nitrogen is the coolant of choice, even though the closeness between the melting and boiling points creates a problem. When a tissue is immersed into liquid nitrogen, the nitrogen boils, which serves to insulate the tissue from the coolant. This problem can, in part, be avoided by vigorous stirring of the tissue, which removes the gaseous phase from the surface of the specimen. Both isopentane and Freon have been tried to avoid the boiling seen with liquid nitrogen, since the differences between their melting and boiling points should minimize this problem. However, both of these freezing agents have to be chilled to their freezing points, which requires the use of another coolant. In addition, the use of isopentane should be avoided because of the possibility of explosion. Although other coolants might match or slightly exceed the freezing rate elicited by liquid nitrogen, the easy accessibility and relatively low cost of this agent are two additional reasons for its selection.

2.2. Rates of Freezing

In this section, the ability of liquid nitrogen and Freon to reduce the temperature of the brain will be considered. The damages encountered with liquid oxygen, liquid air, and isopentane preclude their use in these types of studies and will not be discussed. Swaab (1971) demonstrated that the skin, not the hair or skull, dramatically slowed the rate of freezing in a decapitated rat head and, further, that the rate of freezing was essentially the same in a mouse head whether frozen by liquid nitrogen or Freon. In contrast, liquid nitrogen was found to be more effective than Freon

Table 1
Physical Constants of Coolants Commonly Used in Brain Fixation

Agent	Melting point, °C	Boiling point, °C
Liquid nitrogen	−209.9	−195.8
Liquid helium	<272.2	−269.0
Liquid air	−215.0	−194.4
Isopentane	−160.0	+ 28.0
Freon	−155.0	− 29

in reducing the deeper structures of the mouse brain to 0°C both in intact and severed heads (Ferrendelli et al., 1972). Such studies have led to the preferential use of liquid nitrogen despite the problem with boiling.

Brain temperature measurements have been made during funnel freezing in the gerbil and rat, and the results are presented in Fig 1. The values for both the rat and the gerbil fit the formula for a logarithmic curve with a correlation coefficient greater than 0.95. There is a significant delay before the superficial regions of the cortex reach 0°C, which is similar to that observed in the submersed animals (Ferrendelli et al., 1972). The delay in reaching 0°C in the superficial regions of the cortex was greater in the rat than in either the gerbil or mouse, which may be attributed to differences in the thickness of the skull or skin. If the delayed onset of freezing is disregarded, then the initial rates of freezing of the gerbil, rat, and mouse in liquid nitrogen range between 0.2–0.3 mm/s. Several other laboratories have found similar results (Swaab, 1971; Ferrendelli et al., 1972). As shown in Fig. 1, the rate of freezing in both the gerbil and rat drops by more than 50% in deeper regions of the brain. Another important aspect of the freezing process is that there is a period of hypothermia prior to freezing and that the duration of hypothermia increases in deeper regions of the brain (Fig. 2).

Other methods of determining fixation rates have been tried and generally confirm the results derived from temperature measurements. Freezing times have been approximated in the cat brain using enhanced NADH fluorescence to demark the location of the freezing front when cerebral circulation was arrested during funnel freezing (Welsh and Rieder, 1978; Welsh, 1980). After 1 and 5 min of freezing, the boundary of enhanced fluorescence was 5

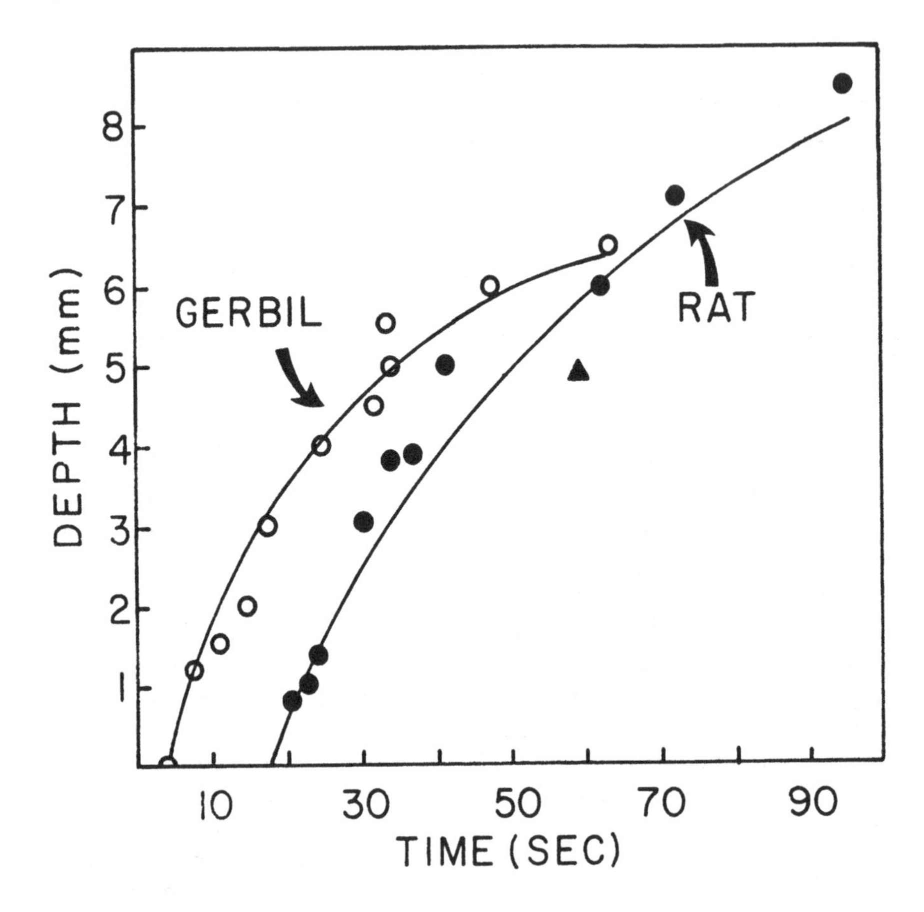

Fig. 1. Freezing times at various depths of the rat, cat, and gerbil brain following *in situ* fixation. Individual values for the rat, cat, and gerbil were taken from articles by Ratcheson (1980); Welsh (1980), and Lust et al., (1980), respectively. Thermistors were placed at various depths of the forebrain in either the gerbil (○) or rat (●) brain, and the time in seconds to reach 0°C following funnel freezing with liquid nitrogen was determined. In contrast, the freezing time was approximated in the cat (▲) by the increase in NADH fluorescence induced by cardiac arrest at 1 and 5 min after the onset of freezing. The temperature results best fit the formula for a logarithmic curve, $y = a + b \ln x$, where a equalled –3.86 and –14.35, and b equalled 2.47 and 4.91 for the gerbil and rat, respectively. The correlation coefficient for both curves was greater than 0.95. For discussion, *see* Sections 2.2. and 2.6.3.

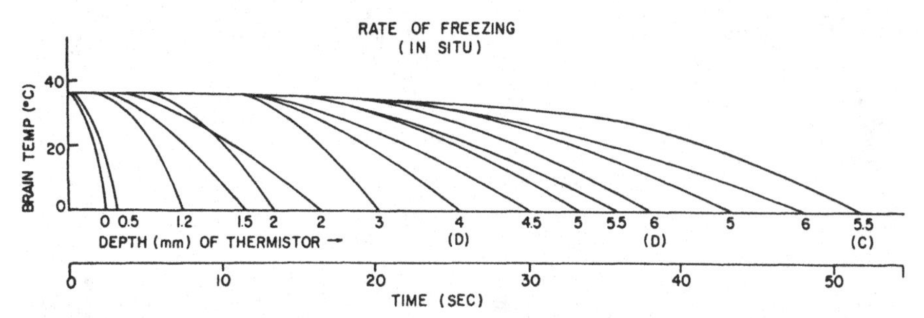

Fig. 2. Representative temperature tracings at different depths of the gerbil brain during funnel freezing. The depth of the thermistor in mm from the surface and the time in s are shown on the abscissa. The temperature was measured in the cerebellum (C) in one animal and in the forebrain of two decapitated heads (D). The remaining measurements were performed in the forebrain of intact gerbils. The brain temperatures in °C are presented on the ordinate. Note the longer periods of hypothermia (< 37°C and > 0°C) at increasing depths of the brain. This figure is reproduced with permission from Williams and Wilkins.

and 15 mm from the surface of the brain, respectively. The relationship of freezing rate to the depth of cat brain is more closely predicted by the constants derived for the rat, rather than those for the gerbil (Fig. 1).

In another set of experiments, the rats were infused with Evans blue to determine the extent of freezing (Fig. 3). The presence of the dye indicates those areas that continued to receive blood flow at 20, 45, and 60 s after the onset of freezing. Although the values were essentially superimposable on those derived from temperature measurements in the rat brain, the photographs graphically illustrate the somewhat irregular pattern of freezing. Even after 20 s of freezing, blood flow is evident in the superficial region served by the anterior cerebral circulation, but not that served by the middle cerebral artery (Fig. 3a). This finding is similar to that observed by Ratcheson (1980), who showed that blood flow of the superficial penetrating vessels ceases before the tissue perfused by these vessels has been frozen. The importance of these observations, in terms of cerebral energy metabolism, will become more apparent in the descriptions of the individual freezing techniques. To iterate:

1. Liquid nitrogen is an acceptable coolant for fixation despite the problem of boiling

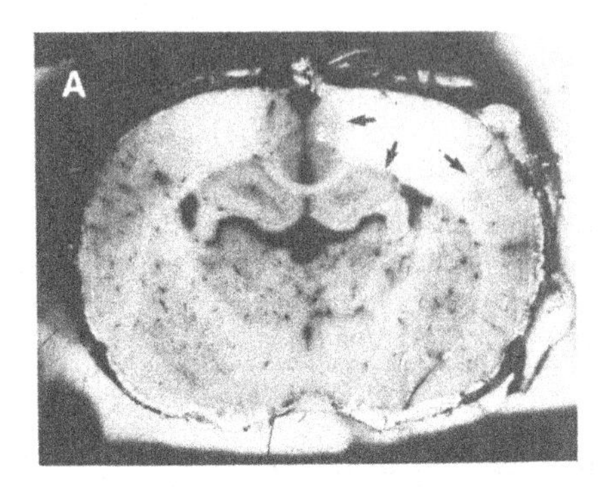

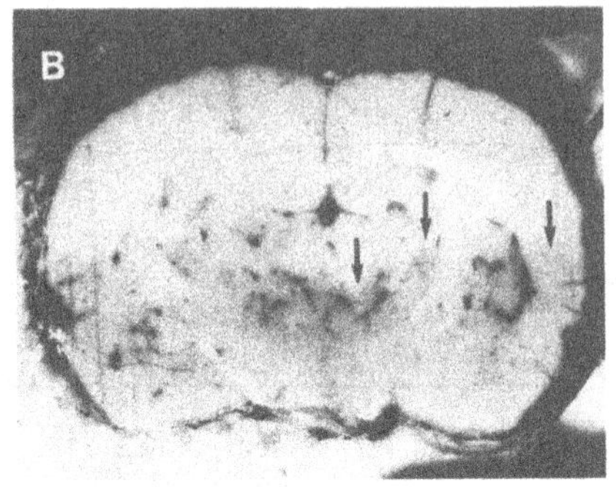

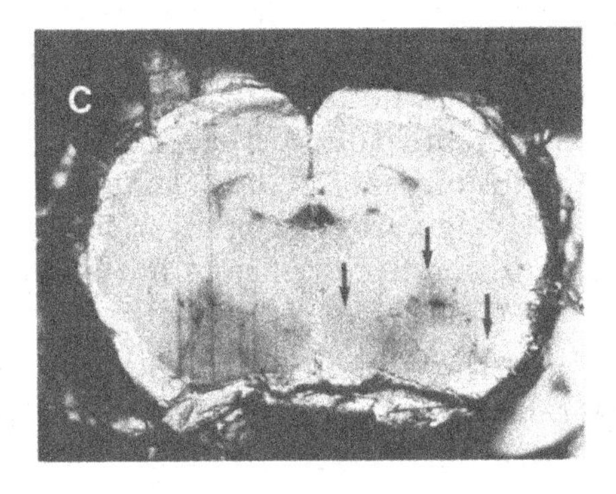

Fig. 3. Freezing rate during funnel freezing determined by the infusion of Evans blue. Rats were anesthetized, the femoral vein was cannulated, and the rats prepared for funnel freezing as described in Section 2.6.3. A 1% solution of Evans blue in 5% bovine serum albumin was infused starting at 20 (A), 45 (B), or 60 (C) s after the onset of freezing. After 3 min, the animals were plunged into liquid nitrogen, and the brains were removed and sectioned. Photographs were taken from the block face of the brains in a refrigerated microtome. The boundaries between areas of perfusion (with dye) and those of no flow (absence of dye) are designated by the arrows shown in one hemisphere. The (a) denotes the area of the superficial cortex served by the anterior cerebral circulation.

2. The rate of freezing in deeper regions of the brain is much slower than in superficial regions and
3. Fixation in this manner is associated with a period of hypothermia prior to freezing.

2.3. Comparison of Brain Metabolite Concentrations

Just as rapid fixation of the brain is important in the study of certain neurotransmitters, neuromodulators, and neuromediators, it is equally critical in the study of metabolites concerned with cerebral energy metabolism. Although some of the fixation techniques described below have evolved from studies of neurotransmitters such as acetylcholine and GABA (Schmidt et al., 1972b; Stavinoha et al., 1973; Knieriem et al., 1977), or of neuromediators like the cyclic nucleotides (Schmidt et al., 1971; Schmidt et al., 1972a), eventually they have all relied, to some degree, on the measurement of energy-related metabolites to establish the validity of the methods. Indeed, the lability of certain metabolites to anoxia/ischemia have made them useful in determining the extent of blood flow disruption prior to fixing the tissue. The changes in energy-related metabolites during ischemia have been described in some detail by Lowry et al. (1964). The high-energy phosphates glucose and glycogen are essentially depleted, whereas lactate and 5'AMP increase manyfold in the first minute of ischemia. Therefore, the presence of high ATP and *P*-creatine along with low lactate and 5'AMP in a control tissue indicates that rapid fixation of the tissue has been achieved. Without knowing the "true" internal concentration of the cerebral metabolites, the efficacy of a fixation technique cannot be precisely assessed. Some in vivo concentrations can be approximated from MRI studies; however, these studies also have limitations. Unfortunately, there is no absolute standard for cerebral energy metabolites in vivo, but as the fixation techniques have achieved quicker inactivation of metabolism, certain approximations of the concentrations in vivo have gained a greater degree of acceptance.

Siesjö (1978) listed other factors that may confound the direct comparisons of data on cerebral energy metabolites. These include experimental differences in species, brain regions, extraction, and analyses. Miller and Shanlon (1977) have also shown that the age of the animal may have an effect on fixation. There also may be unknown effects arising from fixation techniques that either disrupt membranes or that heat the tissue. In addition, certain metabolites such as lactate and glucose are not only a function of endogenous metabolism in the brain, but also of their concentrations in the blood. Many of these pitfalls have been avoided by comparing the various fixation techniques in one laboratory (Ponten et al., 1973b).

Veech et al. (1973) reported that the most sensitive indicators of anoxia were P-creatine and α-oxoglutarate, but that the derived $NAD^+/NADH_2$ ratio was even more sensitive than the individual metabolites. When the data is available, these metabolite concentrations will be presented as a measure of the efficacy of the method. The lactate/pyruvate (L/P) ratio is inversely proportional to the $NAD^+/NADH_2$ ratio according to the lactate dehydrogenase reaction shown in Eq. (1) and these results rather than the ratio of pyridine nucleotides, will be used to compare the efficacy of fixation.

$$\text{Lactate} + \text{NAD}^+ \xleftrightarrow{\text{Lactate dehydrogenase}} \text{Pyruvate} + \text{NADH} + \text{H}^+ \quad (1)$$

Although the measurement of any one of these indicators might seem to establish the efficacy of the freezing method, it is advised that a profile containing many of the more labile metabolites provides a better indication of proper fixation. For example, the concentration of P-creatine in the tissue is a function of pH [*see* Eq. (2)]. A decrease in these levels may merely reflect a change in intracellular pH, and not necessarily a fixation artifact.

$$\text{P-creatine} + \text{ADP} + \text{H}^+ \xleftrightarrow{\text{Creatine kinase}} \text{Creatine} + \text{ATP} \quad (2)$$

Cyclic AMP is another cerebral metabolite that is extremely sensitive to anoxia (Breckenridge, 1964; Steiner et al., 1972). The levels of cyclic AMP in microwaved tissue were also shown to be comparable to those obtained by freezing, and yet subsequent studies on tissue inactivated with a low power microwave oven demonstrated that tissue cerebral energy metabolites were not appropriately fixed (Schmidt et al., 1971; Lust et al., 1973; Lust et al., 1980). Although these studies would tend to support the use of metabolite profile rather than a single metabolite to evaluate a fixation method, it also raises another important question concerning fixation. Low power microwave irradiation has been shown to

Table 2
Concentration of Energy-Related Metabolites in the Cortex and Hypothalamus from Intact Mice (I) or Severed Heads (D) Frozen in Liquid Nitrogen.*

Brain region	Metabolites, mmol/kg wet wt		
	P-creatine	ATP	Lactate
Cerebral cortex (I)	4.30 ± 0.16	2.97 ± 0.04	1.36 ± 0.10
Cerebral cortex (D)	2.69 ± 0.12	3.01 ± 0.03	1.77 ± 0.07
Hypothalamus (I)	2.66 ± 0.15	2.59 ± 0.10	1.80 ± 0.05
Hypothalamus (D)	1.82 ± 0.12	2.69 ± 0.07	3.13 ± 0.11

*Results reported by Ferrendelli et al. (1972). The metabolite values are the mean ± SEM for 6–16 determinations from mouse brains either frozen intact (I) or frozen following decapitation (D).

inactivate the metabolic pathway (i.e., adenylate cyclase and phosphodiesterase) for cyclic AMP (Schmidt et al., 1972b). If the fixation technique allows the accurate measurement of endogenous cyclic AMP, can this procedure then be used when it is totally inadequate by other criteria? Other relatively stable metabolites such as glutamate, isocitrate, and norepinephrine have been measured in brains from decapitated animals. The values are similar to those from frozen brains (Goldberg et al., 1966; Harik et al., 1982); however, subtle differences have been noted (Sharpless and Brown, 1978). It would seem that the success of these studies, using less optimal fixation techniques, supports such an approach, but the investigator must be aware of the possibility that artifacts may arise under certain experimental circumstances.

Metabolite values in Tables 2–4 permit a comparison of the various techniques, but as noted above, there are no absolute reference standards for endogenous metabolites to establish one method to be better than another. Factors other than fixation can be responsible for differences in cerebral metabolite concentrations. Anesthesia, as mentioned, or alterations in pCO_2 and pO_2 can have a profound effect on metabolite levels in the CNS (Siesjö and Nilsson, 1971; Veech, 1980). Inability to monitor physiological

Table 3
The Effect of Fixation Techniques on the Levels of Energy-Related Compounds in Frozen Rat Brain*

	mmol/kg wet wt							
Group	P-creatine	ATP	5'AMP	Glucose	Lactate	Pyruvate	α-OG	L/P
(a) Decapitated nonanesthetized	0.95 ±0.10	2.43 ±0.13	0.23 ±0.06	0.48 ±0.15	5.46 ±0.16	0.109 ±0.012	0.030 ±0.006	52.1 ±6.6
(b) Immersed intact nonanesthetized	3.26 ±0.22	2.97 ±0.13	0.038 ±0.003	1.31 ±0.19	2.50 ±0.14	0.127 ±0.003	0.093 ±0.009	19.7 ±1.0
(c) Immersed intact anesthetized, paralyzed	4.32 ±0.06	2.99 ±0.02	0.041 ±0.002	5.66 ±0.40	1.65 ±0.008	0.105 ±0.002	0.118 ±0.008	15.8 ±1.0
(d) Funnel freezing anesthetized, paralyzed	4.71 ±0.08	3.00 ±0.04	0.040 ±0.001	4.97 ±0.35	1.64 ±0.05	0.121 ±0.005	0.117 ±0.005	13.6 ±0.5
(e) Freeze-blowing	4.05 ±0.07	2.45 ±0.05	0.041 ±0.001	0.962 ±0.084	1.23 ±0.07	0.091 ±0.001	0.213 ±0.001	13.5 –
(f) Freeze clamping	3.79 ±0.14	3.32 ±0.20	0.044 ±0.005	– –	0.72 –	0.079 –	– –	9.1 –
(g) Freeze chopping	– –	3.58 ±0.17	0.40 ±0.05	1.52 ±0.09	0.62 ±0.04	0.051 ±0.013	– –	12.2 –

*Results for groups (a–d) reported by Ponten et al. (1973a), for group (e) by Veech et al. (1973), for group (f) by Quistorff (1975), and for group (g) by McCandless et al. (1976). The anesthesia used in groups (c) and (d) was 70% nitrous oxide and 30% oxygen and paralyzed signifies that the animals were paralyzed with tubocurarine chloride. The metabolite data with the exception of the lactate/pyruvate ratio are expressed in mmol/kg wet wt, and each value represents the mean ± SEM. The abbrevations are as follows: α -OG, α -oxoglutarate, and L/P, lactate/pyruvate. The metabolite values from group (e) were from the forebrain, and those in the remaining groups were from the cerebral cortex.

Table 4
Metabolite Levels from Superficial to Deeper Layers of the Gerbil Brain Following Either Freezing Intact (I) or Funnel Freezing (F)*

	mmol/mg wet wt							
	P-creatine		Lactate		Pyruvate		L/P	
Depth of sample mm	F	I	F	I	F	I	F	I
0–1.2	4.97	4.62	2.50	2.02	0.161	0.196	15.6	10.3
	±0.20	±0.07	±0.34	±0.15	±0.018	±0.047	–	–
1.2–2.4	5.04	4.75	2.01	1.80	0.089	0.144	22.6	12.5
	±0.30	±0.52	±0.30	±0.52	±0.011	±0.041	–	–
2.4–3.6	4.55	4.09	1.43	1.63	0.073	0.097	19.7	16.8
	±0.17	±0.41	±0.16	±0.11	±0.009	±0.008	–	–
3.6–4.8	4.81	3.66	1.42	2.16	0.079	0.091	17.9	23.6
	±0.21	±0.36	±0.13	±0.23	±0.018	±0.017	–	–
4.8–6.0	4.93	2.59	2.01	2.70	0.087	0.086	23.2	31.4
	±0.12	±0.34	±0.31	±0.19	±0.014	±0.011	–	–

*After either in situ or immersion fixation in liquid nitrogen, a dorsoventral plug of brain from the parietal region was removed in a cryostat maintained at –20°C and divided into five wafers approximately 1.2 mm in thickness. Each value represents the mean ± SEM for 4–8 determinations. The metabolite levels are expressed in mmoles/kg wet wt and the L/P is the ratio of lactate to pyruvate.

parameters in smaller animals, with or without anesthesia, has made evaluation of the fixation procedures even more difficult, and the validation of these techniques is almost solely reliant on the endpoint measurement of control metabolite concentrations. In spite of these problems, the metabolite results do offer a basis for the selection of a fixation technique.

2.4. Animal Welfare Concerns on Euthanasia

The following are excerpts from the report of the AVMA Panel on Euthanasia (*JAVMA,* 1986) and are included to inform the reader of existing *guidelines* for euthanasia of experimental animals. Interpretation of these guidelines will undoubtedly differ depending upon the institution. Investigators are advised to consult with their institutional Animal Welfare Committees for interpretation of the guidelines or rules concerning the use of animals.

Recommendations on the use of decapitation with a guillotine: "Until additional information is available to better ascertain whether guillotined animals perceive pain, the technique should be used only after the animal has been sedated or lightly anesthetized, unless the head will be immediately frozen in liquid nitrogen."

Recommendations on the use of rapid freezing of rodents by the following methods:

1. Immersion of intact animal into liquid nitrogen
2. Decapitation and immediate immersion of head into liquid nitrogen
3. Freeze-blowing
4. *In-situ* freezing and
5. Funnel freezing.

"Animal welfare committees and investigators must ensure that animals are made immediately unconscious either from rapid freezing or prior anesthetic. If a paralyzing drug is given in conjunction with an anesthetic, the amount of anesthetic administered must reflect known dose–effect data for a surgical level of anesthesia. Immersion of unanesthetized animals into liquid nitrogen can be used only in animals weighing less than 40 g. Anesthesia must be employed in animals weighing more than 40 g."

2.5. Effects of Neurotropic Agents on the Fixation Process

The AVMA Panel on euthanasia recommended the use of surgical anesthesia if, in their opinion, there was a question about the humane treatment of the experimental animals. Not all concerned and responsible scientists agreed with these recommendations. Dissenting voices are being raised about the recommendation that decapitation requires the use of anesthesia (Allred and Bernston, 1986). In addition, there are questions concerning the use of paralyzing agents in the absence of surgical anesthesia. Although the neuroscientist has always exhibited concern for the humane treatment of experimental animals, the use of agents of unproven benefit, targeted for the very tissue being investigated, is now mandated. Altering the level of excitability of the CNS with depressants of one form or another shifts the investigator further away from the normal condition of the brain. These arbitrary recommendations do not facilitate studies on cerebral energy metabolism.

In most other tissues, anesthetics would be the answer to the problem of euthanasia; however, general surgical anesthesia reduces cerebral metabolic rate secondary to a depression of the CNS (Lowry et al., 1964; Brunner et al., 1971; Heath et al., 1977; Sokoloff et al., 1977; Mrsulja et al., 1984). In addition, anesthesia alters the steady-state levels of a number of metabolites, and the response is dose dependent, which means that the depth of anesthesia introduces another variable to any experiment (Nahrwald et al., 1977). The only anesthetic tested to date that does not affect the metabolism of the CNS to any great degree is nitrous oxide (Ratcheson et al., 1977). Although this agent has long provided satisfactory anesthesia for humans, its potency in rodents has been questioned (Steffey and Eger, 1985). Evaluation of the various techniques for fixation will include inherent pitfalls that may arise when various agents, by necessity, are included in a protocol.

2.6. Protocols for Freezing Methods

2.6.1. Decapitation into Liquid Nitrogen

Freezing the decapitated head in liquid nitrogen or chilled Freon has been used with success over the last 50 yr. Although this method was used in the classic study performed by Lowry et al. (1964) on the metabolic changes that occur following decapitation in the mouse, use of this technique generally has decreased in popularity.

The following equipment is needed for the procedure:

1. 2-L Dewar flask for liquid nitrogen
2. Guillotine or large shears

3. Liquid nitrogen
4. Large forceps for stirring
5. A freezer maintained at –70°C.

Since freezing the tissue does not inactivate the enzymes irreversibly, a freezer maintained at –70°C is needed to prevent warming of the tissue prior to extraction. This is an additional cost for most freezing methods and should be considered when selecting a method of fixation.

Procedure: The flask is filled with liquid nitrogen within 2 in of the top. The experimental animal is decapitated, and the head is dropped directly into the Dewar flask. The liquid nitrogen is vigorously stirred for at least 1 min, and the head is kept in the flask for an additional 2 min to insure that the tissue is completely frozen. Splattering of liquid nitrogen should be avoided, even though it boils off when in contact with the skin. The studies by Swaab (1971) indicate that faster freezing would occur with the removal of the scalp, but this, at a minimum would require use of a local anesthetic. This procedure has, in general, been limited to the following species: rats, mice, gerbils, and hamsters.

This technique offers several advantages that make it quite useful depending on the substance to be measured. It is both fast and inexpensive, requiring neither specialized equipment, surgery, nor anesthesia. The method is compatible with histological evaluation of the tissue and the measurement of enzyme activities. There are several disadvantages to decapitation and then freezing. It has been shown that the rate of freezing in brains of decapitated rats is a function of weight (Jongkind and Bruntink, 1970). Therefore, the severed head would be expected to freeze faster than the whole animal, and the metabolite levels in the brains of decapitated animals should exhibit less autolytic artifact than those in brains from animals frozen intact. However, Ferrendelli et al. (1972) demonstrated a 40% lower concentration of *P*-creatine in the decapitated mouse brain (Table 2). The metabolite differences between immersion freezing and decapitation freezing are even more pronounced in the rat cerebral cortex (Table 3). The levels of *P*-creatine and lactate in the decapitated cortex were 30 and 220%, respectively, of those in the cortex from the rat frozen intact. It is generally accepted that the differences between metabolite values in the decapitation and intact groups arise from stimulation of the CNS resulting in greater energy demands.

A similar effect has been observed in the rabbit brain, when the cerebral cortex was scooped out of the brain and frozen in liquid air (Thorn et al., 1958). In deeper regions of the brain, such as the hypothalamus, there is a marked derangement in all the metabolites examined, indicating that this fixation procedure cannot be used for regional studies of energy metabolism (Table 2). The delayed fixation of deeper regions also precludes this method from the investigation of rapidly occurring events, owing to the relative slowness of the freezing front. In spite of these problems, there are circumstances where this method still may be used for studies on cerebral energy metabolism. Justification for the use of this procedure would, undoubtedly, be required.

2.6.2. Freezing the Intact Animal

Whole animal freezing evolved when it became apparent that the decapitated head did not freeze significantly faster than the entire body (Ferrendelli et al., 1972) and, further, that severing the cord apparently stimulated the brain. Although freezing intact appeared to provide a better approximation of the in vivo situation, many of the disadvantages listed under decapitation similarly apply to this procedure. Necessary supplies are the same as those shown for decapitation, except a guillotine is not required.

Procedure: The Dewar flask is filled with liquid nitrogen. The nape of the neck of the experimental animal is grabbed with forceps, the tail is held with the other hand, and the animal is plunged head first into the liquid nitrogen. The liquid nitrogen is vigorously stirred for 1 min, and the animal removed after a total of 3 min. The animal is stored at –70°C until the brain is removed. In the past, this procedure has been used primarily on rats, gerbils, and mice; however, the new guidelines limit its use to nonanesthetized animals weighing 40 g or less.

Like decapitation with freezing, the advantages of freezing the animal intact are that it is simple, inexpensive, and devoid of any specialized equipment. A large number of animals can be fixed in a comparatively short period of time, and the metabolites in the superficial cortex of the mouse are clearly in the acceptable range, even though the elevated lactate and L/P ratio indicate some anoxia (Table 2; Ponten et al., 1973b). The disadvantages of this procedure would probably be tolerable in the absence of a better method, but such a technique does exist. Regional studies cannot be performed, because of marked changes in the deeper regions of the brain

(Tables 2 and 4). As soon as the animals are plunged into liquid nitrogen, hypoxemia occurs, and it is apparent from the data that there is sufficient time between the onset of freezing and fixation of the deeper brain regions to cause an anoxic response characterized by a 45% loss of high-energy phosphates, a 35% increase in lactate, and a threefold elevation of the L/P ratio. Even when the scalp, mandibles, and tongue are removed, the rate of fixation is not sufficient to obliterate the postmortem changes observed in the deeper regions (Swaab, 1971). By limiting the use of intact fixation by immersion to nonanesthetized animals weighing 40 g or less, this procedure, more than any other presented, will only have limited application in the future.

2.6.3. Funnel Freezing

In situ freezing or funnel freezing permits surface freezing of the animal head, while maintaining oxygenation and blood flow during the process. This avoids the hypoxemia and consequent anoxia that is induced by either freezing intact or decapitation freezing. The technique was first described by Kerr in 1935, and later modified and improved by Ponten et al. (1973a). For a number of reasons, this technique is far superior to the aforementioned methods, although it, too, has some of the disadvantages that arise with freezing.

Although no specialized equipment is necessary, the monitoring of various physiological parameters during the process increases the number of needed supplies. The major equipment includes: a blood gas analyzer, a machine to control the concentration of volatile anesthetics, a rodent respirator, a recording unit for measuring blood pressure, and temperature monitoring equipment. In addition, the following supplies are required: a funnel whose bottom approximates the size of the cranial surface, PE 50 tubing for cannulation, a Dewar flask for liquid nitrogen, liquid nitrogen, and a head holder to position the brain for fixation.

Procedure: Surgical anesthesia is induced either by ether, halothane, or short-acting barbiturates, and the animal is tracheostomized. The animals are ventilated with 70% nitrous oxide and 30% oxygen once the surgery has been completed. The femoral artery is cannulated with polyethylene tubing for monitoring blood pressure and sampling arterial blood for gases. An arterial blood sample is routinely taken for the meaurement of pO_2, pCO_2, and pH. When an agent is to be injected, a femoral vein is also cannulated. Body temperature is monitored by a rectal thermometer, and maintained at 37°C with a heating pad or heat lamp for the entire procedure. The skull is secured in a head holder insuring that the airway is not compromised. The scalp is infiltrated with a local anesthetic, and a midline incision from the level of the eyes to the occiput is made. A funnel is fitted over the skull with the posterior lip approximating the lambdoidal suture. Stopcock grease is applied to the bottom of the funnel to form a seal and prevent leakage of the coolant. The funnel is secured with sutures by pulling the skin edges up around the base of the funnel.

Just prior to freezing, the blood gases and blood pressure should be checked. Acceptable ranges for each of these parameters can be found in the report by Ponten et al. (1973a). The liquid nitrogen is poured into the funnel and replenished as required. Care should be taken to keep the liquid nitrogen away from the airway. Freezing is continued for a period of 3 min, after which the entire rat is plunged into a flask of liquid nitrogen. The animal is stored in a –70°C freezer prior to dissection of the brain. Although this method has been predominantly used on the rat, it also has been performed with some success on the gerbil, cat, and dog (Kerr and Ghantus, 1935; Welsh, 1980; Arai et al, 1986).

This procedure is easily adapted for the measurement of EEG, and the collection of either CSF from the cisterna magna or venous blood from the superior saggital sinus. EEG can be monitored by placing sharpened 22-gage needle electrodes into the skull. Venous blood is sampled with a finely drawn glass capillary tube through a small hole drilled over the superior saggital sinus. CSF samples can be obtained by penetrating the atlanto-occipital membrane with a drawn capillary tube after the occipital muscles are stripped away from the membrane.

This fixation method has gained the widest acceptance for the study of cerebral energy metabolism, and the advantages of the procedure outweigh the disadvantages as shown in Table 5. If the blood flow to the tissue is maintained to provide oxygen and remove carbon dioxide during the freezing period, there is little evidence of autolytic changes in the energy-related metabolites in either the cerebral cortex of rats (Table 3) or deeper regions of the gerbil brain (Table 4). With recognition of the significance of regional changes in energy metabolism, the ability to fix the entire brain without autolytic artifact has gained in importance. For example, metabolic studies on the selective vulnerability of hippocampal CA 1 neurons to ischemia require the sampling of discrete areas (Lust et al., 1985; Arai et al., 1986). Without funnel freezing,

Table 5
Advantages and Disadvantages of Funnel Freezing Technique*

Advantages	Disadvantages
1. Provides optimal levels of metabolites free of autolytic artifact	1. Regional variation in freezing complicates studies requiring exact temporal sampling
2. Allows monitoring and manipulation of physiological parameters	2. Delayed freezing and period of hypothermia may cause regional artifact
3. Allows regional analysis	3. Incompatible with repeat studies
4. Provides measurements in a verified steady state	4. Requires use of anesthetic
5. Allows performance of special sampling procedures (e.g., CSF and venous blood)	5. Elevated tissue glucose levels
6. Readily adapted to larger and smaller laboratory animals	6. Time required for preparation of each animal
7. Tissue obtained can be used for most biochemical determinations	
8. Allows maintenance of blood flow and oxygenation to unfixed tissue	
9. Tissue can be used for histological and immunocytochemical studies	

*Table is adapted from Ratcheson (1980).

these experiments could not be performed. Nevertheless, the reader should be aware of certain pitfalls that might arise during the process. As noted before, the loss of blood flow to regions of the brain prior to their freezing could lead to an autolytic artifact in a region least expected to be improperly fixed, the superficial cerebral cortex. Other problems that arise during funnel freezing center on the effects of the anesthesia and hypothermia (Fig. 2). Both conditions are known to decrease the energy demands of the tissue, which could lead to an artifactual increase in energy reserves (Michenfelder and Theye, 1968; Brunner et al., 1971, Carlsson et al., 1976; Lust and Passonneau, 1976). The glucose levels in the tissue are significantly elevated (Table 3). The problem with anesthesia has been minimized by the use of nitrous oxide, which has been reported to have only minimal effects on metabolic rate (Ratcheson et al., 1977). However, the use of nitrous oxide may be prohibited under the new Animal Welfare guidelines, and alternative anesthetics may have to be considered. As long as the extent of anesthesia and hypothermia cannot be accurately determined, these potentially confounding factors must be considered in the interpretation of the experimental data. In addition, the loss of afferent input from the cortex to deeper structures of the brain during freezing may affect the level of activity, and secondarily, the metabolite profile in the deeper regions. Nevertheless, this approach is recommended because of its proven efficacy and broad application to a variety of projects.

2.6.4. Freeze-Blowing

The next three methods were developed to increase the speed of fixation by fragmenting the brain into smaller pieces. All three procedures have achieved a more rapid fixation of the brain, but in doing so, have limited their usefulness. Freeze-blowing removes and freezes the entire supratentorial portion of the brain within 1 s (Veech et al., 1973). Because of the rapidity of the method, the metabolite values determined from freeze-blown tissue are considered to be a reliable reference for evaluating other methods of fixation.

The brain blower is a specialized piece of equipment fabricated by Precision Medical Industries, Shalimar, FL. The apparatus is described in detail in the paper by Veech et al. (1973). The only additional supplies needed are compressed air and liquid nitrogen.

Procedure: A rat is placed in a restrainer and the head secured by a device that fixes the incisors, mandible, and nose in place without compromising respiration. The procedure commences by shooting two sharp hollow probes into the cranial vault. The left probe enters the head approximately 4 mm posterior to, and level with, the juncture of the eyelids, and the right one enters 4 mm anterior to, and level with, the external auditory canal. Once the probes have penetrated the cranial vault, air at a pressure of 25 lb/in^2 is released into one probe, forcing the brain substance out the other probe into a disc precooled in liquid nitrogen. The entire process is completed in less than 1 s and approximately 1 g of brain tissue is frozen.

The metabolite values from the freeze-blown brain reflect the rapidity of the fixation process (Table 3). Generally, the metabolite results are as good as those for the cerebral cortex from funnel frozen rats. Other advantages of the method are the size of the brain sample, the absence of anesthesia, the ability to study rapidly occurring events, and the relatively short period of time required for the fixation of each animal. A major disadvantage of the procedure is the expenditure for this specialized piece of equipment. In addition, the method cannot be used for either regional studies or histology, and the apparatus is limited to animals weighing between 100 and 400 g. Since the rats are awake, only limited physiological monitoring is possible. If the apparatus were readily available, however, this method would be an invaluable tool for studying energy metabolism.

2.6.5. Freeze Clamping

Freeze clamping is another technique requiring specialized equipment to fix the brain (Bolwig and Quistorff, 1973; Quistorff, 1975; Quistorff and Pederson, 1976). Unlike freeze blowing, which reduces brain size in one step for freezing, this procedure first cuts a portion of the brain into a cross-sectional slab, and then compresses the slab with a precooled press, in a manner reminiscent of the Wollenberger clamps (Wollenberger et al., 1960).

The guillotine clamping device is a specialized apparatus fabricated for the rapid sampling of either brain or liver. The availability of this instrumentation is not known, but the plans for the apparatus were published in the Quistorff paper. A coolant is needed for cooling the clamps.

Procedure: A rat is placed in a restraining tube, and two coronal cuts are simultaneously made through the head with two mechanically driven rotating knives. The resulting brain slab is held in place and clamped between two aluminum blocks precooled in liquid nitrogen. The 13-mm slab is pneumatically compressed to 10.5 mm during the clamping procedure. The entire procedure takes less than 0.1 s.

The metabolite profile for the freeze clamped tissue is similar to that for freeze blowing and funnel freezing (Table 3). The procedure does have the potential for studying regional changes in cerebral metabolites in the unanesthetized rat, but there is apparently an anoxic core within the slab (Quistorff, 1980). The need for specialized equipment makes it unlikely that this method will be used extensively.

2.6.6. Brain Chopping

Like the other invasive fixation techniques, brain chopping was a system devised in the mid-70s for rapidly sampling deeper structures of the brain (McCandless et al., 1976; McCandless and Roseberg, 1980). The principle of the procedure is to cut the brain into a series of cross-sectional slabs, using an air-driven multi-bladed guillotine. The slabs are immediately frozen in chilled 2-methylbutane. As with the two previous methods, the usefulness of this procedure has been hampered by the need for specialized equipment. Besides the brain chopper, the only additional supplies needed are 2-methylbutane and liquid nitrogen.

Procedure: The unrestrained animal is placed on a platform containing a small trap door. Once the head is in position, an air driven piston with a multibladed guillotine is triggered at the pressure of 125 lb/in^2. As the blades are driven through the head, the trap door opens, and the 6-mm thick coronal wafers are stripped from the knife into precooled 2-methylbutane. The cutting process takes 0.2 s, and the freezing time for the wafers is about 4 s. This procedure has only been tried on rats.

Although the metabolite results in frozen chopped tissue are not as convincing as for brain blowing and freeze clamping, the low lactate and L/P ratio suggests that the technique does work (Group e, Table 3). Again, the specialized equipment is not commercially available, and this method has not gained much acceptance. All three of these groups should be commended for not only addressing such a difficult problem as brain fixation, but also for having achieved some degree of success.

2.6.7. Miscellaneous Freezing Methods

a. Cryoplating is a method for surface freezing the cerebral cortex in the unrestrained and nonanesthetized animal (Skinner et al., 1978). The procedure involves subdural implantation of the cryoplate in the region of the parietal cortex. Methanol cooled to –79°C is circulated to cool the system, after which it is shunted into the cryoplate. Freezing in the adjacent 1-mm cortex occurs within 0.2 s. Cerebral energy metabolite levels have not been determined in tissue fixed by cryoplating, but the freezing data is rather convincing.

b. Several laboratories have described a procedure that involves spooning out the brain of unanesthetized animals whose skull had previously been removed under surgical anesthesia (Thorn et al., 1958; Swaab and Boer, 1972). The scooped tissue is immediately frozen in Freon precooled in liquid nitrogen or in liquid air. Although the data from these brains were in the acceptable range, there was some evidence of autolytic artifact.

3. Microwave Irradiation

It has almost been 20 years since the first experimental animal was fixed by microwave irradiation. The progress made in this methodology has been extraordinary. There were periods when the future for microwave irradiation as a fixation method for the investigation of cerebral energy metabolism did not appear promising. When this technique was compared to other freezing methods in several studies reported only 14 yr ago (Table 6), the conclusion regarding the use of microwave irradiation for brain fixation was unfavorable (Lust et al., 1973; Veech et al., 1973). However, the power supply for the microwave apparatus was improved, and within two years, the metabolite profile in the mouse brain closely ressembled that following immersion freezing (Medina et al., 1975). The perserverence of Stavinoha and his colleagues in upgrading this procedure from rather crude beginnings has resulted in a method that now yields metabolite values for the rat comparable to those obtained by funnel freezing (Medina and Stavinoha, 1977; Table 6).

3.1. Animal Welfare Concerns and Microwave Irradiation

Recommendations of the Panel on Euthanasia (*JAVMA*, 1986): "Microwave irradiation is a humane method to euthanatize small laboratory rodents if instruments that induce immediate unconsciousness are used. Only instruments that provide appropriate kilowattage and directed microwaves can be use. *Microwave ovens designed for domestic and institutional kitchens are absolutely condemned for euthanasia use."*

3.2. Microwave Radiation

Although microwave irradiation also stops metabolism by altering the temperature of the brain, it acts by heating rather than freezing the tissue. Microwave irradiation is a form of electromagnetic radiation in the frequency range of 0.2–24 GHz, bounded by the far infrared and radar regions. It has been known for years that the electromagnetic energy of microwave radiation in an aqueous

Table 6
Comparison of Metabolite Levels in Frozen and Microwaved Brains*

Group	mmol/kg wet wt							
	P-creatine	ATP	5'AMP	Glucose	Lactate	Pyruvate	α -OG	L/P
a. Microwave 1.2 kW	1.69	1.69	0.399	–	1.20	0.061	0.100	19.7
2450 MHz, 14 s	±0.08	±0.02	±0.011	–	±0.15	±0.002	±0.007	–
b. Freeze-blowing	4.05	2.45	0.041	0.962	1.23	0.091	0.213	13.5
	±0.07	±0.05	±0.001	±0.084	±0.07	±0.001	±0.001	–
c. Microwave 17 kW	3.99	2.70	0.11	2.01	0.91	0.21	0.29	4.3
915 MHz, 1 s	±0.34	±0.28	±0.012	±0.19	±0.06	±0.04	±0.02	–
d. Funnel freezing	4.71	3.00	0.040	4.97	1.64	0.121	0.117	13.6
anesth., paral.	±0.08	±0.04	±0.001	±0.35	±0.05	±0.005	±0.005	±0.5

*Results for groups (a) and (b) are from the forebrain of 200–250 g rats as reported by Veech et al. (1973), and those for groups (c) and (d) are from the cerebral cortex of rats weighing from 250–400 g as reported by Medina et al. (1980) and Ponten et al. (1973a), respectively. In groups (a) and (c), the power of the microwave apparatus is given in kW, the frequency of the radiation of MHz, and the duration of exposure in seconds. The anesthesia used in group (d) was 70% nitrous oxide and 30% oxygen, and paral. signifies that the animals were paralyzed with tubocurarine chloride. The other groups were not anesthetized. The metabolite data with the exception of the lactate/pyruvate ratio are expressed in mmole/kg wet weight, and each value represents the mean ± SEM. The abbreviations are as follows: α-OG, α-oxoglutarate, and L/P, lactate/pyruvate. For discussion, *see* text.

object is converted to thermal energy. The explanation for this phenomenon is that the electric dipole moment of water causes the molecule to rotate in alignment with the energy field induced by the oscillating radiation. The resistance to the movement of the water molecules generates heat that is transferred to the environment and denatures the tissue proteins. A number of laboratories have shown that an increase of 50°C denatures most of the proteins in the core of the brain. It should be noted that the energy of microwave radiation is in the range of 1 cal/mol, which is insufficient to break hydrogen bonds. Therefore, the denaturation is limited to the destruction of the secondary and tertiary structure of the proteins.

3.3. Microwave Apparatus

The success of microwave irradiation as a fixation tool has followed a logical sequence of adaptations from one of the first experimental ovens, which fixed an entire rat by a 30 s-exposure in a 1.25 kW oven with a frequency of 2.47 GHz (Schmidt et al., 1972b). Although many laboratories have pursued their interest in perfecting microwave irradiation technology for fixation, only one laboratory has extensively applied microwave irradiation to the investigation of cerebral energy metabolism (for reviews, *see* Medina et al., 1980; Jones and Stavinoha, 1979). The first modifications to the original ovens were to focus the energy with a waveguide and to increase the power output to 6 kW. This improved the metabolite results for a mouse brain exposed for 0.4 s. at a frequency of 2.45 GHz (Medina et al., 1975). Subsequently, a 915 MHz unit was designed on the basis of animal size, chamber size, and orientation of the head for use on larger animals. This oven was able to increase the temperature of the brain 50°C in 1 s. with 25 kW output and a 915 MHz frequency. An E-H tuner, ferrite circulators, and a directional coupler have been added to increase the efficiency of the system. Experience has shown that a reasonable standard to insure proper fixation in commercially available ovens is a 50°C increment in core brain temperature in 0.325 and 1 s. for the mouse and rat, respectively. It should be noted, however, that reasonable concentrations of energy-related metabolites have been achieved both in a rat exposed for 2 s. at 2.45 GHz in a 2 kW oven and in a gerbil exposed for 3 s. at 2.45 GHz in a 1.25 kW oven (Guidotti et al., 1974; Mrsulja et al., 1984).

3.4. Microwave Irradiation Protocol

The only equipment needed for this procedure is a microwave oven with sufficient power to fix the brain in 1 s. The component parts suggested are a transformer, magnetron, waveguide system with ferrite circulator, E-H tuner and directional coupler, applicator, and animal holder.

Procedure: The animal is placed in a glass or plastic holder that is attached to the applicator portion of the microwave oven. The power is turned on for sufficient time to heat the core of the brain to 90°C. The brain is removed and dissected after fixation has been confirmed by temperature determination in the brain tissue.

3.5. Discussion of Microwave Irradiation for Studies on Cerebral Metabolism

Many of the problems with using microwave irradiation for studying metabolism have been corrected, and the disadvantages listed in Table 7 are more concerned with the inherent limitations imposed by the nature of the technique rather than by its deficiencies. The concentrations of the energy-related metabolites in brains fixed in the newer microwave ovens are comparable to those for funnel freezing and freeze blowing. As shown in Group a of Table 6, the low levels of high-energy phosphates and α-oxoglutarate, and high 5' AMP and L/P observed in a brain fixed in a weaker oven have been corrected in Group c by using the 17 kW oven. Since microwave irradiation fixes the entire brain in less than a second, regional studies on acute experimental events can be reliably performed on many animals in a short period of time. Metabolism of the brain is stopped by the increase in temperature and the denaturation of the proteins, suggesting secondary metabolic changes should not occur (*see* Section 4).

There are some significant disadvantages, the cost of acquiring an oven being foremost. Obviously, enzyme activities cannot be measured, and sectioning the tissue for histology or immunocytochemistry would be difficult, even though electron micrographs of microwaved tissue have been published and quantitative histochemical analysis in discrete regions of a microwaved brain has been reported (Medina et al., 1980; McCandless et al., 1984). Occasionally, hot spots will form in the brain causing uneven heating of the tissue. A certain number of animals may have to be discarded for this reason. The possibility of metabolite diffusion in microwaved

Table 7
Advantages and Disadvantages of Brain Fixation by Microwave Irradiation

Advantages	Disadvantages
1. Achieves optimal metabolite levels free of autolytic artifact throughout the brain	1. Incompatible with enzyme activity measurements
2. Ideal fixation method for studies on rapidly occurring events	2. Heat denatured tissue unacceptable for quantitative histochemistry, immunocytochemistry, or histology
3. Allows gross regional analysis	3. Expense of the microwave oven
4. Anesthesia and surgery not required	4. Single sampling technique
5. Minimizes likelihood of postfixation artifacts	5. Speculated problems with metabolite diffusion
6. Time required for preparation of each animal	

brain has been considered (for review, *see* Lenox et al., 1982), but does not appear to be a problem when studying energy-related metabolites (McCandless et al., 1984). Microwave irradiation is currently an acceptable fixation technique for studying cerebral energy metabolism. As shown in Table 7, it has some unique advantages over the best of the freezing methods.

4. Tissue Extraction

It was noted previously that the metabolic machinery in frozen tissue was only inactivated by the rapid drop in temperature. The possibility exists that inadvertent warming of the tissue might restore metabolism. Since proof of the reversibility of metabolism in frozen tissue was not readily evident from the literature, a rat was frozen *in situ* as described in section 2.6.3, and 20 micron sections were cut at –20°C in a refrigerated microtome. The tissue samples were incubated at either –20, 0, +5, or +10°C for various times up to 30 min. At the end of incubation, the slices were frozen

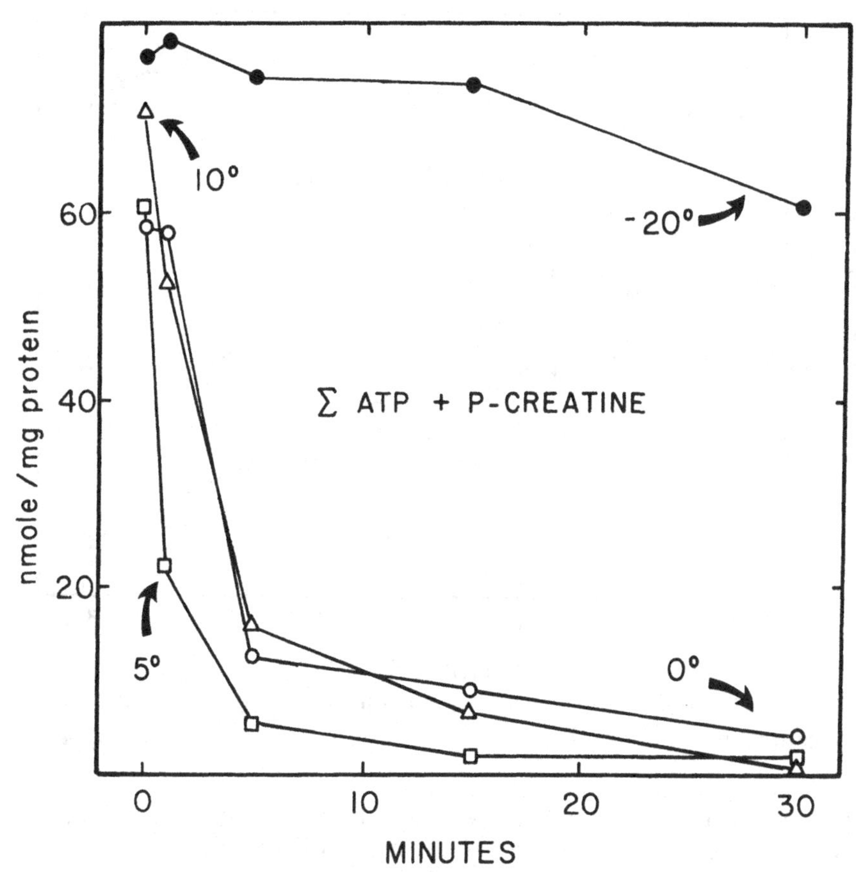

Fig. 4. Effect of warming on the levels of high-energy phosphates (ATP + P-creatine) in frozen brain tissue. A brain from a control rat was frozen *in situ,* and a series of 20 μm sections were cut. The tissue was either kept at –20°C (●) or incubated at 0 (○), 5 (□), or 10 (△) °C for up to 30 min. The tissue was extracted in PCA as described in Section 4.2.1., and the levels of ATP, *P*-creatine, glucose, and glycogen were measured in each sample as described by Mrsulja et al. (1986). The concentration is expressed in nmole/mg protein.

in liquid nitrogen, extracted, and assayed for ATP, P-creatine, glucose, and glycogen. The levels of the sum of ATP and P-creatine (i.e., high-energy phosphates) were maintained at –20°C for up to 30 min, but markedly decreased at the other three incubation temperatures with a mean half-time of approximately 7 min (Fig. 4). The levels of glucose and glycogen (i.e., glucosyl units) were similarly maintained at –20°C, but decreased in the other three

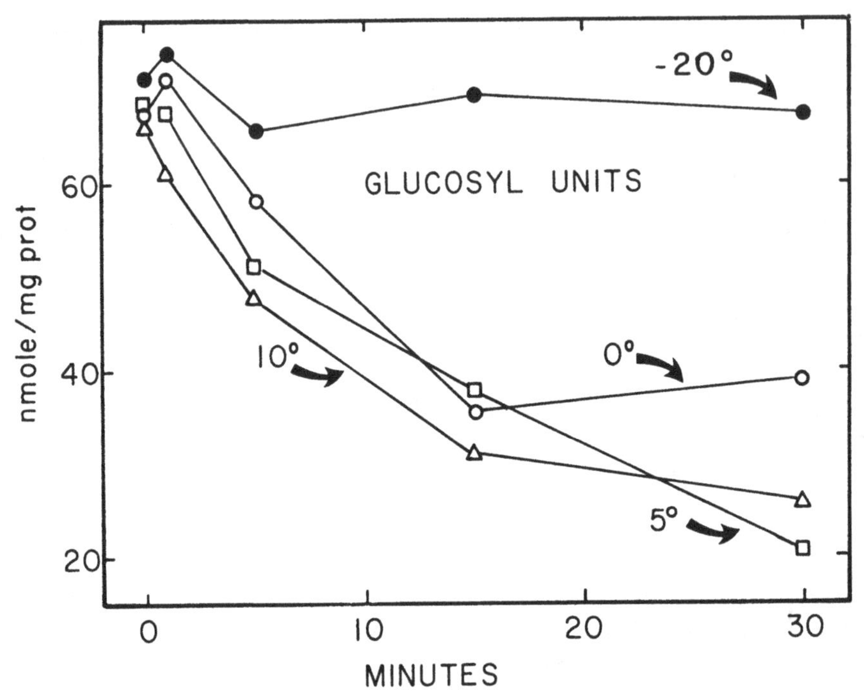

Fig. 5. Effect of warming on the glucosyl (glucose + glycogen) concentration in frozen brain tissue. For details, *see* legend in Fig. 4.

groups with a somewhat slower mean half-time of 24 min (Fig. 5). In most respects, the changes in the metabolites induced by warming the tissue were similar to those observed after ischemia, although glucose was never totally depleted. The rate of loss of both high-energy phosphates and glucosyl units was about 5% of that found in normothermic gerbil brain following ischemia (Mrsulja et al., 1986). Although the relative rate of metabolite loss does not compare with that of ischemia in vivo, the results demonstrate that metabolism in frozen tissue is not irreversibly inactivated. Failure to maintain the tissue at –20°C or less could lead to artifacts that would not be distinguishable from those arising from poor fixation.

As noted before, it is generally accepted that most proteins are denatured by the heat produced from microwave irradiation. Because of the heat denaturation, some investigators have opted not to deproteinate and to extract the tissue in a 5 m*M* EDTA (Medina et al., 1975). However, certain enzymes like adenylate kinase [*see* Eq. (3)] are quite stable even at 100°C, and cooling the tissue to 37°C for any length of time could restore the enzyme activity. Although this possibility is speculative, freezing the microwaved tissue and keeping it cold prior to acid extraction seems advisable.

$$\text{ATP} + 5'\text{AMP} \xleftrightarrow{\text{Adenylate kinase}} 2\ \text{ADP} \quad (3)$$

4.1. Preparation of Tissue

Frozen brains should be removed in a refrigerated glove box maintained at –20°C, since the brains tend to shatter during dissection in colder temperatures. Some investigators have used a power saw to remove the brain, but liquid nitrogen must be continually poured on the blade and tissue to avoid heating resulting from friction. Microwaved brains can be rapidly dissected at room temperature and are then ready for extraction. Since most reagents used for extraction will freeze at temperatures close to 0°C, it is important to complete the extraction as soon as possible after the tissue is added to a chilled extracting agent. Reducing the brain size by using small pieces of tissue aids in extraction. Another common approach is to reduce the size of the brain by grinding it in a mortar and pestle under liquid nitrogen. This insures a greater exposure of the brain to the extractant than is possible with larger samples of brain.

4.2. Tissue Extraction

The purpose of extraction is to quantitatively remove an endogenous compound from the tissue into a medium that should have the following characteristics: (1) the metabolite should be soluble and stable in the reagent, and (2) the medium should not interfere with the assay. The stability of the metabolite should also be determined during storing and with freezing and thawing. (For review on extraction, *see* Lowry and Passonneau, 1972; Passonneau et al., 1979.) The completeness of the metabolite extraction from the tissue can be determined by mixed and multiple extractions of the pellet. Using the multiple extraction method, the tissue is extracted and centrifuged, and the resulting protein pellet is reextracted. There will only be trace amounts of the metabolite in the second supernatant if the first extraction recovered most of the compound. The mixed method uses a different reagent for the second extraction to ensure that the first extractant effectively removed the material. Loss of compound during the remaining steps of the extraction procedure can be determined by spiking a blank or an aliquot of a tissue sample with an appropriate standard. If a portion of the standard is lost during the process, the percent recovery can be determined and used to correct the concentration of the metabolite in the tissue samples. This procedure, however, does not discriminate between a loss in volume or a loss resulting from instability of the compound. The latter can be determined in a different set of experiments.

4.2.1. Acid Extraction

a. Most metabolic processes are irreversibly lost by precipitating the enzymes and structural proteins of the tissue. Proteins can be precipitated by a number of heavy anions that bind to certain reactive groups in the proteins. The agents include phosphotungstate, trichloroacetate, perchlorate, picrate, and so on, but the precipitation does not occur unless the protein is in its predominantly cationic form. This is achieved by using the acid form of these anions. The brain powder can be layered on top of three volumes of 3M perchloric acid (PCA) in a tube kept on dry ice. The tubes are allowed to warm to –10°C, and as the acid thaws, the powder sinks and is extracted (Lowry and Passonneau, 1972). For every 0.3 mL of PCA, 1 mL of glass distilled water is added and mixed. The tubes are centrifuged at 5000*g* for 10 min, and the supernatant free of protein is removed. The acid extract is neutralized either with 0.35 mL of 2M $KHCO_3$/mL of extract or 0.29 mL of a combination of 2*N* KOH, 0.4M imidazole base, and 0.4M KCl/mL of extract. The acid is neutralized by the formation of potassium perchlorate, which is only slightly soluble in water. The protein pellet can be dissolved in 1*N* NaOH for analysis. This extraction procedure is appropriate for most acid soluble metabolites, including P-creatine, the adenylates, and many of the glycolytic and TCA cycle intermediates.

b. Alternatively, approximately 100 mg of tissue is added to 0.1 mL of 0.1N HCl in absolute methanol in a –20°C ethanol–dry ice bath and incubated for 20 min, after which 1 mL of aqueous 0.02 *N* HCl is added to the tube and the tissue is homogenized. Aliquots can be taken for the measurement of glucose and glycogen. To 1 mL of the homogenate, add 0.1 mL of 3N PCA containing 10 m*M* EGTA and mix. Centrifuge at 5000*g* for 10 min, and carefully remove the supernatant. Add 0.15 mL of 2N $KHCO_3$ to each 1 mL of supernatant, and mix well. As a precaution, the pH of the neutralized extract should be determined. The potassium perchlorate precipitate can be spun down, and the neutralized supernatant removed, but this is not necessary. If the tissue was not previously weighed, protein determinations are made after the protein pellet is dissolved in 1*N* NaOH. This extraction procedure is good for the same metabolites mentioned above and also for glycogen, which partially precipitates with the protein in PCA.

4.2.2. *Alkaline Extraction*

NAD and NADP are stable in acid and labile in alkali. In contrast, the reduced form of these pyridine nucleotides are stable in alkali and labile in acid. These properties are quite useful for measuring tissue concentrations of the pyridine nucleotides. Hemoglobin can oxidize the reduced forms. However, this can be avoided by the following procedure. A tissue sample is homogenized rapidly at 0°C in 100 volumes of 0.04*N* NaOH containing 0.5 m*M* cysteine. The extracts can be more concentrated if the amount of hemoglobin is relatively low. At this concentration of NaOH, the half-time for the disppearance of the oxidized pyridine nucleotides at 0°C is greater than 8 h. The total pyridine nucleotide pool can be enzymically determined on this sample. A portion of the alkali extract can be heated at 60°C for 10 min to destroy the oxidized cofactors. This sample can be used to measure the reduced pyridine nucleotides. To another portion of the chilled alkali extract is added 0.03 volumes of 1*M* ascorbic acid to prevent oxidation of the reduced pyridine nucleotides in acid. Two equivalents of a solution containing 0.02*N* sulfuric acid and 0.1*N* sodium sulfate are added, and the sample is heated for 30 min at 60°C. This sample contains only the oxidized pyridine nucleotides.

There are other types of extraction, such as trichloroacetic acid and neutral solvent extraction, but these have had only limited use in the investigation of cerebral energy metabolism. The accurate measurement of cerebral metabolites requires that all steps in the process are working properly. Failure in either fixation, storage, dissection, extraction, or the various assays will invalidate the entire process. Although extraction and storage of the specimens have not received as much attention as fixation, these steps, improperly executed, can ruin the best of fixation techniques.

5. Summary

The unique combination of biochemical and anatomical characteristics found in the brain make it the most difficult organ to fix.

An ideal fixation technique for the brain has yet to be found. Clearly, the brains fixed by funnel freezing, microwave irradiation, and freeze-blowing yield metabolite levels more closely approximating the in vivo condition. It is recommended that these methods be used for the investigation of cerebral energy metabolism. The technological limitations imposed by each method will be a major determinant in the final selection process, as will be concerns for the animal's welfare. Clearly, funnel freezing has the broadest application, but fails to be useful when examining rapidly occurring events. Microwave irradiation, although equally as beneficial, is not applicable to enzyme studies or quantitative histochemistry. Ultimately, the experimental objectives will clearly dictate which of the fixation techniques is consistent with and appropriate for the successful completion of the study.

References

Allred J. B. and Bernston G. G. (1986) Is euthanasia of rats by decapitation inhumane? *J. Nutrition* **116,** 1859–1861.

American Veterinary Medical Association Panel on Euthanasia (1986) *J. Am. Veterinary Med. Assoc.* **188,** 252–268.

Arai H., Passonneau J. V., and Lust W. D. (1986) Energy metabolism in delayed neuronal death of CAl neurons of the hippocampus following transient ischemia in the gerbil. *Metabol. Brain Dis.* **1,** 263–278.

Astrup J. (1982) Energy-requiring cell functions in the ischemic brain: Their critical supply and possible inhibition in protective therapy. *J. Neurosurg.* **56,** 482–497.

Avery B. F., Kerr S. E., and Ghantus M. (1935) The lactic acid content of mammalian brain. *J. Biol. Chem.* **110,** 637–642.

Bolwig T. G. and Quistorff B. (1973) *In vivo* concentration of lactate in the brain of conscious rats before and during seizures: A new ultra-rapid technique for the freeze-sampling of brain tissue. *J. Neurochem.* **21,** 1345–1348.

Breckenridge B. McL. (1964) The measurement of cyclic adenylate in tissues. *Proc. Natl. Acad. Sci. USA* **52,** 1580–1586.

Breckenridge B. M. and Norman J. H. (1962) Glycogen phosphorylase in brain. *J. Neurochem.* **9,** 383–392.

Breckenridge B. M. and Norman J. H. (1965) The conversion of phosphorylase b to phosphorylase a in brain. *J. Neurochem.* **12,** 51–57.

Brunner E. A., Passonneau J. V., and Molstad C. (1971) The effect of volatile anaesthetics on level of metabolites and on metabolic rate in brain. *J. Neurochem.* **18,** 2301–2316.

Carlsson C., Hagerdal M., and Siesjö B. K. (1976) The effect of hyperthermia upon oxygen consumption and upon organic phosphates, glycolytic metabolites, citric acid cycle intermediates and associated amino acids in rat cerebral cortex. *J. Neurochem.* **25,** 1001–1006.

Ferrendelli J. A., Gay M. H., Sedgwick W. G., and Chang M. M. (1972) Quick freezing of the murine CNS: Comparison of regional cooling rates and metabolite levels when using liquid nitrogen or Freon 12. *J. Neurochem.* **19,** 979–987.

Goldberg N. D., Passonneau J. V., and Lowry O. H. (1966) Effect of changes in brain metabolism on the levels of citric acid cycle intermediates. *J. Biol. Chem.* **241,** 3997–4003.

Guide for the care and use of laboratory animals (1985) U.S. Department of Health and Human Services, Public Health Service, National Institutes of Health.

Guidotti A., Cheney D. L., Trabucchi M., Doteuchi M., and Wang C. (1974) Focussed microwave radiation: A technique to minimize post mortem changes of cyclic nucleotides, DOPA, and choline and to preserve brain morphology. *Neuropharmacology* **13,** 1115–1122.

Hansen A. J. (1981) Extracellular ion concentrates in cerebral ischemia, in *The Application of Ion-selective Microelectrodes* (Zeuthen, T., ed.), pp. 239–254, Elsevier/North Holland Biomedical Press.

Harik S. I., Busto R., and Martinez E. (1982) Norepinephrine regulation of cerebral glycogen utilization during seizures and ischemia. *J. Neurosci.* **2,** 409–414.

Heath D. F., Frayn K. N., and Rose J. G. (1977) Rates of glucose utilization and glucogenesis in rats in the basal state induced by halothane anaesthesia. *Biochem. J.* **162,** 643–651.

Jones D. J. and Stavinoha W. B. (1979) Microwave inactivation as a tool for studying the neuropharmacology of cyclic nucleotides, in *Neuropharmacology of Cyclic Nucleotides* (Palmer, G. C., ed.), pp. 253–282, Urban and Schwarzenberg, Baltimore.

Jongkind J. F. and Bruntink R. (1970) Forebrain freezing rates and substrate levels in decapitated rat heads. *J. Neurochem.* **17,** 1615–1617.

Kakiuchi S. and Rall T. W. (1968) Studies on adenosine 3', 5'-phosphate in rabbit cerebral cortex. *Mol. Pharmacol.* **4,** 379–388.

Kerr S. E. (1935) Studies on the phosphorus compounds of brain. I. Phosphocreatine. *J. Biol. Chem.* **110,** 625–636.

Kerr S. E. and Ghantus M. (1935) The carbohydrate metabolism of brain. III. On the origin of lactic acid. *J. Biol. Chem.* **116,** 9–20.

Knieriem K. M., Medina M. A., and Stavinoha W. B. (1977) The levels of GABA in mouse brain following tissue inactivation by microwave irradiation. *J. Neurochem.* **28,** 885–886.

Langendorff V. O. (1886–87) Die chemische reaktion der grauen substanz. *Biol. Centrabl.* **6,** 188–190.

Lenox R. H., Kant G. H., and Meyerhoff J. L. (1982) Rapid enzyme inactivation, in *Handbook of Neurochemistry* (Lajtha, A., ed), pp. 77–102, Plenum, New York.

Lowry O. H. and Passonneau J. V. (1972) *A Felxible System of Enzymatic Analysis,* Academic, New York.

Lowry O. H., Passonneau J. V., Hasselberger F. X., and Schulz D. W. (1964) Effect of ischemia on known substrates and cofactors of the glycolyptic pathway in brain. *J. Biol. Chem.* **239,** 18–42.

Lust W. D., Arai H., Yasumoto Y., Whittingham T. S., Djuricic B., Mrsulja B. B., and Passonneau J. V. (1985) Ischemic encephalopathy, in *Cerebral Energy Metabolism and Metabolic Encephalopathy* (McCandless, D. W., ed.), pp. 79–117, Plenum, New York.

Lust W. D., Murakami N., de Azeredo F., and Passonneau J. V. (1980) A comparison of methods for brain fixation, in *Cerebral Metabolism and Neural Function* (Passonneau J. V., Hawkins R. A., Lust W. D., and Welsh F. A., eds.) pp. 10–19, Williams and Wilkins, Baltimore.

Lust W. D. and Passonneau J. V. (1976) Cyclic nucleotides in murine brain: Effect of hypothermia on adenosine 3', 5'-monophosphate, glycogen phosphorylase, glycogen synthase and metabolites following maximal electroshock or decapitation. *J. Neurochem.* **26,** 11–16.

Lust W. D., Passonneau J. V., and Veech R. L. (1973) Cyclic adenosine monophosphate, metabolites, and phosphorylase in neural tissue: A comparison of methods of fixation. *Science* **181,** 280–282.

McCandless D. W., Abel M., and Stavinoha W. (1984) Maintenance of regional chemical integrity for energy metabolites in microwaved mouse brains. *Brain Res. Bull.* **13,** 253–255.

McCandless D. W. and Roseberg N. C. (1980) Brain chopping: A method for exposure and fixation of deep brain structures, in *Cerebral Metabolism and Neural Function* (Passonneau J. V., Hawkins R. A., Lust W. D., and Welsh F. A., eds.), pp. 53–55, Williams and Wilkins, Baltimore.

McCandless D. W., Roseberg N. C., and Cassidy C. E. (1976) Brain chopping: A new method for rapid removal of newborn rat brain. *Experientia* **32,** 669.

Medina M. A., Deam A. P., and Stavinoha W. B. (1980) Inactivation of brain tissue by microwave irradiation, in *Cerebral Metabolism and Neural Function* (Passonneau J. V., Hawkins R. A., Lust W. D., and Welsh F. A., eds.), pp. 56–69, Williams and Wilkins, Baltimore.

Medina M. A., Jones D. J., Stavinoha W. B., and Ross R. H. (1975) The levels of labile intermediary metabolites in mouse brain following rapid tissue fixation with microwave irradiation. *J. Neurochem.* **24,** 223–227.

Medina M. A. and Stavinoha W. B. (1977) Labile intermediary metabolites in rat brain determined after tissue inactivation with microwave irradiation. *Brain Res.* **132,** 149–152.

Michenfelder J. D. and Milde J. H. (1975) Cerebral protection by anaesthetics during ischaemia (a review). *Resuscitation* **4,** 219–233.

Michenfelder J. D. and Theye R. A. (1968) Hypothermia: Effect on canine brain and whole-body metabolism. *Anesthesiology* **29,** 1107–1112.

Miller A. L. and Shamban A. (1977) A comparison of methods for stopping intermediary metabolism of developing rat brain. *J. Neurochem.* **28,** 1327–1334.

Mrsulja B. B., Yasuyuki Y., and Lust W. D. (1986) Regional metabolite profiles in early stages of global ischemia in the gerbil. *Metabol. Brain Dis.* **1,** 205–220.

Mrsulja B. B., Yasuyuki U., Wheaton A., Passonneau J. V., and Lust W. D. (1984) Release of pentobarbital-induced depression of metabolic rat during bilateral ischemia in the gerbil brain. *Brain Res.* **309,** 152–155.

Nahrwald M. L., Lust W. D., and Passonneau J. V. (1977) Halothane-induced alterations of cyclic nucleotide concentrations in three regions of the mouse nervous system. *Anesthesiology,* **47,** 423–427.

Passonneau J. V., Lust W. D., and McCandless D. W. (1979) The preparation of biological samples for analysis of metabolites. *Techniques in Metabolic Res.* **B212,** 1–27.

Ponten U., Ratcheson R. A., Salford L. G., and Siesjö B. K. (1973a) Optimal freezing conditions for cerebral metabolites in rats. *J. Neurochem.* **21,** 1127–1138.

Ponten U., Ratcheson R. A., and Siesjö B. K. (1973b) Metabolic changes in the brains of mice frozen in liquid nitrogen. *J. Neurochem.* **21,** 1121–1126.

Quistorff B. (1975) A mechanical device for the rapid removal and freezing of liver or brain tissue from unanaesthetized and nonparalyzed rats. *Anal. Biochem.* **68,** 102–118.

Quistorff B. (1980) Guillotine freeze clamping of rat brain, in *Cerebral Metabolism and Neural Function* (Passonneau J. V., Hawkins R. A., Lust W. D., and Welsh F. A., eds.), pp. 42–52, Williams and Wilkins, Baltimore.

Quistorff B. and Pederson E. (1976) A new device for freeze-clamping of tissue samples. *Anal. Biochem.* **73,** 236–239.

Ratcheson R. A. (1980) Funnel freezing for preservation of intermediary metabolites in rats, in *Cerebral Metabolism and Neural Function* (Passonneau J. V., Hawkins R. A., Lust W. D., and Welsh F. A., eds.), pp. 20–29, Williams and Wilkins, Baltimore.

Ratcheson R. A., Bilezikjian L., and Ferrendelli J. A. (1977) Effect of nitrous oxide anesthesia upon cerebral energy metabolism. *J. Neurochem.* **28,** 223–225.

Rhodes G. M. (1892) *The Nine Circles of the Hell of the Innocent,* Sonneschein, London.

Schmidt D. E., Speth R. C., Welsch F., and Schmidt M. J. (1972a) The use of microwave radiation in the determination of acetylcholine in the rat brain. *Brain Res.* **38,** 377–389.

Schmidt M. J., Hopkins J. T., Schmidt D. E., and Robison G. A. (1972b) Cyclic AMP in brain areas: Effect of amphetamine and norepinephrine assessed through the use of microwave radiation as a means of tissue fixation. *Brain Res.* **42,** 465–477.

Schmidt M. J., Schmidt D. E., and Robison G. A. (1971) Cyclic adenosine monophosphate in brain areas: Microwave irradiation as a means of tissue fixation. *Science* **173,** 1142–1143.

Sharpless N. S. and Brown L. L. (1978) Use of microwave irradiation to prevent postmortem catecholamine metabolism: Evidence for tissue disruption artifact in a discrete region of rat brain. *Brain Res.* **140,** 171–176.

Siesjö B. K. (1978) *Brain Energy Metabolism,* John Wiley and Sons, Chichester, New York.

Siesjö B. K. and Nilsson L. (1971) The influence of arterial hypoxemia upon labile phosphates and upon extracellular and intracellular lactate and pyruvate concentrations in the rat brain. *Scand. J. Clin. Lab. Invest.* **27,** 83–96.

Skinner J. E., Welch K. M. A., Reed J. C., and Nell J. H. (1978) Psychological stress reduces cyclic 3',5'-adenosine monophosphate levels in the cerebral cortex of conscious rats, as determined by a new cyrogenic method of rapid tissue fixation. *J. Neurochem.* **30,** 691–698.

Sokoloff L. (1977) Relation between physiological function and energy metabolism in the central nervous system. *J. Neurochem.* **29,** 13–26.

Sokoloff L., Reivich M., Kennedy C., Des Rosiers M. H., Patlak C. S., Pettigrew K. D., Sakurada O., and Shinohara M. (1977) The [^{14}C]deoxyglucose method for the measurement of local cerebral glucose utilization: Theory, procedure, and normal values in the conscious and anesthetized albino rat. *J. Neurochem.* **28,** 897–916.

Stavinoha W. B., Pepelko B., and Smith P. (1970) Microwave radiation to inactivate cholinesterase in the rat brain prior to analysis for acetylcholine. *Pharmacologist* **12,** 257.

Stavinoha W. B., Weintraub S. T., and Modak A. T. (1973) The use of microwave heating to inactivate cholinesterase in the rat brain prior to analysis for acetylcholine. *J. Neurochem.* **20,** 361–371.

Steffey E. P. and Eger II E. I. (1985) Nitrous oxide in veterinary practice and animal research, in *Nitrous Oxide*/N_2O (Eger E. I., ed.), Elsevier, New York.

Steiner A. L., Ferrendelli J. A., and Kipnis D. M. (1972) Radioimmunoassay for cyclic nucleotides: Effect of ischemia, changes during development and regional distribution of cyclic AMP and cyclic GMP in mouse brain. *J. Biol. Chem.* **247,** 1121–1124.

Swaab D. F. (1971) Pitfalls in the use of rapid freezing for stopping brain and spinal cord metabolism in rat and mouse. *J. Neurochem.* **18,** 2085–2092.

Swaab D. F. and Boer K. (1972) The presence of biologically labile compounds during ischemia and their relationship to the EEG in rat cerebral cortex and hypothalamus. *J. Neurochem.* **19,** 2843–2853.

Thorn W., Scholl H., Pfleiderer G., and Mueldener B. (1958) Stoffwechselvorgange im gehirn bei normaler und herabgesetzer korpertemperatur unter ischamischer und anoxischer belastung. *J. Neurochem.* **2,** 150–165.

Veech R. L. (1980) Freeze blowing of brain and the interpretation of the meaning of certain metabolite levels, in *Cerebral Metabolism and Neural Function* (Passonneau J. V., Hawkins R. A., Lust W. D., and Welsh F. A., eds.), pp. 34–41, Williams and Wilkins, Baltimore.

Veech R. L., Harris R. L., Veloso D., and Veech E. H. (1973) Freeze-blowing: A new technique for the study of brain *in vivo*. *J. Neurochem.* **20,** 183–188.

Welsh F. A. (1980) *In situ* freezing of cat brain, in *Cerebral Metabolism and Neural Function* (Passonneau J. V., Hawkins R. A., Lust W. D., and Welsh F. A., eds.), pp. 28–33, Williams and Wilkins, Baltimore.

Welsh F. A. and Rieder W. (1978) Evaluation of *in situ* freezing of cat brain by NADH fluorescence. *J. Neurochem.* **31,** 299–309.

Wollenberger A., Ristau O., and Schoffa G. (1960) Eine einfache technik der extrem schnellen abkuhlung grosserer gewebstucke. *Pflugers Arch. Ges. Physiol.* **270,** 399–412.

Isolation and Characterization of Synaptic and Nonsynaptic Mitochondria from Mammalian Brain

James C. K. Lai and John B. Clark

1. Introduction

1.1. Historical Aspects of Brain Mitochondrial Isolation and Metabolism

Until the mid-1960s, studies on the metabolism of brain mitochondria were hampered by the lack of suitable methodologies for brain mitochondrial isolation. In the majority of the studies up to this time, crude mitochondrial preparations were isolated from mammalian brain homogenates using differential centrifugation techniques that were adapted from those that had originally been designed for isolating liver mitochondria. During the early 1960s, two groups of workers [see Whittaker (1969 and 1984) and De Robertis and de Lores Arnaiz (1969) for discussions] independently designed relatively elaborate subcellular fractionation procedures, whereby the crude mitochondrial fraction derived from brain homogenates was subfractionated, using sucrose density gradients, into three or more discrete fractions, including pinched-off nerve-ending particles or "synaptosomes" and myelinated axon fragments (usually referred to as "myelin"), in addition to "free" (i.e., nonsynaptic) mitochondria. Thus, the structural heterogeneity of the crude mitochondrial fraction derived from brain homogenates was revealed. Consequently, many of the metabolic properties (e.g., glycolysis) that were erroneously attributed to isolated brain mitochondria in the studies in that era could be accounted for by the presence of vesicular and other membranous particles (e.g., synaptosomes, myelin, and microsomes) that contain cytosolic material to a greater or lesser extent [*see* Clark and Nicklas (1970) for a discussion].

1.2. Methodological Problems Associated with Early Studies of Brain Mitochondrial Metabolism

As indicated above, it has been a recognized problem to prepare brain mitochondria that are significantly free from contamination by nonmitochondrial material, and are, at the same time, metabolically active and well-coupled. The mitochondria isolated by high density sucrose gradients satisfy the criterion of purity, but not the criterion of metabolic integrity [*see* Clark and Nicklas (1970) and Lai and Clark (1976) for a discussion]. These density gradient procedures usually involve centrifuging mitochondria through high density and hyperosmotic sucrose solutions for long periods, so that significant proportions of the isolated mitochondria appear shrunken. (The degree of structural alteration of the mitochondria after isolation depends critically upon the osmolarity of the gradient material and the duration of the centrifugation.) Consequently, the mitochondrial preparations so isolated are clearly not ideal for detailed structural studies. Nonetheless, these mitochondrial preparations may be adequate for enzymatic and some chemical studies. Moreover, such hypertonic sucrose gradients can be skillfully employed, in conjunction with zonal centrifugation, to elucidate the microheterogeneity of brain mitochondria [*see* Blokhuis and Veldstra (1970), Reijnierse et al. (1975a,b), Lai and Clark (1976), and references cited therein for discussions and experimental details].

Another problem relates to whether or not the isolated mitochondria are representative of the mitochondria in brain in vivo. In this regard, one should be cautious in extrapolating conclusions based on studies using isolated mitochondria to mitochondria in brain in vivo.

1.3. Heterogeneity of Brain Mitochondria and Metabolic Compartmentation

In the last two decades, a good deal of evidence has accumulated that brain mitochondria are heterogeneous not only in their enzyme contents and complement, but also in their biophysical attributes (e.g., size, density, inner membrane/cristae configurations). This biochemical/biophysical heteogeneity is hardly surprising in view of the established morphological heteogeneity of the brain at the regional, cellular, and subcellular levels.

On the other hand, the relations between the morphological heteogeneity and the biochemical/biophysical counterpart have not been fully elucidated. For example, whether or not the various populations of mitochondria (derived from the same whole brain homogenates), separated by density gradient and/or zonal centrifugation, are actually from different regions of the brain, remains to be completely resolved. More recent studies (Lai et al., 1983; Leong et al., 1984) demonstrate that the biochemical/enzymatic activity differences can also be detected in various mitochondrial populations isolated from a *single* defined brain region, suggesting that there is at least some degree of mitochondrial hereogeneity at the subcellular level.

There is ample metabolic evidence supporting the concept of heterogeneity of brain mitochondria. For example, there is solid evidence [*see* Balázs and Cremer (1973), Berl et al. (1975), and articles therein] that brain metabolism of glutamate is compartmented—at least two glutamate compartments have been biochemically differentiated. These compartments are designated "small" and "large" compartments. Linked to these two glutamate compartments are the "small" and the "large" citric acid cycles. Since the citric acid cycle and glutamate-metabolizing enzymes are largely, if not exclusively, localized in mitochondria, the metabolic compartmentation of glutamate and citric acid cycle intermediates suggests that brain mitochondria consist of at least two metabolically distinct subpopulations.

Additionally, metabolic and enzymatic studies [*see* Van den Berg et al. (1975) and references cited therein] led to the proposal that both the "large" and the "small" compartments can be further subdivided. Thus, this proposal implies that brain mitochondria may consist of many more than the two subpopulations that correspond to the originally designated "large" and "small" compartments. Indeed, data from other enzymatic studies on subpopulations of mitochondria isolated (a) from whole brain homogenates by density gradient and zonal centrifugation (Blokhuis and Veldstra, 1970; Reijnierse et al., 1975a,b; Lai and Clark, 1976) and (b) from discrete brain regions by density gradients (Leong et al., 1984; Lai et al., 1983 and 1985) are consistent with this notion.

Metabolic compartmentation can also occur within a mitochondrion (Van Dam, 1973). However, such "microcompartments" have not been investigated in brain mitochondria [*see* Srere (1985) for a general discussion and references].

1.4. Anatomical and Functional Classification of Heterogeneous Populations of Brain Mitochondria

From the anatomical point of view, mitochondrial heterogeneity could have originated from the heterogeneity at the cellular (e.g., mitochondria from neurons vs glia, mitochondria from different types of neurons or glia) and the subcellular (e.g., mitochondria from neuronal cell body vs those from dendrites or nerve-endings) levels.

Although it is theoretically possible to study the structure and function of mitochondria isolated from a particular cell type in brain, in practice this approach suffers from several methodological limitations. For example, this *cellular approach* allows the isolation of mitochondria from bulk-isolated neurons or glia (Hamberger et al., 1970). However, the yields of the bulk-isolated neurons and glia are relatively low (<10% of the starting material). Thus, the yields of the mitochondria derived from the bulk-isolated cells are correspondingly smaller. Consequently, it is very difficult to conduct good metabolic experiments with such small amounts of mitochondrial material. In the past, many workers have been very critical of the actual purity of such cell preparations isolated using techniques developed in the 1960s [*see* Johnston and Roots (1970), Balázs et al. (1975) and references cited therein]. [For example, one concern was that the "glial" preparation(s) could be contaminated with varying amounts of synaptosomal material (J. C. K. Lai, unpublished observations). Thus, the mitochondria subsequently isolated from such "glial" preparations actually contain a proportion of mitochondria of nonglial (actually neuronal) origin.] Additionally, whether or not the mitochondria isolated from the small quantities of bulk-prepared cells are really representative of the mitochondria in these cell types in vivo is still controversial.

Compared with the cellular approach, the *subcellular approach* (which addresses mitochondrial heterogeneity from the intracellular perspective) has one major advantage: the yields of "free" or nonsynaptic mitochondria are substantially higher and are, thus, adequate for metabolic studies (*see* later Sections). There is an added bonus to the subcellular approach: the synaptosomes [derived from the same isolation procedure as for the "free" (i.e., nonsynaptic) mitochondria] and intraterminal mitochondria can also be isolated in purity, metabolic integrity, and yields that are adequate for metabolic studies.

1.5. *The Scope of This Review*

The authors' evaluation of the literature (1965–1986) reveals that the topic of brain mitochondria has not been comprehensively reviewed from the methodological perspective. Nonetheless, because of space limitation, a comprehensive and exhaustive critique of this literature is beyond the scope of the present review. Discussions of older literature have appeared elsewhere [*see* Abood (1969), Moore and Strasberg (1970), Clark and Nicklas (1970), and references cited therein). It should be noted that many of the studies prior to 1970 were complicated by methodological problems that severely "cloud" the interpretations and the significances of the results of those early studies. The present review focuses on the methodological advances based upon the *subcellular approach* and illustrates how these methodological advances have made possible the diverse range of studies.

2. Methods for the Isolation of Relatively Pure and Metabolically Active Fractions of Synaptic and Nonsynaptic Mitochondria

2.1. *Isolation of Nonsynaptic Mitochondria from Whole Brain*

The majority of the existing literature on brain mitochondria concerns the isolation of nonsynaptic mitochondria from whole brain using density gradients.

2.1.1. *Isolation of Nonsynaptic Mitochondria Using Sucrose Density Gradients*

To meet the criterion of mitochondrial purity, the sucrose employed in this kind of preparative procedure is usually hyperosmotic (>0.9*M*); the mitochondria so isolated are metabolically inactive and poorly coupled to oxidative phosphorylation [*see* Clark and Nicklas (1970), Lai and Clark (1976) and Lai et al. (1977) for a full discussion]. Thus, the details of the sucrose density gradient procedures will not be elaborated upon, as the present review primarily focuses on methods that yield mitochondria that meet *both* the criteria of *purity* and *metabolic integrity*.

2.1.2. Isolation of Nonsynaptic Mitochondria Using Ficoll Density Gradients

The development of density gradient material that can achieve high density without the concomitant high osmolarity represents a milestone in the compartively short (less than 50 yr) history of subcellular fractionation. Ficoll is a good example of such a density gradient material.

Although there were reports of isolation of nonsynaptic mitochondria with Ficoll gradients prior to 1970 [*see* Abood (1969), Moore and Strasberg (1970) and references cited therein], these early reported attempts at using Ficoll gradients suffered from several disadvantages. These included:

1. the need to use high Ficoll concentrations
2. the isolated mitochondria showed absolute requirement for bovine serum albumin (which apparently "preserves" the metabolic integrity of mitochondria during and subsequent to the isolation procedure) to retain metabolic integrity and
3. long isolation procedures.

Clark and Nicklas (1970) were the first to arrive at a procedure without these disadvantages. Subsequently, Lai and Clark (1976, 1979) devised procedures whereby metabolically active and relatively pure synaptic, as well as nonsynaptic, mitochondria could be routinely isolated from brain homogenates in yields large enough for metabolic studies.

2.1.2.1. The Clark and Nicklas (1970) Method

2.1.2.1.1. Reagents. *Ficoll 400* can be obtained from Pharmacia Ltd., Uppsala, Sweden, or Sigma Chemical Co., St. Louis, MO, USA. A 40% (w/v, i.e., 40 g in a final volume of 100 mL) solution of Ficoll in glass-distilled water is dialyzed against glass-distilled water (to remove low molecular weight impurities) for at least 4 h, with two changes of water. To accurately determine the concentration of Ficoll, the density of the dialyzed Ficoll is then measured using a specific gravity bottle. With the Pharmacia Chart, which plots the density of the Ficoll solution vs the Ficoll % (w/w), the Ficoll % (usually between 25–30) corresponding to the density of the dialyzed solution can be determined. The dialyzed Ficoll is then diluted to a concentration of 20% (w/w) and kept as a stock solution. Other Ficoll solutions are prepared by suitable dilutions of this 20% (w/w) stock solution (*see* later Sections).

Analytical grade chemicals should be used. All reagents-media are prepared with glass-distilled (preferably double-distilled or distilled and deionized) water.

Isolation medium. This contains 0.25*M* sucrose, 0.5 m*M* EDTA (K salt), and 10 m*M* Tris-HCl pH 7.4. In the original protocol, Tris was used: in the authors' experience, other buffers [e.g., 5 m*M* HEPES (*N*-2-hydroxyethylpiperazine-*N*'-2-ethanesulfonic acid) or MOPS (3-(*N*-morpholino)propanesulfonic acid)] can also substitute for Tris [*see* Lai et al. (1980 and 1985) and Lai and Blass (1984) for details]. However, the final medium *pH is critical* and should be kept between 7.3–7.5. Similarly, the final pH of the Ficoll media *(see below)* should also be kept within these limits, irrespective of the choice of the buffer.

6% (w/w) Ficoll medium. This consists of 6% (w/w) Ficoll [diluted from the 20% (w/w) stock], 0.24*M* mannitol, 60 m*M* sucrose, 0.05 m*M* EDTA (K salt), and 10 m*M* Tris-HCl, pH 7.4.

3% (w/w) Ficoll medium. This is the 6% (w/w) Ficoll medium diluted 1:1 with glass-distilled water and the final pH adjusted to 7.4 with buffer.

2.1.2.1.2. Procedure. The following steps are carried out at 4°C. It is important to follow these steps with minimum delays in order to obtain well-coupled brain mitochondria consistently.

Brain dissection. Eight young adult rats (usually male, preferably between 150 and 180 g in body weight) are used in each experiment. After decapitation, the top of the skull is removed to expose the brain. The brain is then transsected at the level of the inferior/superior colliculi. The part of the brain rostral to this transsection (except the olfactory bulbs) is removed and put into ice-cold isolation medium in a 50 mL beaker on ice.

Preparation of brain homogenates. The chilled brains, immersed in ice-cold isolation medium in a 50 mL beaker, are chopped finely into small cubes with scissors. The chopped tissue is allowed to sediment to the bottom of the beaker. The isolation medium is then carefully, but rapidly decanted. Fresh, ice-cold isolation medium is poured onto the chopped tissue in the beaker to wash away as much blood as possible. Again, the tissue cubes are allowed to sediment to the bottom of the beaker, and the isolation medium is decanted. This "washing" procedure is repeated until there is no visible sign of blood.

The washed tissue cubes are transferred to a glass Dounce homogenizer with a glass pestle (total clearance = 0.1 mm). The tissue is manually homogenized with 30 mL of ice-cold isolation

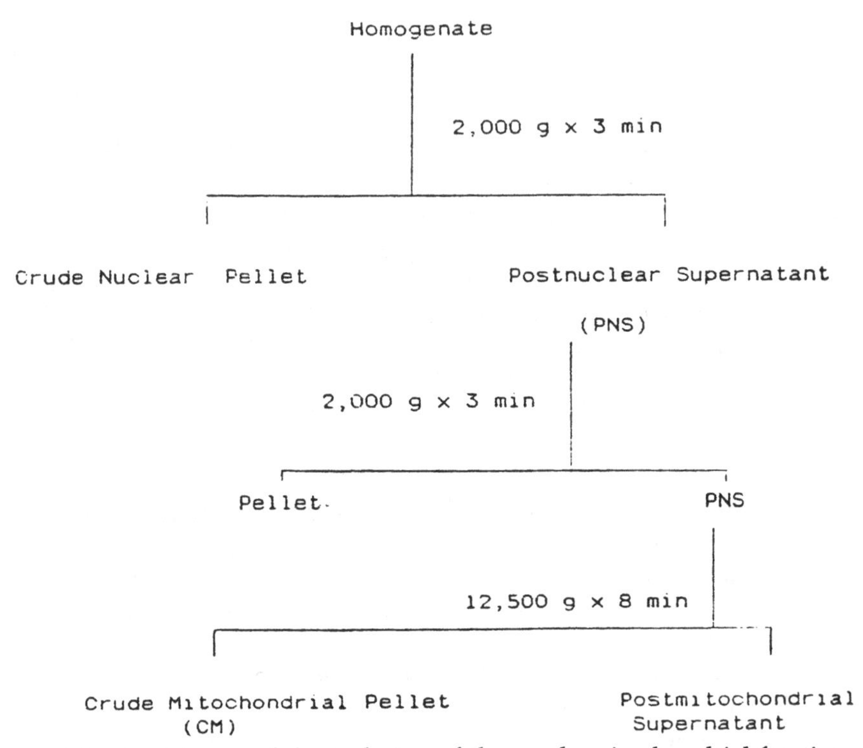

Fig. 1. Scheme of the isolation of the crude mitochondrial fraction from rat brain homogenates by the Clark and Nicklas method.

medium with 12 up-and-down strokes. This homogenate is diluted with cold isolation medium to a final volume of 60 mL.

Isolation of the crude mitochondrial fraction (CM) (Fig. 1). The homogenate is centrifuged (in two tubes) at 2000 g for 3 min. The supernatant (the postnuclear supernatant, PNS) is carefully decanted (making sure that the "opaque, white" crude nuclear pellet is not disturbed) and recentrifuged at 2000 g for 3 min (to ensure that there is no "carry-over" of the crude nuclear fraction from the previous step). The resulting supernatant (i.e., the PNS) is then centrifuged at 12,500*g* for 8 min. The supernatant (i.e., the postmitochondrial supernatant) is sharply but carefully decanted, leaving the dark brown crude mitochondrial pellet (i.e., the CM) undisturbed.

Purification of the crude mitochondrial fraction (CM) using a discontinuous Ficoll gradient. The CM pellet is resuspended with 12 mL of the 3% (w/w) Ficoll medium; 6 mL of the resulting suspension is then layered onto 25 mL of the 6% (w/w) Ficoll medium. The gradient is centrifuged at 11,500*g* for 30 min. After the gradient centrifugation, the supernatant is sharply decanted to remove the loose, fluffy, white top layer of the pellet. The remaining brown pellet is suspended with 5 mL of isolation medium; this suspension is centrifuged at 11,500*g* for 10 min. At the end of the centrifugation, the supernatant is again sharply decanted to get rid of the remaining, small amount of loose, white fluffy top layer of the pellet. The brown pellet is then resuspended with a small volume (usually 1 mL) of isolation medium and is designated the B fraction of nonsynaptic mitochondria. Aliquots of this preparation are then used for metabolic and enzymatic experiments.

2.1.2.2. THE LAI AND CLARK METHOD. Lai and Clark have designed two methods whereby metabolically active and relatively pure nonsynaptic mitochondria can be routinely isolated using discontinuous Ficoll gradients from forebrain homogenates in yields large enough for metabolic studies (Lai et al., 1975, 1977; Lai and Clark, 1976, 1979). Since method II (Lai et al., 1975, 1977; Lai and Clark, 1979) shows improvements (e.g., increased mitochondrial yield) over method I (Lai and Clark, 1976), only method II will be detailed below.

2.1.2.2.1. Reagents. *Ficoll-400* was obtained and dialyzed as described above (*see* Section 2.1.2.1.1).

Analytical grade chemicals should be used. All reagents/media are prepared with glass-distilled (preferably double-distilled or distilled and deionized) water.

Isolation medium contains: 0.32*M* sucrose, 1 m*M* EDTA (K salt), and 10 m*M* Tris-HCl, pH 7.4. [Alternatively, 5 m*M* HEPES-Tris (or HEPES-KOH) pH 7.4 can substitute for Tris-HCl. *Also see* Section 2.1.2.1.1 above.]

The *7.5% (w/w) and 10% (w/w) Ficoll media* contain 7.5% (w/w) or 10% (w/w) Ficoll (diluted from the 20% (w/w) stock solution; *see* Section 2.1.2.1.1 above) with 0.32*M* sucrose, 0.05 m*M* EDTA (K salt), and 10 m*M* Tris-HCl pH 7.4. [Alternatively, 5 m*M* HEPES-KOH (or -Tris), pH 7.4, may be used instead of Tris-HCl; *also see* Section 2.1.2.1.1 above.]

The *bovine plasma albumin (BPA) medium* consists of 1 mL of BPA (containing 10 mg fatty acid-free BPA/mL glass-distilled water) plus 19 mL of isolation medium.

2.1.2.2.2. Procedure. The following steps are carried out at 4°C. Again, it is important to go through the procedure with minimum delays in order to obtain well-coupled brain mitochondria consistently.

Brain dissection. Four adult rats (usually male, preferably between 150 and 180 g in body weight) are used in each experiment. The rest of the procedure is as detailed above (*see* Section 2.1.2.1.2).

Preparation of brain homogenates. This step is essentially the same as that detailed above in Section 2.1.2.1.2, except that the isolation medium is different from the one used in the Clark and Nicklas (1970) method. The washed, chopped brain tissue is manually homogenized (in a glass Dounce homogenizer with a glass pestle; total clearance = 0.1 mm) with 12 up-and-down strokes. The homogenate is then diluted with cold isolation medium to a final volume of 60 mL.

Isolation of the crude mitochondrial fraction (CM) by differential centrifugation (Fig. 2). The homogenate (in two tubes) is centrifuged at 1300 g for 3 min, and then the bulk of the supernatant is carefully decanted, leaving behind approximately 5 mL (in each tube) of the supernatant, which covers the crude nuclear (CN) pellet. The CN pellets are gently loosened with the remaining supernatant using a Teflon rod and poured back into the all-glass Dounce homogenizer; 10 mL of cold isolation medium is pipetted into the homogenizer. The CN pellets are then manually homogenized with six up-and-down strokes. The homogenate is centrifuged at 1300 g for 3 min. After centrifugation, the supernatant is carefully decanted (making sure that the "off-white" nuclear pellet remains undisturbed), and combined with the supernatant from the first centrifugation step (Fig. 2). The combined postnuclear supernatants are then mixed (by gently stirring with a Teflon rod) and centrifuged (in two tubes) at 17,000*g* for 10 min to obtain the crude mitochondrial pellet (CM) (Fig. 2).

Purification of the crude mitochondrial fraction (CM) using a discontinuous Ficoll gradient (Fig. 2). The CM pellets are resuspended with a total of 15 mL of cold isolation medium by manually homogenizing in a homogenizer with a loose pestle (total clearance approximately 0.25 mm) with four up-and-down strokes. A third of this suspension is layered onto 7 mL of the 7.5% (w/w) Ficoll medium on top of 7 mL of 10% (w/w) Ficoll medium (in each of three centrifuge tubes) and centrifuged at 99,000*g* for 30 min in a 3 × 23 mL swing-out rotor in a MSE 65 ultracentrifuge. [Note that the 7.5% (w/w)/10% (w/w) Ficoll gradient is usually prepared sometime (10–20 min) prior to this centrifugation step by carefully (usually slowly) layering 7 mL of the 7.5% (w/w) Ficoll medium onto 7 mL of the 10% (w/w) Ficoll medium in a centrifuge tube cooled on ice. This is an important step: it requires some practice to

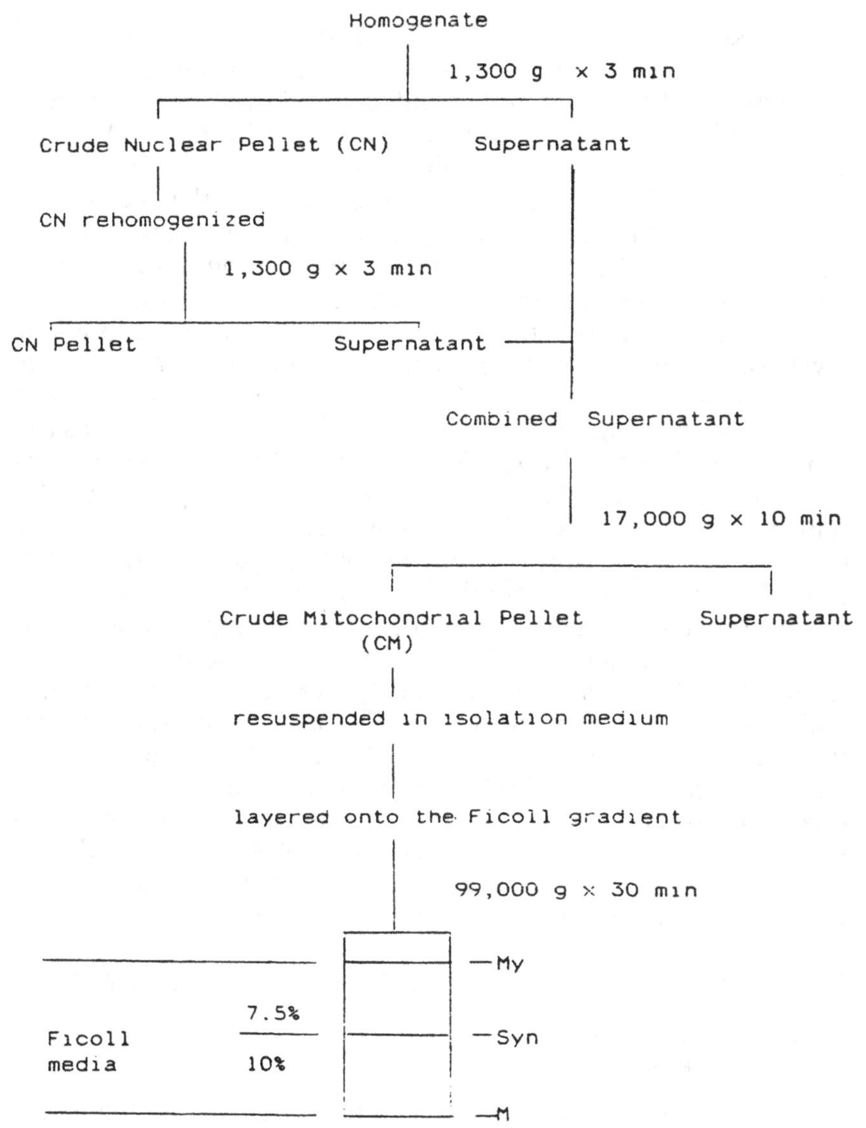

Fig. 2. Scheme of the isolation of nonsynaptic mitochondria from rat brain homogenates by the Lai and Clark method.

obtain a discontinuous density gradient with "sharp" interphases.] At the end of the centrifugation, the myelin fraction (My) bands at the interphase between the isolation medium and the 7.5% (w/w) Ficoll medium. The bulk of the synaptosomal fraction (Syn) bands at the interphase between the 7.5% (w/w) and the 10% (w/w)

Ficoll media, although some of the "heavier" synaptosomes have sedimented through to the 10% (w/w) Ficoll layer, but remained above the clearly defined, brown mitochondrial pellet (M). There is usually only a very small, white, fluffy layer (~1 mm thick) on top of the brown mitochondrial pellet. (If the white, fluffy layer is much thicker than 2 mm or if the mitochondrial pellet is not uniformly brown, the mitochondrial preparation should be discarded, because these are the indications that the mitochondrial fraction is heavily contaminated with synaptosomes and other membranous material, probably as a result of density gradient inversion.) The myelin (My) fraction is removed with a Pasteur pipet until almost all of the 7.5% (w/w) Ficoll medium is removed. Then, the remaining 10% (w/w) Ficoll medium (which contains the synaptosomal fraction) is quickly dumped into another clean centrifuge tube. Five mL of isolation medium is gently pipeted into the centrifuge tube, such that the isolation medium is allowed to flow from the side of the tube onto the top of the mitochondrial pellet. The centrifuge tube is gently rotated such that the isolation medium dislodges the white, fluffy layer on top of the brown mitochondrial pellet. The isolation medium, together with the white, fluffy layer, is then decanted. (After this step, if the fluffy, white layer on top of the brown mitochondrial pellet is still visible, the "rinsing" procedure is repeated.) The brown mitochondrial pellet is suspended with 5 mL of the BPA medium. The suspension is centrifuged at 9800*g* for 10 min. After centrifugation, the supernatant is sharply decanted. The brown pellet is resuspended with isolation medium, usually to a final volume of 1 mL, and is designated the A fraction of nonsynaptic (i.e., "free") mitochondria.

2.2. Isolation of Synaptic Mitochondria from Whole Brain

2.2.1. The Nonsynaptosomal Contaminants in Synaptosomal Preparations

The success of the methodology for the isolation of synaptic mitochondria depends on an initial isolation of a relatively pure preparation of synaptosomes. However, a 100% pure synaptosomal preparation is impossible to obtain in practice. Thus, it is important to realize that maximum purity of the synaptosomal preparation isolated using the state-of-the-art techniques rarely, if ever, exceeds 80%. Moreover, within the context of the isolation of intraterminal mitochondria, it is also important to determine the sources that may contribute nonsynaptosomal mitochondria to the isolated fraction(s) that actually contain the intraterminal mitochondria. At least three sources are known:

1. "vesicular bodies" derived from pinched-off dendrites
2. "gliosomes" and
3. "free" mitochondria that cosediment with synaptosomes in density gradients.

In addition, microsomes and membrane fragments of unknown origin(s) may appear as contaminants. [The extent of the contamination with these four kinds of organelles critically depends on the density gradient(s) and the centrifugation procedure(s) employed.]

Little information is available regarding the contamination of the synaptosomal fraction with "vesicular bodies" derived from "pinched-off" (during the initial homogenization process) dendrites. Systematic quantitative electron-microscopic investigation of the synaptosomal fraction isolated using different density gradients and centrifugation conditions may elucidate the degree of this contamination. (These studies have yet to be carried out.) However, one positive aspect of this contamination is that at least such "vesicular bodies" derived from dendrites ultimately yield mitochondria of *neuronal* origin.

The contamination of the synaptosomal fraction with "gliosomes" present a more serious problem, since "gliosomes" will lead to the isolation of *glial* rather than *neuronal* mitochondria. Although workers in this field are generally aware of this problem, they have not arrived at a systematic quantitative assessment of the "gliosomal" contamination of the synaptosomal fraction. However, there is some indication that, by manipulating the density gradient(s) employed for the purification of the synaptosomal fraction, the contamination of the synaptosomal fraction with "gliosomes" can be minimized [*see* Dennis et al. (1980) for a discussion and references]. Obviously, other studies are needed to resolve the problem of "gliosomal" contamination.

Significant presence (>10% of the total synaptosomal protein) of "free" mitochondria in the synaptosomal fraction does complicate the interpretation of the results of studies on the mitochondria isolated from the synaptosomal fraction [*see* Lai et al. (1975) for a discussion]. Some attempts have been made to address this problem by using marker enzyme assays (Lai et al., 1975, 1977). By this approach, the contamination with "free" mitochondria in the synaptosomal fraction isolated by the Lai and Clark method (Lai et al., 1975, 1977; Lai and Clark, 1979) or by methods modified therefrom (Booth and Clark, 1978a; Lai et al., 1978, 1980) has been estimated to be between 3 and 7% on a protein basis.

The contamination of synaptosomal preparations with microsomes and membrane fragments of unknown origin(s) can be minimized by (1) suitable design(s) of the differential centrifugation steps (*see* Fig. 2) such that a minimum amount of these contaminants cosediments with the crude mitochondrial fraction, and (2) the separation of microsomes and other membrane fragments from synaptosomes using a density gradient. (Subsequent to the osmotic lysis of the synaptosomal fraction, the microsomes and other membrane fragments originally present in the synaptosomal fraction are well separated from intraterminal mitochondria during the differential and density gradient centrifugation. *See* Section 2.2.4 below and Fig. 3.)

2.2.2. Disruption of the Synaptosomal Plasma Membrane to Release Intraterminal Organelles and Macromolecules

In order to isolate intraterminal organelles (e.g., mitochondria, synaptic vesicles) and synaptic plasma membrane from synaptosomes, the synaptosomal plasma membrane has to be disrupted under reasonably controlled conditions. Classically, this kind of membrane disruption can be achieved with physical [e.g., ultrasonication, freezing and thawing, forcing through a narrow slit (using a Hughes press) or orifice (using a French press)], physicochemical [e.g., osmotic lysis, shrinking-swelling cycle(s)] and chemical [treatment with detergent(s) and other membrane-disruptive agent(s) (e.g., digitonin)] methods. [*See* Hughes et al. (1971) and Lai (1975) for detailed discussions of the various methods.] Irrespective of the choice of the disruptive method, the primary concern is that the synaptosomal plasma membrane be disrupted in such a way that the intraterminal organelles and macromolecules are released from within the confines of the synaptosomal plasma membrane without structural or functional damage.

Although theoretically any one of the three kinds of methods (namely physical, physicochemical, and chemical methods) may be feasible for the membrane disruption, in practice, very few of the various disruptive methods have been systematically and critically evaluated [*see* Lai (1975) and Whittaker (1984) for a discussion]. Even in the case of osmotic lysis, the most popular method

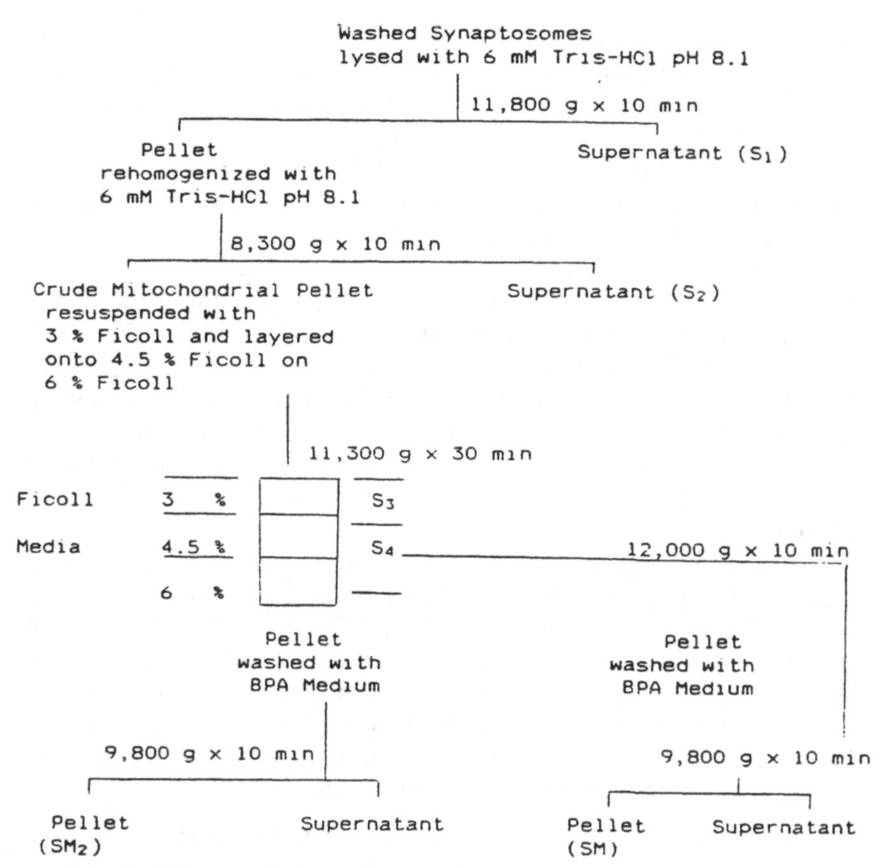

Fig. 3. Scheme of the isolation of synaptic mitochondria (fractions SM and SM_2) from rat forebrain homogenates by the Lai and Clark method.

amongst investigators, the disruptive efficiency and the underlying technical and mechanistic details have not been extensively investigated.

One important factor known to affect the efficiency of the osmotic lysis process is pH. Osmotic lysis using distilled water (pH < 6.5) alone does not completely disrupt the synaptosomal plasma membrane [J. C. K. Lai and J. B. Clark, unpublished data; *also see* Whittaker (1984) for an additional discussion]. On the other hand, Cotman and Matthews (1971) reported that an alkaline hypotonic buffer is much better than distilled water in attaining complete, or near complete, osmotic lysis of the synaptosomal plasma membrane. Their approach led to the successful isolation of relatively pure preparations of synaptic plasma membranes (Cotman and Matthews, 1971). The approach of Cotman and Matthews (1971) was therefore adapted (Lai, 1975; Lai and Clark, 1976; Lai et al., 1975, 1977) for the "controlled" osmotic lysis of synaptosomes, such that relatively pure and metabolically active synaptic mitochondria can be routinely isolated in quantities adequate for metabolic studies (*see* Section 2.2.4 below).

Alternative methods have also been examined for the "controlled" disruption of the synaptosomal plasma membrane. For example, preliminary studies using ultrasonication at high frequency (20 MHz) suggest that ultrasonication is a potentially useful method, although more systematic studies on parameters, such as the relationships between the number of cavitation events and the amplitude and/or the irradiation time (or pulse), will be helpful in designing optimum conditions for sonication [*see* Lai (1975) for a discussion].

Another approach that showed some promise even in initial pilot studies [*see* Lai (1975) for a discussion and experimental details] is the use of digitonin. Booth and Clark (1979) subsequently extended the studies of Lai (1975) and arrived at a procedure whereby the synaptosomal plasma membrane could be disrupted with digitonin under "controlled" conditions. Independently, Nicholls (1978) also devised a method for disrupting the synaptosomal plasma membrane using digitonin.

2.2.3. Isolation of Synaptic Mitochondria Using Sucrose Density Gradient(s)

The early and pioneering work from the laboratories of Whittaker (Whittaker, 1969, 1984) and De Robertis (De Robertis and de Lores Arnaiz, 1969) established the possibility that, upon osmotic lysis of the synaptosomal plasma membrane, relatively pure preparations of synaptosomally derived mitochondria can be isolated using discontinuous sucrose density gradients in swing-out rotors. Subsequently, other workers have also shown that synaptosomally derived mitochondria can be purified and subfractionated on discontinuous sucrose gradients in zonal centrifugation (Lai, 1975; Lai and Clark, 1976; J. C. K. Lai and J. B. Clark, unpublished observations). It is important to note that, although sucrose density gradients used in conjunction with zonal rotors have been powerful tools for elucidating the heterogeneity of brain mitochondria in regard to enzyme contents [*see* Lai (1975),

Reijnierse et al. (1975a,b), and Lai and Clark (1976) for discussions], very hypertonic (>1.0*M*) sucrose gradients are required in these procedures. However, brain mitochondria isolated using similarly hypertonic sucrose gradients are metabolically inactive and poorly coupled (Lai and Clark, 1976), rendering such sucrose density gradient procedures unsuitable for the isolation of metabolically competent mitochondria from lysed synaptosomes. Nonetheless, these procedures may be used to isolate synaptic mitochondria and plasma membranes for the enyzmatic and chemical studies in which the metabolic integrity of the isolated fractions is not an absolute requirement [for procedural details, *see* Whittaker (1969 and 1984), De Robertis and de Lores Arnaiz (1969) and references cited therein].

2.2.4. Isolation of Synaptic Mitochondria Using Ficoll Density Gradient

Since hypertonic sucrose density gradients are unsuitable for routine isolation of metabolically active synaptic mitochondria, alternative methods are needed. To this end, Lai and Clark (1976) successfully adapted the Clark and Nicklas (1970) discontinuous Ficoll gradient for isolating a metabolically active and relatively pure population of synaptic mitochondria. Since synaptosomally derived mitochondria are heterogeneous [*see* Lai (1975) and Lai and Clark (1976) for discussions], Lai and Clark (Lai, 1975; Lai et al., 1975, 1977; Lai and Clark, 1979) designed another discontinuous Ficoll gradient (Fig. 3), whereby two metabolically active and relative pure populations of synaptosomally derived mitochondria (designated SM and SM_2) can be routinely isolated in quantities large enough for metabolic studies.

2.2.4.1. Reagents. *Ficoll-400* was obtained and dialyzed as described above (*see* Section 2.1.2.1.1).

Analytical grade chemicals should be used. All reagents/media are prepared with glass-distilled (preferably double-distilled or deionized and distilled) water.

Isolation medium contains: 0.32*M* sucrose, 1 m*M* EDTA (K salt), and 10 m*M* Tris-HCl, pH 7.4. [Alternatively, 5 m*M* HEPES-Tris (or -KOH) pH 7.4 may be used as the buffer. *Also see* Section 2.1.2.1.1 above.]

7.5% (w/w) and 10% (w/w) Ficoll media contain 7.5% (w/w) or 10% (w/w) Ficoll (diluted from the 20% (w/w) stock solution; *see* Section 2.1.2.1.1 above), with 0.32*M* sucrose, 0.05 m*M* EDTA (K salt), and 10 m*M* Tris-HCl, pH 7.4. [5 m*M* HEPES-Tris (or -KOH) pH 7.4 can substitute for Tris-HCl as the buffer; *also see* Section 2.1.2.1.1 above.]

4.5% (w/w) and 6% (w/w) Ficoll media contain, in addition to Ficoll, 0.24*M* mannitol, 60 m*M* sucrose, 0.05 m*M* EDTA (K salt), and 10 m*M* Tris-HCl pH 7.4. (Again, other buffers can substitute for Tris; *see* Section 2.1.2.1.1 above.)

3% (w/w) Ficoll medium is the 6% (w/w) Ficoll medium diluted 1:1 with glass-distilled water and pH adjusted to 7.4 with buffer.

Bovine plasma albumin (BPA) medium consists of 1 mL of BPA (containing 10 mg fatty acid-free BPA/mL of glass-distilled water) and 19 mL of isolation medium.

2.2.4.1. PROCEDURE. The following steps are carried out at 4°C. (*Also see* Section 2.1.2.1.2 above.)

Brain dissection. In each experiment, four young adult rats (usually male, preferably between 150–180 g in body weight) are employed. The dissection procedure is as detailed above (*see* Section 2.1.2.1.2).

Preparation of brain homogenates. The procedure is the same as that detailed above in Section 2.1.2.2.2.

Isolation of the crude mitochondrial fraction (CM) by differential centrifugation. The procedure is the same as that described in Section 2.1.2.2.2 above (Fig. 2).

Isolation of the synaptosomal fraction (Syn) by a discontinuous Ficoll gradient (Fig. 2). The fractionation procedure is identical to that shown in Fig. 2. After the Ficoll gradient centrifugation, the Syn fraction is removed as described in Section 2.1.2.2.2 above, and washed by diluting with three volumes of isolation medium (usually about 50 mL) and centrifuging at 18,500*g* for 20 min. After this centrifugation, the pellet is designated the washed Syn fraction.

Alternatively, the Syn fraction can be isolated from the CM by the "floatation" centrifugation technique (Booth and Clark, 1978a; Lai et al., 1978, 1980). The "floatation" technique yields slightly more pure Syn fraction in that the Syn fraction so isolated contains slightly fewer "free" mitochondria as a contaminant. However, the overall yield of synaptosomes by the "floatation" technique is lower than that achieved through the more conventional centrifugation (Fig. 2). This decrease in yield is because some of the more dense synaptosomes sediment onto the "free" mitochondrial pellet during the "floatation" procedure (J. C. K. Lai, unpublished observations). Consequently, the nonsynaptic mitochondrial fraction isolated by the "floatation" method may be slightly more contaminated with synaptosomes.

Osmotic lysis of the synaptosomal fraction (Syn) using hypotonic alkaline buffer. The washed synaptosomes (Syn) are gently and manually homogenized in an all-glass Dounce homogenizer (total clearance = 0.25 mm) with 10 mL of 6 m*M* Tris-HCl (or Tris-HEPES), pH 8.1, with four up-and-down strokes. Then another 20 mL of 6 m*M* Tris-HCl, pH 8.1, is added and mixed. The lysed synaptosomes are immediately centrifuged at 11,800*g* for 10 min (Fig. 3). The supernatant (S_1) is carefully decanted without disturbing the pellet. The pellet is manually rehomogenized in 10 mL of 6 m*M* Tris-HCl, pH 8.1, using the all-glass Dounce homogenizer (total clearance = 0.25 mm) with four up-and-down strokes. The resultant suspension is centrifuged at 8300*g* for 10 min (Fig. 3). The supernatant (S_2) is then sharply decanted.

Isolation of two populations of synaptic mitochondria (fractions SM and SM_2) from the lysed synaptosomal fraction using a discontinuous Ficoll gradient. The pellet from the previous centrifugation step (Fig. 3) is suspended with 10 mL of 3% (w/w) Ficoll medium by sucking up and down four times in a 10 mL pipet. A third of the suspension is layered onto 5 mL of 4.5% (w/w) Ficoll medium on 10 mL of 6% (w/w) Ficoll medium and centrifuged at 11,300*g* for 30 min in a 3 × 23 mL swing-out rotor in a MSE 65 ultracentrifuge (Fig. 3). After this centrifugation step, the S_3 fraction is sucked out with a Pasteur pipet. The S_4 fraction is decanted into another centrifuge tube, diluted with an equal volume of isolation medium, and centrifuged at 12,000*g* for 10 min (Fig. 3). The pellet is washed with 5 mL of BPA medium by gently and manually homogenizing in an all-glass Dounce homogenizer (total clearance = 0.25 mm) with four up-and-down strokes. The suspension is centrifuged at 9800*g* for 10 min. After this centrifugation step, the supernatant is sharply decanted; the pellet is resuspended in a small amount of isolation medium, usually to a final volume of 1 mL. This fraction is designated the SM population of synaptic mitochondria (Fig. 3).

The pellet below the 6% (w/w) Ficoll (Fig. 3) is similarly washed with 5 mL of BPA medium and centrifuged at 9800*g* for 10 min. After this centrifugation, the supernatant is sharply decanted. The pellet is resuspended with a small amount of isolation medium, usually to a final volume of 1 mL. The fraction is designated the SM_2 population of synaptic mitochondria (Fig. 3).

2.3. Isolation of Synaptic and Nonsynaptic Mitochondria from Discrete Brain Regions

The Ficoll density gradient method (Figs. 2 and 3) of Lai and Clark (1979) has been adapted for the isolation of synaptic and nonsynaptic mitochondria from several rat brain regions (Lai et al., 1981; Dagani et al., 1983; Leong et al., 1984; J. C. K. Lai, unpublished observations). More recently, Lai and coworkers have demonstrated that metabolically active and relatively pure nonsynaptic mitochondria can be routinely isolated from both large (e.g., cerebral cortex) and small (e.g., olfactory bulbs) brain regions (Lai et al., 1985; Sheu et al., 1985a; Lai and Sheu, 1985).

2.4. Isolation of Mitochondria from the Developing and Aging Brain

Land and Clark (Land and Clark, 1975; Clark and Land, 1975; Land et al., 1976) have demonstrated that metabolically competent and relatively pure nonsynaptic mitochondria can be routinely isolated from the developing postnatal rat brain. Other workers have also isolated nonsynaptic mitochondria from the neonatal rat brain using Ficoll [*see* Holtzman and Moore (1973), O'Neill and Holtzman (1985), and references cited therein] or sucrose (Dienel et al., 1977) gradient procedure.

More recently, Patel and coworkers (Deshmukh et al., 1980; Deshmukh and Patel, 1982) have shown that metabolically active synaptic and nonsynaptic mitochondria can be isolated from the forebrains of old rats using the earlier (Lai and Clark, 1976) of the two Ficoll density gradient methods of Lai and Clark.

3. Characterization of Preparations of Synaptic and Nonsynaptic Brain Mitochondria

3.1. The Purity of Various Preparations

The purity of various mitochondrial preparations should ideally be assessed both by the appropriate marker enzyme assays and quantitative electron-microscopy. (However, it is technically difficult and exceedingly time-consuming to do quantitative electron-microscopy: usually, the compromise of a semi-quantitative approach is adopted.)

In practice, the assessment of the purity of different mitochondrial preparations isolated by "standard" method(s) is rarely reported. Perhaps workers deem it unnecessary to undertake such "quality control" studies, since "standard" method(s) is/are employed. However, in the authors' experience, these "quality controls" are imperative for comparative purposes. Moreover, "standard" methods are often modified to suit the equipment (e.g., centrifuges and rotors, gradient generator) available in a particular researcher's laboratory. Consequently, the lack of such "quality control" data makes it difficult to compare results from various laboratories, even though the same "standard" method(s) has/have been employed. [For detailed discussions of the assessment of the purity of mitochondrial preparations, consult Clark and Nicklas (1970 and 1984), Lai et al. (1975, 1977, 1985), Lai and Clark (1976, 1979), and references cited therein.] Nonetheless, it should be pointed out that this kind of "quality control" study only provides information regarding nonmitochondrial contaminants, but sheds little light on the question of the heterogeneity of brain mitochondria (*see* Sections 3.5 and 3.6 below).

Judging from results from studies using electron-microscopy (EM) and marker enzyme assays (Clark and Nicklas, 1970; Lai and Clark, 1976, 1979; Lai et al., 1975, 1977), nonsynaptic and synaptic mitochondria isolated from forebrain homogenates using Ficoll density gradient procedures are relatively free from cytosolic and nonmitochondrial membranous contamination. For example, in these mitochondrial preparations, the lactate dehydrogenase (a marker for cytosolic and synaptoplasmic material) and acetylcholinesterase (a general marker for synaptic and other plasma membranes) levels are usually less than 0.5% and 0.8%, respectively, of the homogenate values (Table 1). In addition, the EM studies indicate that 90–95% of the subcellular particles present in these mitochondrial preparations can be identified as mitochondria [*see* Lai and Clark (1979) and Clark and Nicklas (1984) for detailed discussions].

3.2. The Structural and Metabolic Integrity of Different Mitochondrial Preparations

Usually, EM studies are most revealing regarding the structural integrity of isolated mitochondria. However, it is important to complement the structural studies using EM with biochemical assessment of metabolic integrity, because it is possible that a

Table 1
Recoveries of Activities of Lactate Dehydrogenase and Acetylcholinesterase and Protein in Synaptic and Nonsynaptic Mitochondria Derived from Rat Brain[a]

	Protein		LDH		AChE	
Fraction	Total (mg)	%	Sp. act.	%	Sp. act.	%
Homogenates	957* or 1888**	100	681	100	60	100
Synaptic mitochondria						
SM	15	1.6	92	0.2	26	0.7
SM_2	23	2.4	59	0.2	15	0.6
Nonsynaptic mitochondria						
A	25	2.7	105	0.4	13	0.5
B	19	1.0	79	0.1	25	0.4

[a]Values are Means of three or more experiments and are from Lai and Clark (1976) and Lai et al. (1977). Nonsynaptic mitochondria (fraction B) were isolated by the Clark and Nicklas (1970)** method (*see* Section 2.1.2.1); the other mitochondria were isolated by the Lai and Clark method* (*see* Sections 2.1.2.2 and 2.2.4). The abbreviations are: LDH, lactate dehydrogenase; AChE, acetylcholinesterase; Sp.act., specific activity (in mU/mg protein); mU, milliunit (nmol/min) of enzymatic activity.

preparation of mitochondria may *appear* to be *structurally intact* as determined by EM and, yet, turn out to be *metabolically inactive* and *uncoupled*.

The synaptic and nonsynaptic mitochondria isolated by the Ficoll gradient procedures described above are usually structurally intact, with well-defined outer and inner membranes (Clark and Nicklas, 1970; Lai and Clark, 1976; Lai et al., 1977). The overall appearances in cross-section of the isolated nonsynaptic mitochondria are usually slightly better than those of isolated synaptic mitochondria. In general, within a preparation, there is some variability in the mitochondrial sizes, shapes, and cristal structures (e.g., condensed or otherwise) (Clark and Nicklas, 1970; Lai and Clark, 1976; Lai et al., 1977; J. C. K. Lai and J. B. Clark, unpublished data).

It is generally accepted that the measurement of the respiratory control ratios (RCRs: the ratio of state 3 respiratory rate to state 4 respiratory rate; *see* Chance and Williams, 1956) in a mitochondrial preparation gives a good indication as to how metabolically intact the preparation is [*see* Tzagoloff (1982) for a discussion]. Alternatively, the P/O ratios [see Tzagoloff (1982) for a discussion] can be used as an index of metabolic integrity, although even mitochondrial preparations that exhibit almost ideal P/O ratios may show poor RCRs.

As shown in Table 2, the various populations of synaptic and nonsynaptic mitochondria isolated by the Ficoll gradient procedures are metabolically active and well-coupled. For example, with pyruvate and malate as substrates, the RCRs in the different preparations are routinely higher than 5.

3.3. Respiratory Properties of Isolated Mitochondria

One property that is common to all kinds of brain mitochondria so far studied is the respiratory response to raised potassium ions [*see* Clark and Nicklas (1970), Nicklas et al. (1971), Lai and Clark (1976, 1979) and Clark and Nicklas (1984) for detailed discussions]. For example, when the potassium ion concentration in the respiratory medium is increased from 5 to 100 m*M* (Table 2), the state 3 respiratory rates are increased to a greater (200%) or lesser (20%) extent depending on the substrate(s) employed. In respect to potassium stimulation of substrate-supported oxygen uptake, pyruvate (in the presence of malate) is consistently the best substrate, whereas 3-hydroxybutyrate is the poorest (Table 2).

Concerning the potassium stimulation of pyruvate-supported oxygen uptake by brain mitochondria, Nicklas et al. (1971) proposed that the most likely underlying mechanism is a direct stimulation by potassium ions of the pyruvate dehydrogenase complex (PDHC). Recently, using cerebro-cortical nonsynaptic mitochondria, Lai and Sheu (1985) have obtained direct evidence that elevated levels of potassium ions *do raise* the *activation state* of the pyruvate dehydrogenase complex (PDHC).

In general, pyruvate (in the presence of malate) is the best respiratory substrate. Succinate is almost as good a substrate as pyruvate, although the RCRs using succinate are usually quite low (Table 2). In the presence of malate, 2-oxoglutarate, citrate, and

Table 2
Substrate-Dependent Oxygen Uptake by Synaptic and Nonsynaptic Mitochondria from Rat Brain[a]

[S]	$[K^+]$,mM	O_2 Uptake rate, natom/min/mg protein							
		Synaptic mitochondria				Nonsynaptic mitochondria			
		SM		SM_2		A		B	
		State 3	RCR	State 3	RCR	State 3	RCR	State 3	RCR
5 m*M* Pyr	5	73 ± 9	8.6	55 ± 15	6.7	66 ± 15	5.8	93 ± 10	6.7
	100	161 ± 6	12.8	135 ± 15	11.6	158 ± 15	5.4	183 ± 13	8.0
10 m*M* Glut	5	54 ± 12	5.8	54 ± 6	9.3	40 ±	14.5	97 ± 8	4.8
	100	89 ± 15	7.7	67 ± 11	5.2	107 ± 20	5.3	158 ± 11	4.1
10 m*M* Glutamine	5	—	—	—	—	—	—	47 ± 3	2.6
	100	—	—	43 ± 3	—	—	—	55 ± 2	2.4
5 m*M* DL-BHB	5	57 ± 5	2.5	46 ± 1	2.5	30 ± 3	3.0	30 ± 6	3.0
	100	66 ± 1	1.7	55 ± 1	2.0	53 ± 5	2.5	60 ± 7	3.0
5 m*M* Cit	5	59 ± 1	3.7	52 ± 1	3.6	54 ± 1	10.0	66 ± 5	3.2
	100	131 ± 1	2.4	105 ± 1	2.7	60 ± 12	11.0	92 ± 9	2.0
5 m*M* KG	5	46 ± 4	4.0	52 ± 9	2.7	66 ± 6	9.4	46 ± 5	3.0
	100	135 ± 8	3.8	113 ± 17	3.9	135 ± 9	5.3	95 ± 10	2.8
10 m*M* Succ	5	106 ± 2	2.7	80 ± 8	2.3	93 ± 3	1.8	122 ± 10	3.3
	100	135 ± 2	2.9	140 ± 17	2.6	138 ± 8	2.7	150 ± 10	2.0

[a]Results are Means ± SD of three or more experiments, carried out at 28°C and in the presence of 2.5 m*M* malate. Synaptic mitochondria and the A preparation of nonsynaptic mitochondria were prepared by the method of Lai and Clark (*see* Sections 2.1.2.2 and 2.2.4). The B preparation of nonsynaptic mitochondria was isolated by the Clark and Nicklas (1970) method (*see* Section 2.1.2.1). State 3 is the substrate-dependent oxygen uptake rate in the presence of ADP, and the respiratory control ratio (RCR) is defined as the substrate-dependent oxygen uptake rate in the presence of ADP divided by the substrate-dependent oxygen uptake rate in the absence of ADP. The data are compiled from those of Lai et al. (1977), Land (1974), Dennis et al. (1977), and from unpublished data from the authors' laboratories. The abbreviations are: [S], substrate concentration; $[K^+]$, potassium concentration in the medium; Pyr, pyruvate; Glut, glutamate; DL-BHB, DL-3-hydroxybutyrate; Cit, citrate; KG, 2-oxoglutarate; Succ, succinate.

glutamate are all moderately good respiratory substrates for brain mitochondria (Table 2). Other substrates, including glutamine, GABA, and 3-hydroxybutyrate, are also oxidized, but at much lower rates. [*See* Clark and Nicklas (1970) and Lai and Clark (1976 and 1979) for detailed discussions concerning the oxidation of these and other substrates.]

The rate of pyruvate-supported oxygen uptake by the SM_2 fraction of synaptic mitochondria is consistently and significantly lower (by approximately 30%) than the corresponding rates by other synaptic or nonsynaptic mitochondria (Table 2). The observation correlates quite closely with the distribution of the activities of pyruvate dehydrogenase complex (PDHC), being also lowest in the SM_2 fraction of synaptic mitochondria (Table 4).

The rates of glutamate-supported oxygen uptake by both populations (SM and SM_2) of synaptic mitochondria are lower than the corresponding rates in the two populations of nonsynaptic mitochondria: the contrast in the rates between the SM_2 fraction of synaptic mitochondria (the lowest) and the B fraction of nonsynaptic mitochondria (the highest) are especially noticeable (Table 2). The differences in the glutamate-supported oxygen uptake rates between the synaptic and nonsynaptic mitochondria correlate quite well with the differences in the activities of aspartate aminotransferase (AAT) in synaptic (lower AAT activity) and nonsynaptic (higher AAT activity) mitochondria. The good correlation between the glutamate-supported oxygen uptake rates (Table 2) and the AAT activities (Table 4) in the various mitochondrial preparations is consistent with the earlier proposal that conversion of exogeneously supplied glutamate to 2-oxoglutarate by synaptic as well as nonsynaptic mitochondria are mainly through the AAT-catalyzed reaction rather than the glutamate dehydrogenase (GDH)-catalyzed reaction (Dennis et al., 1977).

3.4. The Cytochromes in Synaptic and Nonsynaptic Mitochondria

The various cytochromes normally found in mammalian mitochondria can also be found in the different populations of synaptic and nonsynaptic mitochondria (Table 3). The concentrations of cytochrome c and b are, respectively, the highest and the lowest among the various cytochromes in all types of brain mitochondria (Table 3). In general, the levels (on a protein basis) of all the cytochromes in nonsynaptic mitochondria are higher than

Table 3
Contents of Cytochromes in Synaptic and Nonsynaptic Rat Brain Mitochondria[a]

Cytochrome content, nmol/mg protein	Synaptic mitochondria		Nonsynaptic mitochondria	
	SM	SM_2	A	B
b	0.14	0.13	0.19	0.20
c_1	0.27	0.23	0.31	0.31
c	0.45	0.46	0.54	0.51
a	0.28	0.27	0.33	0.36
a_3	0.20	0.21	0.24	0.23

[a]Values are Means of at least two separate determinations. The SM and SM_2 preparations of synaptic mitochondria and the A preparation of nonsynaptic mitochondria were isolated by the method of Lai and Clark (*see* Sections 2.1.2.2 and 2.2.4); the cytochrome contents of these mitochondrial preparations are from Dennis (1976). The B preparation on nonsynaptic mitochondria were isolated by the Clark and Nicklas (1970) method (*see* Section 2.1.2.1), and the cytochrome contents of this preparation are from Booth (1978).

those in synaptic mitochondria. However, further studies are required to establish the significance of this trend.

3.5. Enzymatic Properties of Different Mitochondrial Populations from Whole Brain and from Discrete Regions

3.5.1. Enzymatic Properties of Mitochondrial Populations from Whole Brain

The specific activities of citrate synthase (Cit Syn), 2-oxoglutarate dehydrogenase complex (KGDHC), fumarase (Fum), aspartate aminotransferase (AAT), creatine kinase (CK), and hexokinase (HK) are generally higher in nonsynaptic mitochondria than in synaptic mitochondria (Table 4). By contrast, the specific activities of 3-hydroxybutyrate dehydrogenase (BOBDH) are generally higher in synaptic mitochondria than in nonsynaptic mitochondria (Table 4).

The specific activities of citrate synthase (Cit Syn), NADP-isocitrate dehydrogenase (NAD-ICDH), succinic semialdehyde dehydrogenase (SSADH), acetoacetyl-CoA thiolase (A-CoA Th), 3-hydroxy-3-methylglutaryl-CoA synthase (HMG-CoA Syn), and acetyl-CoA synthase (A-CoA Syn) in the lighter fraction (SM) of synaptic mitochondria are lower than the corresponding specific activities of these enzymes in the other three mitochondrial populations (Table 4). On the other hand, the specific activities of 3-oxo-acid:CoA transferase (OCoA Tr) in this mitochondrial fraction (SM) are higher than those in the other 3 mitochondrial populations (Table 4).

In the heavier fraction (SM_2) of synaptic mitochondria, the specific activities of the following enzymes are lower than the corresponding specific activities in the other three mitochondrial populations: pyruvate dehydrogenase complex (PDHC), NAD-isocitrate dehydrogenase (NAD-ICDH), 2-oxoglutarate dehydrogenase complex (KGDHC), fumarase (Fum), and aspartate aminotransferase (AAT) (Table 4). On the other hand, the specific activities of the following enzymes are higher in this fraction (SM_2) than those in other mitochondrial populations: NADP-isocitrate dehydrogenase (NADP-ICDH), NAD-glutamate dehydrogenases (NAD-GDH), NADP-glutamate dehydrogenase (NADP-GDH), 3-hydroxybutyrate dehydrogenase (BOBDH), acetoacetyl-CoA thiolase (A-CoA Th), and acetyl-CoA synthase (A-CoA Syn) (Table 4).

The specific activities of citrate synthase (Cit Syn), NAD-isocitrate dehydrogenase (NAD-ICDH), 2-oxoglutarate dehydrogenase complex (KGDHC), and NAD-malate dehydrogenase (NAD-MDH) in fraction A of nonsynaptic mitochondria are higher than the corresponding values in the other three mitochondrial populations (Table 4).

The specific activities of 4-aminobutyrate transaminase (GABA T) and succinic semialdehyde dehydrogenase (SSADH) in the B fraction of nonsynaptic mitochondria are higher than the corresponding values in the other three mitochondrial populations (Table 4).

The specific activities of type A (serotonin-oxidizing) (MAO-A) and type B (phenylethylamine-oxidizing) (MAO-B) monoamine oxidases (MAO) are much higher in synaptic mitochondria than in nonsynaptic mitochondria (Table 4). In addition, the ratios of the specific activity of MAO-A to the specific activity of MAO-B in the lighter fraction (SM) of synaptic mitochondria are significantly

Table 4
Distribution of Enzymatic Activities in Rat Brain Mitochondria[a]

Enzyme	Specific activity, mU/mg protein			
	Synaptic mitochondria		Nonsynaptic mitochondria	
	SM	SM_2	A	B
PDHC	73 ± 5(3)	57 ± 6(3)	72 ± 12(3)	72 ± 6(14)
Cit Syn	621 ± 0(3)	771 ± 0(3)	1242 ± 0(3)	1070 ± 104(14)
NAD-ICDH	142 ± 12(6)	127 ± 6(6)	161 ± 18(3)	141 ± 13(4)
NADP-ICDH	24 ± 6(7)	47 ± 8(7)	28 ± 2(3)	34 ± 7(4)
KGDHC	45 ± 9(3)	36 ± 3(3)	62 ± 4(3)	53 ± 3(3)
Fum	323 ± 38(4)	281 ± 42(4)	364 ± 37(5)	371 ± 24(3)
NAD-MDH	8625 ± 1143(5)	7955 ± 720(5)	10357 ± 950(3)	7919 ± 1055(4)
NAD-GDH	550 ± 38(6)	992 ± 99(6)	534 ± 71(7)	578 ± 44(9)
NADP-GDH	469 ± 66(3)	786 ± 41(3)	454 ± 34(3)	492 ± 39(4)
AAT	1592 ± 141(7)	1216 ± 168(7)	1794 ± 142(6)	1752 ± 83(4)
PDG	—	294 ± 37(3)	—	384 ± 65(3)
PIG	—	70(1)	—	70(1)
GABA T	21 ± 1(4)	26 ± 5(4)	20 ± 1(3)	38 ± 3(4)
SSADH	23 ± 4(7)	29 ± 5(7)	27 ± 2(3)	57 ± 1(7)

CK	208 ± 18(3)	215 ± 13(3)	774 ± 43(3)	—
HK	138 ± 9(3)	131 ± 6(3)	300 ± 21(3)	—
BOBDH	26 ± 3(3)	46 ± 3(3)	19 ± 1(3)	16 ± 1(5)
OCoA Tr	180 ± 12(50)	98 ± 5(3)	99 ± 8(3)	98 ± 6(5)
A-CoA Th	53 ± 3(3)	69 ± 3(3)	66 ± 4(3)	66 ± 3(5)
HMG-CoA Syn	14 ± 1(3)	18 ± 1(3)	18 ± 1(3)	18 ± 1(3)
A-CoA Syn	2 ± 1(3)	7 ± 1(3)	3 ± 0(3)	3 ± 0(3)
MAO-A	4 ± 0(8)	4 ± 0(8)	2 ± 1(8)	—
MAO-B	1 ± 0(8)	2 ± 0(8)	1 ± 0(8)	—

[a]Values are Means ± SD with the number of determinations in parenthesis. Synaptic (SM, SM_2) and nonsynaptic (population A) mitochondria were prepared by the method of Lai and Clark (*see* Sections 2.1.2.2 and 2.2.4) and nonsynaptic mitochondria (population B) by the method of Clark and Nicklas (1970) (*see* Section 2.1.2.1). The abbreviations are: PDHC, pyruvate dehydrogenase complex; Cit Syn, citrate synthase; NAD- and NADP-ICDH, NAD- and NADP-linked isocitrate dehydrogenase; KGDHC, 2-oxoglutarate dehydrogenase complex; Fum, fumarase; NAD-MDH, NAD-linked malate dehydrogenase; NAD- and NADP-GDH, NAD- and NADP-linked glutamate dehydrogenase; AAT, aspartate aminotransferase; PDG, phosphate-dependent glutaminase; PIG, phosphate-independent glutaminase; GABA T, 4-aminobutyrate transaminase; SSADH, succinic semialdehyde dehydrogenase; CK, creatine kinase; HK, hexokinase; BOBDH, 3-hydroxybutyrate dehydrogenase; OCoA Tr, 3-oxoacid: CoA transferase; A-CoA Th, acetoacetyl-CoA thiolase; HMG-CoA Syn, 3-hydroxy-3-methylglutaryl-CoA synthase; A-CoA Syn, acetyl-CoA synthase; MAO-A, type A monoamine oxidase; MAO-B, type B monoamine oxidase; mU, milli-unit (of enzymatic activity) (i.e., nmol of substrate used or product formed/min). The data are compiled from those of Lai and Clark (1979) (PDHC; NAD- and NADP-ICDH; KGDHC; Fum; NAD-MDH; NAD- and NADP-GDH; AAT; GABA T; SSADH; A-CoA Syn), Land (1974) PDHC; (Cit Syn), Booth (1978) (Cit Syn; CK; HK), Lai (1975) (PDG; PIG), Patel (1978) (BOBDH; OCoA Tr; A-CoA Th; HMG-CoA Syn), and Owen et al. (1977) (MAO-A and -B).

higher than the corresponding ratios in the heavier fraction (SM_2) of synaptic mitochondria and in the A fraction of nonsynaptic mitochondria (Table 4). This difference in MAO A/B between synaptic and nonsynaptic mitochondria has been confirmed in studies employing the selective inhibitors of the two forms of MAO [*see* Owen et al. (1977) and Lai et al. (1983) for a discussion].

3.5.2. Enzymatic Properties of Mitochondrial Populations from Discrete Regions

3.5.2.1. ENZYMATIC PROPERTIES OF MITOCHONDRIA ISOLATED FROM CEREBRAL CORTEX, STRIATUM AND PONS AND MEDULLA. With the exception of hexokinase (HK) in the cerebral cortex, the specific activities of the citric acid cycle and related enzymes (Table 5) are generally significantly higher in the A fraction of nonsynaptic mitochondria than in the SM_2 fraction of synaptic mitochondria irrespective of the brain region (Leong et al., 1984). In the case of hexokinase (HK) in the cerebral cortex (Leong et al., 1984), the specific activities of this enzyme associated with nonsynaptic (A) and synaptic (SM_2) mitochondria do not differ significantly (Table 5).

Regarding the regional distribution of enzymatic activities in nonsynaptic mitochondria (A), the specific activities of pyruvate dehydrogenase complex (PDHC), NAD-isocitrate dehydrogenase (NAD-ICDH), and hexokinase (HK) are higher in cerebral cortex than in striatum and pons and medulla (Table 5). In contrast, the specific activities of fumarase (Fum) are higher in pons and medulla than in the other regions (Table 5), whereas the specific activities of NADP-isocitrate dehydrogenase (NADP-ICDH) are highest in pons and medulla, intermediate in cerebral cortex and lowest in striatum (Table 5). However, the specific activities of citrate synthase (Cit Syn), NAD-malate dehydrogenase (NAD-MDH), 3-hydroxybutyrate dehydrogenase (BOBDH), and creatine kinase (CK) do not show significant regional variations (Table 5).

The regional distribution of enzymatic activities differ between synaptic (SM_2) and nonsynaptic (A) mitochondria (Table 5). For example, in the cases of synaptic mitochondria (SM_2), the specific activities of citrate synthase (Cit Syn) and fumarase (Fum) are lower in the cerebral cortex than in the striatum and pons and medulla (Table 5). On the other hand, the specific activities of

Table 5
Distribution of Enzymatic Activities in Synaptic (SM_2)
and Nonsynaptic (A) Mitochondria Isolated from Selected Brain Regions[a]

	Specific activity, mU/mg protein					
	Cerebral cortex		Striatum		Pons and medulla	
Enzyme	SM_2	A	SM_2	A	SM_2	A
PDHC	20 ± 3	81 ± 6	14 ± 1	61 ± 9	13 ± 4	65 ± 8
Cit Syn	429 ± 83	1210 ± 189	604 ± 131	1057 ± 141	605 ± 40	1188 ± 197
NAD-ICDH	50 ± 11	190 ± 36	42 ± 4	103 ± 27	37 ± 5	135 ± 19
NADP-ICDH	32 ± 5	43 ± 2	24 ± 7	35 ± 3	29 ± 8	62 ± 5
Fum	134 ± 38	494 ± 33	200 ± 48	409 ± 53	234 ± 28	656 ± 90
NAD-MDH	5204 ± 235	14970 ± 1100	5314 ± 628	12850 ± 2470	6214 ± 94	14100 ± 370
BOBDH	3 ± 1	10 ± 2	3 ± 1	7 ± 2	3 ± 0	7 ± 2
HK	471 ± 130	408 ± 19	219 ± 76	301 ± 31	158 ± 76	281 ± 40
CK	205 ± 81	575 ± 162	256 ± 70	672 ± 8	244 ± 61	654 ± 3

[a]Values are Means ± SD of three or more experiments and are from Leong et al. (1984). Synaptic (population SM_2) and nonsynaptic (population A) mitochondria were isolated from the respective regions of the rat brain by the method of Lai and Clark [*see* Sections 2.1.2.2 and 2.2.4, and Leong et al. (1984) for details]. The abbreviations are the same as those in Table 4. In the studies of Leong et al. (1984), the hippocampus was also included in the cerebrocortical tissue.

pyruvate dehydrogenase complex (PDHC) and hexokinase (HK) are higher in the cerebral cortex than in the other two regions (Table 5). The specific activities of NAD- and NADP-isocitrate dehydrogenases (NAD-ICDH and NADP-ICDH), 3-hydroxybutyrate dehydrogenase (BOBDH), and creatine kinase (CK) do not differ between the three regions (Table 5). Nonetheless, the specific activity of NAD-malate dehydrogenase (NAD-MDH) is slightly higher in the pons and medulla than in the other two regions (Table 5).

The results of the enzyme distribution studies (Table 5) strongly suggest that the mitochondrial enzyme activities in a particular population of mitochondria, be they synaptic or otherwise, differ significantly from those in another population of mitochondria derived from either the same or another brain region[*see* Leong et al. (1984) for a detailed discussion]. Moreover, the results from studies on the heterogeneity of MAO—another mitochondrial enzyme—in synaptic and nonsynaptic mitochondria from selected brain regions also support such a conclusion (Lai et al., 1981, 1983).

3.5.2.2. Enzymatic Properties of Mitochondria Isolated from the Olfactory Bulb. Recent studies on the brain regional distribution of several enzymes of intermediary metabolism indicate that the olfactory bulb is unusual in that the specific activities of pyruvate dehydrogenase complex (PDHC) (Sheu et al., 1984a, 1985a), pyruvate dehydrogenase kinase (Sheu et al., 1984a; Lai et al., 1985) and 2-oxyglutarate dehydrogenase complex (KGDHC) (Lai and Cooper, 1986) are lowest among all the brain regions studied. Studies were therefore initiated to determine the specific activities of a number of enzymes associated with energy metabolism in metabolically active nonsynaptic mitochondria isolated from the olfactory bulb and compare the enzymatic activities with those of the nonsynaptic mitochondria isolated from the cerebral cortex (Table 6).

The specific activities of pyruvate dehydrogenase complex (PDHC), pyruvate dehydrogenase kinase, NAD-isocitrate dehydrogenase (NAD-ICDH), and hexokinase (HK) are significantly lower in the olfactory bulb nonsynaptic mitochondria than in cerebral cortex nonsynaptic mitochondria (Lai et al., 1985) (Table 6). By contrast, the specific activities of NAD- and NADP-glutamate dehydrogenase (NAD-GDH and NADP-GDH) and NADP-isocitrate dehydrogenase (NADP-ICDH) are significantly higher in olfactory

Table 6
Specific Activities of Enzymes in Nonsynaptic (Population A) Mitochondria from Rat Cerebral Cortex and Olfactory Bulb[a]

Enzyme	Specific activity, mU/mg protein	
	Cerebral cortex	Olfactory bulb
PDHC	109 ± 17 (4)	41 ± 12 (4)
NAD-ICDH	135 ± 22 (7)	95 ± 20 (5)
NADP-ICDH	36 ± 3 (7)	44 ± 4 (5)
NAD-GDH	518 ± 55 (7)	908 ± 193 (5)
NADP-GDH	495 ± 71 (7)	846 ± 158 (5)
HK	234 ± 35 (7)	125 ± 21 (5)

[a]Values are Means ± SD, with the number of experiments in parenthesis. The data are from Lai et al. (1985): in these studies, only cerebro-cortical or olfactory bulb tissue was used for preparation of homogenates. The abbreviations are the same as those employed in Table 4. The A population of nonsynaptic mitochondria were isolated from cerebral cortex and olfactory by the Lai and Clark method (*see* Section 2.1.2.2).

bulb nonsynaptic mitochondria than in cerebral cortex nonsynaptic mitochondria (Lai et al., 1985) (Table 6).

3.6. The Relations Between the Heterogeneity of Brain Mitochondria and Metabolic Compartmentation

It is well established that brain metabolism of the tricarboxylic acid cycle and related metabolites (e.g., glutamate, glutamine, GABA) is compartmented [*see* Balázs and Cremer (1973) and Berl et al. (1975) and articles therein for detailed discussions]. The concept of metabolic compartmentation in the brain entails the hypothetical construction of a two-compartment—the "large" and the "small" compartments—model. The "large" compartment contains relatively large pools of tricarboxylic acid cycle intermediates that exchange with a large pool of glutamate and a small pool of glutamine. The "small" compartment consists of relatively small pools of citric acid cycle intermediates that exchange rapidly with a small pool of glutamate that is, in turn, in equilibrium with a large pool of glutamine [*see* Balázs and Cremer (1973) and articles therein].

Some findings of the heterogeneous distributions of enzymatic activities in brain mitochondria have been correlated with the different metabolic compartments (Van den Berg, 1973; Van den Berg et al., 1975). Since ammonia, acetate, and GABA are precursors of the "small" pool of glutamate, and the activities of glutamate dehydrogenase (GDH), acetyl-CoA synthase (A-CoA Syn), and GABA transaminase (GABA T) (i.e., enzymes that are involved in the metabolism of ammonia, acetate, and GABA, respectively) show very similar distributions in sucrose density gradients, these three enzymes were postulated to be localized in the mitochondria in the "small" compartment (Van den Berg, 1973; Van den Berg et al., 1975). Using a similar argument, the similar distribution patterns of other mitochondrial enzymes [e.g., fumarase (Fum), citrate synthase (Cit Syn), NAD-isocitrate dehydrogenase (NAD-ICDH), and aspartate aminotransferase (AAT)] were interpreted as being indicative of the presence of these enzymes in the mitochondria of the "large" compartment, Based upon the same kind of argument, Balázs et al. (1973) have proposed a three-compartmental model of glutamate and GABA metabolism in the brain.

Other studies by Van den Berg and coworkers (Reijnierse et al., 1975a,b; Van den Berg et al., 1975) on the sedimentation profiles of the activities of various enzymes [including glutamate dehydrogenase (GDH), acetyl-CoA synthase (A-CoA Syn), NAD-isocitrate dehydrogenase (NAD-ICDH), and butyryl- and propionyl-CoA synthases] in sucrose density gradients separated in a zonal rotor indicate that none of the distributions of the activities of these enzymes show complete overlap. Van den Berg and his associates interpreted these data to imply that the mitochondria within the "small" compartment are also heterogeneous, since most of the enzymes showing dissimilar sedimentation profile are involved in the metabolism of the "small" compartment (Reijnierse et al., 1975a,b; Van den Berg et al., 1975). This consideration (and others) led Van den Berg and coworkers (Van den Berg et al., 1975) to propose that both the "small" and the "large" compartments may consist of several subcompartments.

Until the mid-1970s, this approach of Van den Berg and coworkers represented the state-of-the-art approach (Reijnierse et al., 1975a,b; Van den Berg et al., 1975). However, two limitations are apparent in this and similar approaches. First, until the mid-1970s, studies on the heterogeneity of brain mitochondria involved the fractionation on a density gradient of the crude mitochondrial fraction derived from the homogenates of the whole brain or the forebrain (Salganicoff and De Robertis, 1965; Van Kempen et al., 1965; Balázs et al., 1966: Salganicoff, and Koeppe, 1968; Neidle et al., 1969; Blokhuis and Veldstra, 1970; Reijnierse et al., 1975a,b; Lai and Clark, 1976; Lai et al., 1975,1977). Thus, it remains possible (and likely) that the apparent heterogeneous populations of mitochondria found in the whole brain could have originated from different regions of the whole brain or forebrain [*see* Leong et al. (1984) for a discussion]. Secondly, most of these early studies involved subjecting mitochondria to hyperosmotic sucrose gradient for prolonged periods; consequently, despite the apparent purity based on the morphological criterion, the isolated heterogeneous mitochondrial populations are metabolically inactive [*see* Clark and Nicklas (1970, 1984) and Lai and Clark (1976, 1979) for discussions].

In regard to the first limitation, evidence is accumulating that the heterogeneity of brain mitochondria also occurs at the regional level (Wilkin et al., 1979; Lai et al., 1981, 1983, 1985: Leong et al., 1984). Thus, these findings of brain regional mitochondrial hereogeneity are consistent with the notion that at least some aspects of the mitochondrial heterogeneity in the whole brain may be accounted for by the mitochondrial hereogeneity at the regional level (Lai et al., 1983, 1985; Leong et al., 1984).

Concerning the second limitation, Ficoll gradient procedures have been designed such that metabolically active and relatively pure but heterogeneous populations of mitochondria can be routinely isolated from either the whole brain (Clark and Nicklas, 1970; Lai and Clark, 1976, 1979; Lai et al., 1977) or particular brain regions (Leong et al., 1984; Sheu et al., 1985a; Lai et al., 1985).

However, a more fundamental issue is whether or not the apparent V_{max}-type of enzymatic activity determinations can yield quantitative data that are directly relevant to the fluxes through the reactions catalyzed by the enzymes under study. For enzymes that catalyze "near-equilibrium" reactions, the apparent V_{max}-type of enzymatic activity is only, at best, a very rough indicator of the potential flux through the enzyme-catalyzed reaction [*see* Newsholme and Start (1973) and Newsholme and Leech (1983) for a discussion].

On the other hand, for enzymes that catalyze nonequilibrium and/or rate-limiting reactions, the apparent V_{max}-type of enzymatic activity is a reasonable indicator of the potential flux through the enzyme-catalyzed reaction [*see* Newsholme and Start (1973) for a discussion]. Nonetheless, it is important to note that other determinants of fluxes (e.g., metabolite concentrations and transport, activator and inhibitor concentrations, positive and negative regulator concentrations, and coenzyme availability) may be just as important as, if not more important than, the apparent V_{max}-type activity potentials [*see* Van Dam (1973), Lai et al. (1975) and Lai and Clark (1976) for detailed discussions].

In view of the above considerations, further metabolic experiments using the heterogeneous populations of brain mitochondria are needed to elucidate the more precise relations between the heterogeneity of brain mitochondria (with respect to the mitochondrial enzyme contents) and metabolic compartmentation.

4. Integrated Metabolic Studies with Various Intact Mitochondrial Populations

The availability of methods whereby metabolically intact and relatively pure preparations of mitochondria can be isolated from the whole brain or a particular brain region renders it possible to address discrete issues pertaining to the regulation of intermediary metabolism in intact brain mitochondria.

4.1. Metabolism of Pyruvate and Ketone Bodies

The importance of aerobic glycolysis as a primary source of energy for the adult brain can be determined by the fact that, under "normal circumstances," pyruvate is the best substrate that supports oxygen uptake by brain mitochondria (*see* Table 2). During development or prolonged starvation in adulthood, the brain may be able to derive its energy from the aerobic oxidation of alternative substrates such as ketone bodies. In the following sections, the discussions of some of the studies on the transport and metabolism of pyruvate and ketone bodies by brain mitochondria serve to illustrate the variety of studies made possible by recent methodological advances.

4.1.1. Transport

The monocarboxylate carrier, which transports both pyruvate and the ketone bodies (i.e., 3-hydroxybutyrate and acetoacetate) across the inner mitochondrial membrane, has been extensively characterized using peripheral tissue mitochondria [*see* Halestrap (1975, 1978) and Denton and Halestrap (1979) for discussions]. However, although this carrier has not been so extensively studied in brain mitochondria, some of its properties that are known to occur in peripheral tissue mitochondria are also found in brain mitochondria. For example, alpha-cyanocinnamate, a known inhibitor of the monocarboxylate translocase (Halestrap, 1975, 1978; Denton and Halestrap, 1979), also inhibits both the oxidation (i.e., substrate-dependent oxygen uptake) (Land and Clark, 1974; Patel et al., 1977) and uptake of pyruvate and ketone bodies (Land et al., 1976) by nonsynaptic mitochondria isolated from the rat forebrain.

The importance of the monocarboxylate carrier in the regulation of brain energy metabolism is twofold. On the one hand, as suggested by Land et al. (1976), the capacity of nonsynaptic rat brain mitochondria to accumulate pyruvate against a concentration gradient may allow pyruvate oxidation to occur at a rate high enough to account for the observed rate of glycolysis in the rat brain. On the other hand, the mitochondrial monocarboxylate translocase may be one of the sites that differentially regulates the brain oxidation (i.e., substrate-dependent oxygen uptake) of pyruvate and ketone bodies (Land et al., 1976). However, although synaptic mitochondria derived from the rat forebrain can oxidize ketone bodies (Lai and Clark, 1976; *also see* Table 2) (and, thus, the presence therein of the carrier can be inferred), the properties of the translocase in synaptic mitochondria remain to be evaluated.

4.1.2. Metabolism and Its Regulation

4.1.2.1. Pyruvate Metabolism and Its Regulation

4.1.2.1.1. Pyruvate Metabolism via Pyruvate Carboxylase. Patel (Patel and Tilghman, 1973; Patel, 1974) has demonstrated the presence of pyruvate carboxylase in nonsynaptic mitochondria isolated from the forebrain. However, the quantitative contribution of the carboxylase to the flux of pyruvate-C into the citric acid cycle in brain remains to be fully elucidated. There is a concensus that the bulk of the pyruvate oxidation (i.e., pyruvate-supported oxygen uptake) in the adult brain is mediated by the flux through the pyruvate dehydrogenase complex (PDHC).

4.1.2.1.2. Pyruvate Metabolism via the Pyruvate Dehydrogenase Complex. *Pyruvate dehydrogenase complex (PDHC)* is a multienzyme complex consisting of pyruvate dehydrogenase (E_1), dihydrolipoyl transacetylase (E_2), and dihydrolipoyl dehydrogenase (E_3) [*see* Reed (1981) for a detailed discussion]. Associated with the PDHC are the two regulatory enzymes, pyruvate dehydrogenase kinase and pyruvate dehydrogenase phosphate phosphatase [*see* Randle (1981) and Reed (1981) for a discussion]. The kinase phosphorylates the α-peptide of E_1 and inactivates the PDHC; the phosphatase dephosphorylates the phosphorylated α-peptide of E_1 and activates the PDHC [*see* Randle (1981) and Reed (1981) for a discussion].

The enzymatic properties of the pyruvate dehydrogenase phosphate phosphatase (Sheu et al., 1983) and pyruvate dehydrogenase kinase (Sheu et al., 1984a) in rat brain have been found to be similar to the properties of these enzymes in peripheral tissues (Randle, 1981; Reed, 1981). Bovine (Sheu and Kim, 1984), rat (Sheu et al., 1985a; Malloch et al., 1986) and human (Sheu et al., 1985c) brain PDHCs have been characterized immunochemically using antibodies against purified preparations of PDHC.

Brain mitochondrial pyruvate-supported oxygen uptake and PDHC activities. The state 3 pyruvate-supported oxygen uptake rates by the SM_2 fraction of synaptic mitochondria isolated from the rat forebrain are lower than the corresponding rates in other fractions of synaptic (SM) and nonsynaptic (A and B) mitochondria from the rat forebrain (Table 2). The oxygen uptake rates by various fractions of synaptic and nonsynaptic mitochondria from the rat forebrain (Table 2) correlate well with the distribution of PDHC activities in the same mitochondrial fractions: the PDHC activity in the SM_2 fraction of synaptic mitochondria is lower than the corresponding values in the other mitochondrial fractions (Table 4).

Regulation of pyruvate metabolism in brain mitochondria and the activation state of PDHC. The relation between the activation state of PDHC and pyruvate metabolism in brain mitochondria has not been fully elucidated. It is well known that elevated levels (>20 m*M*) of potassium ions stimulate pyruvate-supported state 3 oxygen uptake by synaptic and nonsynaptic mitochondria (Clark and Nicklas, 1970; Nicklas et al., 1971; Lai and Clark, 1976, 1979; *also see* Table 2). Studies on the pyruvate-supported oxygen uptake (Clark and Nicklas, 1970; Nicklas et al., 1971) and pyruvate disappearance from the incubation medium (Nicklas et al., 1971) using one population (B) of nonsynaptic mitochondria from the rat forebrain led Nicklas et al. (1971) to hypothesize that the potassium stimulation of pyruvate oxidation by brain mitochondria is mediated through a direct activation of PDHC by potassium. Subsequently, in studies in which the flux through the PDHC (as determined by [1-^{14}C]pyruvate decarboxylation) and the activation state of the PDHC were determined using a fraction of nonsynaptic mitochondria isolated from the cerebral cortex, Lai and Sheu (1985) have provided direct evidence that the potassium stimulation of pyruvate-supported oxygen uptake in brain mitochondria is achieved through the direct elevation of the activation state of the PDHC.

Other earlier studies on the factors that may regulate pyruvate metabolism in nonsynaptic brain mitochondria have implicated a role for adenine nucleotides (Jope and Blass, 1975; Booth and Clark, 1978b). More recent studies indicate that no single class of "regulators" (e.g., adenine nucleotides) predominate in the control of pyruvate metabolism in brain mitochondria (Sheu et al. 1984b, 1985b; Sheu and Lai, 1985; Lai and Sheu, 1985, 1987). However, it is likely that a number of factors (including adenine nucleotides, alternative substrates, redox potential, and divalent cations) may be important depending on the metabolic state(s) of the mitochondria (Lai and Sheu, 1985, 1987; Sheu and Lai, 1985).

4.1.2.2. METABOLISM OF KETONE BODIES. It is well established that ketone bodies are oxidized by both synaptic and nonsynaptic mitochondria (*see* Table 2 and references cited therein). Consistent with these observations are the findings that the activities of several ketone body-metabolizing enzymes [e.g., 3-hydroxybutyrate dehydrogenase (BOBDH), 3-oxo-acid: CoA transferase (OCoA Tr), and acetoacetyl-CoA thiolase (A-CoA Th)] are detectable in all types of brain mitochondria so far investigated (*see* Table 4 and references cited therein).

As proposed by Land and Clark (Land and Clark, 1975; Clark and Land, 1975), the monocarboxylate translocase may play an important role in the regulation of brain mitochondrial ketone body metabolism.

4.2. Metabolism of Glutamate and GABA

4.2.1. Transport

4.2.1.1. TRANSPORT OF GLUTAMATE. Brain mitochondrial glutamate transport has been studied so far primarily with nonsynaptic mitochondria isolated from whole brain or forebrain. Although several mitochondrial glutamate transporters (e.g., glutamate-hydroxyl antiporter, glutamate-aspartate antiporter, glutamine-glutamate antiporter) have been recognized [*see* Dennis et al. (1976) for a discussion], an early study (Brand and Chappell, 1974) reported the presence of only the glutamate-aspartate antiporter in nonsynaptic brain mitochondria. By contrast, Dennis et al. (1976) and Minn and coworkers (Minn et al., 1975; Minn and Gayet, 1977) have provided some evidence that the glutamate-hydroxyl antiporter also exists in nonsynaptic brain mitochondria, although probably operating at rates slower than those of the glutamate-aspartate antiporter. However, the glutamine-glutamate antiporter is reportedly absent in nonsynaptic brain mitochondria (Brand and Chappell, 1974; Dennis et al., 1976).

4.2.1.2. TRANSPORT OF GABA. No specific system for GABA transport has been found in synaptic (Walsh and Clark, 1976) or nonsynaptic (Brand and Chappell, 1974; Walsh and Clark, 1976) brain mitochondria. It is likely that GABA enters brain mitochondria as a neutral species (Brand and Chappell, 1974; Walsh and Clark, 1976).

4.2.2. Metabolism

4.2.2.1. METABOLISM OF GLUTAMATE. Evidence from studies using isolated synaptic and nonsynaptic brain mitochondria is in favor of glutamate metabolism (i.e., conversion to 2-oxoglutarate) primarily through transamination [by aspartate aminotransferase (AAT)] (Dennis et al., 1977; Dennis and Clark, 1977, 1978). That transamination is the preferred pathway may be an indication of the glutamate-aspartate antiporter in these mitochondria operating at a significantly higher rate than the glutamate-hydroxyl antiporter (Dennis et al., 1976). Moreover, in regard to the importance of the transamination pathway, it is pertinent to note that distribution of AAT activities in various populations of synaptic (i.e., SM and SM_2) and nonsynaptic (A and B) mitochondria isolated from the forebrain correlates quite closely with the rates of glutamate-supported oxygen uptake by the different mitochondrial populations (*see* Tables 2 and 4). For example, in comparison with other mitochondrial populations, the activity of AAT and the rate of glutamate-supported oxygen uptake in the SM_2 fraction of synaptic mitochondria are significantly lower (Tables 2 and 4).

4.2.2.2. METABOLISM OF GABA. GABA-supported oxygen uptake by nonsynaptic brain mitochondria is accompanied by the formation of glutamate and aspartate (Walsh and Clark, 1976; Cunningham et al., 1980). This observation is consistent with the notion that GABA-transaminase (GABA-T) is primarily responsible for the intramitochondrial metabolism of GABA. That GABA oxidation decreases brain mitochondrial substrate-level phosphorylation rates is indicative of a direct interaction between GABA oxidation and the fluxes through the tricarboxylic acid cycle in brain mitochondria (Rodichok and Albers, 1980).

Glutamate production from GABA oxidation is 50% higher in nonsynaptic (fraction B; *see* Table 4) forebrain mitochondria than in synaptic (fraction SM_2; *see* Table 4) forebrain mitochondria (Walsh and Clark, 1976). This finding of lower glutamate production from GABA oxidation by synaptic mitochondria is consistent with the observation that, in general, the activities of both GABA-transaminase (GABA T) and succinic semialdehyde dehydrogenase (SSADH) are lower in synaptic than in nonsynaptic mitochondria (*see* Table 4 and references cited therein).

4.3. Mitochondrial-Cytosolic Transport and Shuttle Systems

In the majority of the studies in this area, nonsynaptic mitochondria isolated from the whole brain or the forebrain have been employed. Thus, whether or not results obtained using one kind of brain mitochondria can be generalized to all kinds of brain mitrochondria remains to be determined.

4.3.1. Metabolite Transport

In addition to the transport systems discussed above, a citrate-malate translocase has been demonstrated in nonsynaptic (Patel, 1975; Patel and Clark, 1981) and synaptic (Patel and Clark, 1981) forebrain mitochondria. There is some evidence to suggest the presence of the malate-phosphate and succinate-phosphate translocases in nonsynaptic brain mitochondria (Minn et al., 1975).

4.3.2. Redox Shuttles

Since the inner mitochondrial membrane is impermeable to NADH, various shuttle mechanisms that mediate the transfer of reducing equivalents across the inner mitochondrial membrane have been proposed [*see* Meijer and Van Dam (1974) and La Noue and Schoolwerth (1984) for comprehensive reviews of the literature]. Of these mechanisms, the malate-aspartate and the alpha-glycerophosphate shuttles are probably both operative in brain

(Brand and Chappell, 1974; Dennis et al, 1977; Minn and Gayet, 1977; Clark and Nicklas, 1984). Evidence is accumulating that the malate-aspartate shuttle may play an important role in the regulation of brain energy metabolism (Dennis et al., 1977; Dennis and Clark, 1978; Cooper et al., 1983; Fitzpatrick et al., 1983; Lai et al., 1986).

4.3.3. *Transfer of Acetyl Moiety for Acetylcholine Synthesis*

Because acetyl-CoA, one of the precursors for acetylcholine synthesis, does not readily cross the inner mitochondrial membrane and acetyl-CoA derived from aerobic glycolysis is a precursor source for acetylcholine synthesis (Tucek and Cheng, 1974), some transport mechanisms have to be invoked to account for the exit of acetyl-CoA from mitochondria to the cytosol where the acetyl-CoA joins the precursor pool for acetylcholine synthesis [*see* Clark and Nicklas (1984) for a discussion]. Export of citrate from the mitochondria into the cytosol and the regeneration of acetyl-CoA via the cytosolic citrate lyase appears to be one of the primary mechanisms [*see* Hafalowska and Ksiezak (1976), Sterling and O'Neill (1978), Clark and Nicklas (1984), and references cited therein]. *N*-Acetyl-L-aspartate exported from mitochondria into cytosol via the dicarboxylate translocase may also be a source of the acetyl moiety for acetylcholine synthesis (D'Adamo et al., 1968; Patel and Clark, 1979). Nonetheless, the relative importance of the origins of the acetyl moiety that regulates acetylcholine synthesis in brain remains to be fully elucidated.

5. Pathophysiological and Toxicological Studies with Brain Mitochondria

In view of space limitations, it is impossible to cover the diverse range of interesting studies relevant to these two important areas of research. Consequently, the authors have chosen to discuss only some of studies that serve to illustrate how the mitochondrial preparative methodology can be usefully exploited to address mechanistic questions that are of pathophysiological and/or toxicological relevance. [The following reviews should be consulted for areas not discussed by the authors: Land and Clark (1979), Siesjö (1981). Nowicki et al. (1982), Lai et al. (1983), and O'Neill and Holtzman (1985)].

5.1. Pathophysiological Studies

5.1.1. Studies in Relation to Phenylketonuria and Maple Syrup Urine Disease

Land and Clark (Land and Clark, 1974; Clark and Land 1974; Land et al., 1976) have demonstrated the usefulness of nonsynaptic brain mitochondria as an in vitro model system to investigate some of the pathophysiological (and perhaps also the pathogenetic) mechansims implicated in the mental retardation commonly found in patients who suffer from phenylketonuria and maple syrup urine disease. It is known that phenylpyruvate and 2-oxo-4-methylpentanoic acid (alpha-ketoisocaproate, KIC) are the toxic compounds that accumulate in individuals with phenylketonuria and maple syrup urine disease, respectively. Employing nonsynaptic mitochondria isolated from forebrains of adult and 22-day-old rats, Land and Clark demonstrated that both phenylpyruvate and KIC, at pathophysiological levels, inhibit pyruvate- and 3-hydroxybutyrate-supported state-3 oxygen uptake by these mitochondria (Land and Clark, 1974; Clark and Land, 1974, 1975; Land et al., 1976). Moreover, at concentrations that are effective in inhibiting pyruvate- and 3-hydroxybutyrate-dependent oxygen uptake by adult and developing forebrain nonsynaptic mitochondria, neither phenylpyruvate nor KIC has any significant inhibitory effect on brain pyruvate dehydrogenase complex (PDHC) and 3-hydroxybutyrate dehydrogenase (BOBDH) activities [*see* Land et al. (1976) for a discussion]. These observations led Land and Clark to propose that "a primary lesion in phenylketonuria and maple syrup urine disease is the inhibition of pyruvate and 3-hydroxybutyrate utilization by the inhibition of the transport of these substrates across the mitochondrial membrane" (Land and Clark, 1974). Indeed, their subsequent study (Land et al., 1976) has provided direct evidence that both phenylpyruvate and KIC do inhibit pyruvate and 3-hydroxybutyrate transport across the inner membrane of nonsynaptic mitochondria.

5.1.2. Studies Related to Aging and Dementia

It is known that, in aging and senile dementia, brain energy metabolism is compromised [*see* Lai and Blass (1984), Sun et al., (1985), and references cited therein]. Since mitochondrial metabolism constitutes a major aspect of energy metabolism, brain mitochondrial preparations may be one of the ideal in vitro model systems for elucidating the mechanisms underlying the disturbances in energy metabolism noted in the brain in vivo in aging and in senile dementia. Moreover, in support of the efficacy of this approach is the observation that, in different regions of the aging rat brain, the activities of several energy-metabolizing enzymes (known to be localized in brain mitochondria) are altered [Ryder, 1980; Leong et al., 1981; Lai et al., 1982; *also see* the chapter by Clark and Lai in this volume].

Patel and coworkers (Deshmukh et al., 1980; Deshmukh and Patel, 1982) employed Ficoll density gradients (*see* Section 2.4. above) to isolate synaptic and nonsynaptic mitochondria from the forebrains of 3-, 12-, and 24-mo-old rats. These workers found that the pyruvate-supported state-3 oxygen uptake rates of these two types of mitochondrial preparations from the old (>12-mo-old) rats were significantly lower than corresponding rates of the mitochondria isolated from the young (i.e., 3-mo-old) rats. However, the activities of the "fully activated" PDHC in these mitochondrial preparations isolated from young and old rats did not significantly differ (Deshmukh et al., 1980). Subsequently, Deshmukh and Patel (1982) provided some evidence that the lower state 3 pyruvate-supported oxygen uptake by synaptic and nonsynaptic mitochondria isolated from the forebrains of aged (i.e., 24-mo-old) rats (as compared to the rates in mitochondrial preparations isolated from the forebrains of young (i.e., 3-mo-old) rats can be accounted for by the age-dependent decreases in pyruvate uptake into these two types of brain mitochondria.

5.1.3. Studies Related to Ischemia and Anoxia

Using nonsynaptic mitochondrial preparations, several groups of researchers have demonstrated alterations of mitochondrial metabolism (measured ex vivo) in ischemia and anoxia [*see* Hillered et al. (1984), Sims et al. (1986), Wagner and Myers (1986), and references cited therein for detailed discussions]. It is interesting that Sims et al. (1986) recently showed that derangements of mitochondrial oxidative metabolism in ischemia can also be detected in brain homogenates using experimental approaches (Sims and Blass, 1986) derived from mitochondrial studies. For example, brain homogenates from rats subjected to 30 min of postdecapitation ischemia exhibited large decreases (~50%) of oxygen uptake measured in a high-potassium medium in the presence of ADP or an uncoupler of oxidative phosphorylation (Sims et al., 1986).

5.2. Toxicological Studies

5.2.1. Neurotoxic Effects of Metals

Studies with nonsynaptic mitochondria isolated from cerebella of control and lead-treated rat pups indicate that chronic lead poisoning interferes with cerebellar oxidative metabolism. For example, in the mitochondria isolated from lead-treated animals, the state 4 glutamate-supported oxygen uptake is increased in the initial phase of lead exposure, whereas the state 3 glutamate-supported oxygen uptake and the respiratory control ratio are decreased in the later phase of lead exposure [*see* O'Neill and Holtzman (1985) and references cited therein].

High and toxic levels of calcium decrease [1-^{14}C]pyruvate decarboxylation by cerebro-cortical nonsynaptic mitochondria in vitro and suppress the activation state of the pyruvate dehydrogenase complex (PDHC) in these mitochondria (Sheu et al., 1985b; J. C. K. Lai and K.-F. R. Sheu, unpublished data).

5.2.2. Neurotoxicity of MPTP

Administration of MPTP (1-methyl-4-phenyl-1,2,3,6-tetrahydropyridine) induces a selective lesion of the central dopaminergic neurons in monkeys and mice, and a Parkinsonian-like syndrome in man [*see* Nicklas et al. (1985), Vyas et al. (1986), and references cited therein]. Thus, the neurotoxic effects of MPTP in experimental animals have been employed as model systems to stimulate the underlying pathophysiology and pathogenesis of Parkinson's disease. By using nonsynaptic forebrain mitochondria, Nicklas and co-workers (Nicklas et al., 1985; Vyas et al., 1986) were able to demonstrate that MPP$^+$ [1-methyl-4-phenylpyridinium, a product of the action of type B monoamine oxidase (MAO-B) on MPTP] preferentially inhibits the oxidation of NAD-linked substrates (e.g., pyruvate and glutamate) by these mitochondria. The results of these studies led Vyas et al. (1986) to propose that the "compromise of mitochondrial function and its metabolic sequelae within dopaminergic neurons could be an important factor in the neurotoxicity observed after MPTP administration." Furthermore, studies using synaptosomes have shown that MPP$^+$ and other inhibitors of mitochondrial NAD-linked substrate-dependent oxygen uptake may have marked inhibitory effects on acetylcholine synthesis (Cheeseman and Clark, 1987).

6. Concluding Remarks and Prospects for Future Studies

6.1. Concluding Remarks

Major advances in our understanding of the regulation of metabolism in heterogeneous populations of brain mitochondria have been made in the last 15 yr resulting largely from new and improved methodology. However, in comparison with mitochondria from peripheral tissues, our knowledge of brain mitochondrial metabolism and its regulation is still rudimentary and incomplete. Thus, the following remark that the present authors made some 10 yr ago still applies: "the availability of metabolically active, tightly coupled and relatively pure brain mitochondrial preparations such as those described in the present paper renders it practicable to study this multitude of control points with the ultimate view of clarifying the connexion between the heterogeneity of brain mitochondria and metabolic compartmentation" (Lai and Clark, 1976).

6.2. Prospects for Future Studies

It is impossible to discuss comprehensively the various major areas that merit future studies, although several potentially fruitful areas for investigation have been indicated in the previous sections. However, in view of the recent methodological advances, the following two areas certainly deserve more immediate and serious considerations.

1. Metabolic studies can be undertaken to elucidate the properties of synaptic and nonsynaptic mitochondria isolated from the developing brain using existing or new methodology. Studies from this perspective will undoubtedly improve our understanding of the regulation of oxidative metabolism during brain development.
2. The recent advances in the techniques of primary cultures of various neural cell types make it possible to evaluate the regulation of mitochondrial metabolism from a *cellular perspective*. For example, it would be revealing to extend the present brain mitochondrial isolation techniques to preparing metabolically active and relatively pure fractions of

mitochondria from primary cultures of astrocytes, and elucidating the regulation of oxidative metabolism in the mitochondrial fractions therefrom.

7. Summary

The historical aspect of the development of brain mitochondrial isolation methodology is briefly reviewed. The Ficoll density gradient procedures for the isolation of metabolically active and relatively pure preparations of synaptic and nonsynaptic mitochondria from the rat forebrain (or brain regions) are detailed. Some of the enzymatic and respiratory properties of the various synaptic and nonsynaptic mitochondrial populations are reviewed. A brief discussion on the different aspects of brain mitochondrial metabolism is focused on the "theme" of how the various mitochondrial preparations may be skillfully exploited for metabolic, pathophysiological, pathological, and toxicological studies. The conclusion touches on the need to apply and extend the brain mitochondrial preparative methodology to elucidating the regulation of mitochondrial metabolism during brain development on the one hand, and in mitochondrial populations isolated from primary cultures of different neural cell types on the other.

Acknowledgments

The authors are grateful to Dr. Arthur J. L. Cooper for his critical reading of the present review and helpful comments and to the authors' past and present collaborators for their contributions over the years. J. C. K. Lai's research was supported, in part, by the Medical Research Council (UK), the Worshipful Company of Pewterers (UK), and the National Institutes of Health (USA). J. B. Clark's research was supported, in part, by the Medical and Science and Engineering Research Councils (UK) British Diabetic Association, and NATO.

References

Abood L. G. (1969) Brain Mitochondria, in *Handbook of Neurochemistry*, Vol. 2 (1st Ed.) (Lajtha A., ed.). pp. 303–326. Plenum, New York.

Balázs R. and Cremer J. E. (1973) *Metabolic Compartmentation in the Brain.* MacMillan, London.

Balázs R., Dahl D., and Harwood J. R. (1966) Subcellular distribution of enzymes of glutamate metabolism in rat brain. *J. Neurochem.* **13,** 897–905.

Balázs R., Machiyama Y., and Patel A. J. (1973) Compartmentation and the metabolism of gamma-aminobutyrate, in *Metabolic Compartmentation in the Brain* (Balázs R. and Cremer J. E., eds.), pp. 57–70. MacMillan, London.

Baláza R., Wilkin G. P., Wilson J. E., Cohen J., and Dutton G. R. (1975) Biochemical dissection of the cerebellum—isolation of perikarya from the cerebellum with well-preserved ultrastructure, in *Metabolic Compartmentation and Neurotransmission* (Berl S., Clarke D. D., and Schneider D., eds.). pp. 437–448. Plenum, New York.

Berl S., Clarke D. D., and Schneider D. (1975) *Metabolic Compartmentation and Neurotransmission.* Plenum, New York.

Blokhuis G. G. D. and Veldstra H. (1970) Heterogeneity of mitochondria in rat brain. *FEBS Lett.* **11,** 197–199.

Booth R. F. G. (1978) Ph.D. Thesis, University of London, England.

Booth R. F. G. and Clark J. B. (1978a) A rapid method for the preparation of relatively pure metabolically competent synaptosomes from rat brain. *Biochem. J.* **176,** 365–370.

Booth R. F. G. and Clark J. B. (1978b) The control of pyruvate dehydrogenase in isolated brain mitochondria. *J. Neurochem.* **30,** 1003–1008.

Booth R. F. G. and Clark J. B. (1979) A method for the rapid separation of soluble and particulate components of rat brain synaptosomes. *FEBS Lett* **107,** 387–392.

Brand M. D. and Chappell J. B. (1974) Glutamate and aspartate transport in rat brain mitochondria. *Biochem. J.* **140,** 205–210.

Chance B. and Williams G. R. (1956) The respiratory chain and oxidative phosphorylation, in *Advances in Enzymology,* **Vol. 17** (Nord F. F., ed.). pp. 63–134. Interscience, New York.

Cheeseman A. J. and Clark J. B. (1987) Effects of 1-methyl-4-phenyl-1,2,5,6-tetrahydropyridine and its metabolite 1-methyl-4-phenylpyridine on acetylcholine synthesis in synaptosomes from rat forebrain. *J. Neurochem.* **48,** 1209–1214.

Clark J. B. and Lai J. C. K. (1988) Glycolytic, tricarboxylic acid cycle and related enzymes in brain, in *Neuromethods,* **Vol. 11** (Boulton, AA, Baker, G. B. and Butterworth R. F., eds) Humana Press, Clifton, pp. 233–281.

Clark J. B. and Land J. M. (1974) Differential effects of 2-oxo-acids on pyruvate utilization and fatty acid synthesis in rat brain. *Biochem. J.* **140,** 25–29.

Clark J. B. and Land J. M. (1975) Phenylketonuria and maple syrup disease and their association with brain mitochondrial substrate utilization, in *Normal and Pathological Development of Energy Metabolism* (Hommes F. A. and Van den Berg C. J., eds.). pp. 177–191. Academic, London.

Clark J. B. and Nicklas W. J. (1970) The metabolism of rat brain mitochondria. *J. Biol. Chem.* **245,** 4724–4731.

Clark J. B. and Nicklas W. J. (1984) Brain mitochondria, in *Handbook of Neurochemistry* **Vol.** 7 (2nd Ed.) (Lajtha A., ed.). pp. 135–159. Plenum, New York.

Cooper A. J. L., Fitzpatrick S. M., Ginos J. Z., Kaufman C., and Dowd P. (1983) Inhibition of glutamate-aspartate transaminase by beta-methylene-*DL*-aspartate. *Biochem. Pharmacol.* **32,** 679–689.

Cotman C. W. and Matthews D. A. (1971) Synaptic plasma membranes from rat brain synaptosomes: isolation and partial characterization. *Biochim. Biophys. Acta* **249,** 380–394.

Cunningham J., Clarke D. D., and Nicklas W. J. (1980) Oxidative metabolism of 4-aminobutyrate by rat brain mitochondria: inhibition by branched-chain fatty acids. *J. Neurochem.* **34,** 197–202.

D'Adamo Jr. A. F., Gidez L. I., and Yatzu F. M. (1968) Acetyl transport mechanisms. Involvement of *N*-acetyl aspartic acid in *de novo* fatty acid biosynthesis in the developing brain. *Exp. Brain Res.* **5,** 267–273.

Dagani F., Gorini A., Polgatti M., Villa R. F., and Benzi G. (1983) Rat cortex synaptic and nonsynaptic mitochondria: enzymatic characterization and pharmacological effects of naftidrofuryl. *J. Neurosci. Res.* **10,** 135–140.

Dennis S. C. (1976) Ph.D. Thesis, University of London, England.

Dennis S. C. and Clark J. B. (1977) The pathway of glutamate metabolism in rat brain mitochondria. *Biochem. J.* **168,** 521–527.

Dennis S. C. and Clark J. B. (1978) The regulation of glutamate metabolism by tricarboxylic acid-cycle activity in rat brain mitochondria. *Biochem. J.* **172,** 155–162.

Dennis S. C., Lai J. C. K., and Clark J. B. (1977) Comparative studies on glutamate metabolism in synaptic and non-synaptic rat brain mitochondria. *Biochem. J.* **164,** 727–736.

Dennis S. C., Lai J. C. K., and Clark J. B. (1980) The distribution of glutamine synthetase in subcellular fractions of rat brain. *Brain Res.* **197,** 469–475.

Dennis S. C., Land J. M., and Clark J. B. (1976) Glutamate metabolism and transport in rat brain mitochondria. *Biochem. J.* **156,** 323–331.

Denton R. M. and Halstrap A. P. (1979) Regulation of pyruvate metabolism in mammalian tissues, in *Essays in Biochemistry,* **Vol. 15** (Campbell P. N. and Marshall R. D., eds.). pp. 37–77. The Biochemical Society, U.K.

De Robertis E. and de Lores Arnaiz G. R. (1969) Structural components of the synaptic region, in *Handbook of Neurochemistry,* **Vol. 2** (1st Ed.) (Lajtha A., ed.). pp. 365–392. Plenum, New York.

Deshmukh D. R. and Patel M. S. (1982) Age-dependent changes in pyruvate uptake by nonsynaptic and synaptic mitochondria from rat brain. *Mech. Ageing Dev.* **20,** 343–351.

Deshmukh D. R., Owen O. E., and Patel M. S. (1980) Effect of aging on the metabolism of pyruvate and 3-hydroxybutyrate in nonsynaptic and synaptic mitochondria from rat brain. *J. Neurochem.* **34,** 1219–1224.

Dienel G., Ryder E., and Greengard O. (1977) Distribution of mitochondrial enzymes between the perikaryal and synaptic fractions of immature and adult rat brain. *Biochim. Biophys. Acta* **496,** 484–494.

Fitzpatrick S. M., Cooper A. J. L., and Duffy T. E. (1983) Use of β-methylene-*D,L*-aspartate to assess the role of aspartate aminotransferase in cerebral oxidative metabolism. *J. Neurochem.* **41,** 1370–1383.

Hafalowska U. and Ksiezak H. (1976) Subcellular localization of enzymes oxidizing citrate in the rat brain. *J. Neurochem.* **27,** 813–815.

Halestrap A. P. (1975) The mitochondrial pyruvate carrier. Kinetics and specificity for substrates and inhibitors. *Biochem. J.* **148,** 85–96.

Halestrap A. P. (1978) Pyruvate and ketone-body transport across the mitochondrial membrane. Exchange properties, pH-dependence and mechanism of the carrier. *Biochem. J.* **172,** 377–387.

Hamberger A., Blomstrand C., and Lehninger A. L. (1970) Comparative studies on mitochondria isolated from neuron-enriched and glia-enriched fractions of rabbit and beef brain. *J. Cell Biol.* **45,** 221–234.

Hillered L., Siesjö B. K., and Arfors K.-E. (1984) Mitochondrial response to transient forebrain ischemia and recirculation in the rat. *J. Cereb. Blood Flow Metab.* **4,** 438–446.

Holtzman D. and Moore C. L. (1973) Oxidative phosphorylation in immature rat brain mitochondria. *Biol. Neonate* **22,** 230–242.

Hughes D. E., Wimpenny J. W. T., and Lloyd D. (1971) The disintegration of micro-organisms, in *Methods in Microbiology,* **Vol. 5B** (Norris J. R. and Ribbons D. W., eds.). pp. 1–54. Academic, London.

Johnston P. V. and Roots B. I. (1970) Neuronal and glial perikarya preparations: an appraisal of present methods. *Int. Rev. Cytol.* **29,** 265–280.

Jope R. and Blass J. P. (1975) A comparison of the regulation of pyruvate dehydrogenase in mitochondria from rat brain and liver. *Biochem. J.* **150,** 397–403.

Lai J. C. K. (1975) Ph.D. Thesis, University of London, England.

Lai J. C. K. and Blass J. P. (1984) Inhibition of brain glycolysis by aluminum. *J. Neurochem.* **42,** 438–446.

Lai J. C. K. and Clark J. B. (1976) Preparation and properties of mitochondria derived from synaptosomes. *Biochem. J.* **154,** 423–432.

Lai J. C. K. and Clark J. B. (1979) Preparation of synaptic and nonsynaptic mitochondria from mammalian brain, in *Methods in Enzymology* **55,** Pt. F (Fleischer S. and Packer L., eds.). pp. 51–60. Academic, New York.

Lai J. C. K. and Cooper A. J. L. (1986) Brain α-ketoglutarate dehydrogenase complex: kinetic properties, regional distribution, and effects of inhibitors. *J. Neurochem.* **47,** 1376–1386.

Lai J. C. K., Guest J. F., Lim L., and Davison A. N. (1978) The effects of transition-metal ions on rat brain synaptosomal amine-uptake systems. *Biochem. Soc. Trans.* **6,** 1010–1012.

Lai J. C. K., Guest J. F., Leung T. K. C., Lim L., and Davison A. N. (1980) The effects of cadmium, manganese and aluminium on sodium-potassium-activated and magnesium-activated adenosine triphosphatase activity and choline uptake in rat brain synaptosomes. *Biochem. Pharmacol.* **29,** 141–146.

Lai J. C. K., Leung T. K. C., and Lim L. (1981) Monoamine oxidase in synaptic and non-synaptic mitochondria from brain regions. *Proc. 8th Meet Int. Soc. for Neurochem.* p. 291.

Lai J. C. K., Leung T. K. C., and Lim L. (1982) Activities of the mitochondrial NAD-linked isocitric dehydrogenase in different regions of the rat brain. Changes in ageing and the effect of chronic manganese chloride administration. *Gerontology* **28,** 81–85.

Lai J. C. K., Murthy Ch. R. K., Hertz L., and Cooper A. J. L. (1986) NH_3 & beta-methyleneaspartate inhibit neuronal & glial glutamate oxidation. *Trans. Am. Soc. Neurochem.* **17,** 217.

Lai J. C. K. and Sheu K.-F.R. (1985) Relatiosnhip between activation state of pyruvate dehydrogenase complex and rate of pyruvate oxidation in isolated cerebro-cortical mitochondria: effects of potassium ions and adenine nucleotides. *J. Neurochem.* **45,** 1861–1868.

Lai J. C. K. and Sheu K.-F. R. (1987) The effect of 2-oxoglutarate or 3-hydroxybutyrate on pyruvate dehydrogenase complex in isolated cerebro-cortical mitochondria. *Neurochem. Res.* **12,** 715–722.

Lai J. C. K., Sheu K.-F. R. and Carlson Jr. K. C. (1985) Differences in some of the metabolic properties of mitochondria isolated from cerebral cortex and olfactory bulb in the rat. *Brain Res.* **343,** 52–59.

Lai J. C. K., Walsh J. M., Dennis S. C., and Clark J. B. (1975) Compartmentation of citric acid cycle and related enzymes in distinct populations of rat brain mitochondria, in *Metabolic Compartmentation and Neurotransmission* (Berl S., Clarke D. D. and Schneider D., eds.). pp. 487–496, Plenum, New York.

Lai J. C. K., Walsh J. M., Dennis S. C., and Clark J. B. (1977) Synaptic and non-synaptic mitochondria from rat brain: isolation and characterization. *J. Neurochem.* **28,** 625–631.

Lai J. C. K., Wong P. C. L., and Lim L. (1983) Structure and function of synaptosomal and mitochondrial membranes: elucidation using neurotoxic metals and neuromodulatory agents, in *Neural Membranes* (Sun G. Y., Bazan N., Wu J.-Y., Porcellati G., and Sun A. Y., eds.). pp. 355–374, Humana, Clifton.

Land J. M. (1974) Ph.D. Thesis, University of London, England.

Land J. M. and Clark J. B. (1974) Inhibition of pyruvate and beta-hydroxybutyrate oxidation in rat brain mitochondria by phenylpyruvate and alpha-ketoisocaproate. *FEBS Lett.* **44,** 348–351.

Land J. M. and Clark J. B. (1975) The changing pattern of brain mitochondrial substrate utilization during development, in *Normal and Pathological Development of Energy Metabolism* (Hommes F. A. and Van den Berg C. J., eds.). pp. 155–167, Academic, New York–London.

Land J. M. and Clark J. B. (1979) Mitochondrial myopathies. *Biochem. Soc Trans.* **7,** 231–245.

Land J. M., Mowbray J., and Clark J. B. (1976) Control of pyruvate and beta-hydroxybutyrate utilization in rat brain mitochondria and its relevance to phenylketonuria and maple syrup urine disease. *J. Neurochem.* **26,** 823–830.

La Noue K. F. and Schoolwerth A. C. (1984) Metabolite transport in mammalian mitochondria, in *Bioenergetics* (Ernster L., ed.). pp. 221–268, Elsevier, Amsterdam.

Leong S. F., Lai J. C. K., Lim L., and Clark J. B. (1981) Energy-metabolizing enzymes in brain regions of adult and aging rats. *J. Neurochem.* **37,** 1548–1556.

Leong S. F., Lai J. C. K., Lim L., and Clark J. B. (1984) The activities of some energy-metabolizing enzymes in nonsynaptic (free) and synaptic mitochondria derived from selected brain regions. *J. Neurochem.* **42,** 1306–1312.

Malloch G. D. A., Munday L. A., Olson M. S., and Clark J. B. (1986) Comparative development of the pyruvate dehydrogenase complex and citrate synthase in rat brain mitochondria. *Biochem. J.* **238,** 729–736.

Meijer A. J. and Van Dam (1974) The metabolic significance of anion transport in mitochondria. *Biochim. Biophys. Acta* **346,** 213–244.

Minn A. and Gayet J. (1977) Kinetic study of glutamate transport in rat brain mitochondria. *J. Neurochem.* **29,** 873–881.

Minn A., Gayet J., and Delorme P. (1975) The penetration of the membrane of brain mitochondria by anions. *J. Neurochem.* **24,** 149–156.

Moore C. L. and Strasberg P. M. (1970) Cytochromes and oxidative phosphorylation, in *Handbook of Neurochemistry,* Vol. **3** (1st Ed.) (Lajtha A., ed.). pp. 53–85, Plenum, New York.

Neidle A., Van den Berg C. J., and Grynbaum A. (1969) The heterogeneity of rat brain mitochondria isolated on continuous sucrose gradients. *J. Neurochem.* **16,** 225–234.

Newsholme E. A. and Leech A. R. (1983) *Biochemistry for the Medical Sciences,* John Wiley & Sons, Chichester–New York.

Newsholme E. A. and Start C. (1973) *Regulation in Metabolism.* John Wiley & Sons, London.

Nicholls D. G. (1978) Calcium transport and proton electrochemical potential gradient in mitochondria from guinea-pig cerebral cortex and rat heart. *Biochem. J.* **170,** 511–522.

Nicklas W. J., Clark J. B., and Williamson J. R. (1971) Metabolism of rat brain mitochondria. Studies on the potassium ion-stimulated oxidation of pyruvate. *Biochem. J.* **123,** 83–95.

Nicklas W. J., Vyas I., and Heikkila R. E. (1985) Inhibition of NADH-linked oxidation in brain mitochondria by 1-methyl-4-phenylpyridine, a metabolite of the neurotoxin, 1-methyl-4-phenyl-1,2,5,6-tetrahydropyridine. *Life Sci.* **36,** 2503–2508.

Nowicki J.-P., MacKenzie E. T., and Spinnewyn B. (1982) Effects of agents used in the pharmacotherapy of cerebrovascular disease on the oxygen consumption of isolated cerebral mitochondria. *J. Cereb. Blood Flow Metab.* **2,** 33–40.

O'Neill J. J. and Holtzman D. (1985) Heavy metal toxicity and energy metabolism in the development brain: lead as the model, in *Cerebral Energy Metabolism and Metabolic Encephalopathy* (McCandless D. W., ed.). pp. 391–424, Plenum, New York.

Owen, F., Bourne R. C., Lai J. C. K., and Williams R. (1977) The heterogeneity of monoamine oxidase in distinct populations of rat brain mitochondria. *Biochem. Pharacol.* **26,** 289–292.

Patel M. S. (1974) The relative significance of CO_2-fixing enzymes in the metabolism of rat brain. *J. Neurochem.* **22,** 717–724.

Patel M. S. (1975) Citrate transport and oxidation by isolated rat brain mitochondria. *Brain Res.* **98,** 607–611.

Patel M. S. and Tilghman S. M. (1973) Regulation of pyruvate metabolism via pyruvate carboxylase in rat brain mitochondria. *Biochem. J.* **132,** 185–192.

Patel T. B. (1978) Ph.D. Thesis, University of London, England.

Patel T. B., Booth R. F. G., and Clark J. B. (1977) Inhibition of acetoacetate oxidation by brain mitochondria from the suckling rat by phenylpyruvate and alpha-ketoisocaproate. *J. Neurochem.* **29,** 1151–1153.

Patel T. B. and Clark J. B. (1979) Synthesis of *N*-acetyl-*L*-aspartate by rat brain mitochondria and its involvement in mitochondrial/cytosolic carbon transport. *Biochem. J.* **184,** 539–546.

Patel T. B. and Clark J. B. (1981) Mitochondrial/cytosolic carbon transfer in the developing rat brain. *Biochim. Biophys. Acta* **677,** 373–380.

Randle P. J. (1981) Phosphorylation-dephosphorylation cycles and the regulation of fuel selection in mammals, in *Current Topics in Cellular Regulation,* **Vol. 18** (Estabrook R. W. and Srere P., eds.). pp. 107–128, Academic, New York.

Reed L. J. (1981) Regulation of mammalian pyruvate dehydrogenase complex by a phosphorylation-dephosphorylation cycle, in *Current Topics in Cellular Regulation,* **Vol. 18** (Estabrook R. W. and Srere P., eds.). pp. 95–106, Academic, New York.

Reijnierse G. L. A., Veldstra H., and Van den Berg C. J. (1975a) Short-chain fatty acid synthases in brain. Subcellular localization and changes during development. *Biochem. J.* **152,** 477–484.

Reijnierse G. L. A., Veldstra H., and Van den Berg C. J. (1975b) Subcellular localization of gamma-aminobutyrate transaminase and glutamate dehydrogenase in adult rat brain. Evidence for at least two small glutamate compartments in brain. *Biochem. J.* **152,** 469–475.

Rodichok L. D. and Albers R. W. (1980) The effect of gamma-aminobutyric acid on substrate-level phosphorylation in brain mitochondria. *J. Neurochem.* **34,** 808–812.

Ryder E. (1980) Enzymatic profile of mitochondria isolated from selected brain regions of young adult and one-year-old rats. *J. Neurochem.* **34,** 1550–1552.

Salganicoff L. and De Robertis E. (1965) Subcellular distribution of the enzymes of the glutamic acid, glutamine and gamma-aminobutyric acid cycles in rat brain. *J. Neurochem.* **12,** 287–309.

Salganicoff L. and Koeppe R. E. (1968) Subcellular distribution of pyruvate carboxylase, diphosphopyridine nucleotide and triphosphopyridine nucleotide isocitrate dehydrogenases, and malate enzyme in rat brain. *J. Biol. Chem.* **243,** 3416–3420.

Sheu K.-F.R. and Kim Y. T. (1984) Studies on the bovine brain pyruvate dehydrogenase complex using the antibodies against kidney enzyme complex. *J. Neurochem.* **43,** 1352–1358.

Sheu K.-F. R. and Lai J. C. K. (1985) Regulation of pyruvate metabolism in brain mitochondria. *J. Neurochem.* **44** (Suppl), S166D.

Sheu K.-F. R., Lai J. C. K., and Blass J. P. (1983) Pyruvate dehydrogenase phosphate (PDH_b) phosphatase in brain: activity, properties and subcellular localization. *J. Neurochem.* **40,** 1366–1372.

Sheu K.-F. R., Lai J. C. K., and Blass J. P. (1984a) Properties and regional distribution of pyruvate dehydrogenase (PDH_a) kinase in rat brain. *J. Neurochem.* **42,** 230–236.

Sheu K.-F. R., Lai J. C. K., and Dorant G. (1984b) ADP induces pyruvate dehydrogenase inactivation in brain mitochondria. *Trans. Am. Soc. Neurochem.* **15, 188.**

Sheu K.-F.R., Lai J. C. K., Kim Y. T., Bagg J., and Dorant G. (1985a) Immunochemical characterization of pyruvate dehydrogenase complex in rat brain. *J. Neurochem.* **44,** 593–599.

Sheu K.-F. R., Lai J. C. K., DiLorenzo J. C., and Blass J. P. (1985b) Calcium inactivates pyruvate dehydrogenase complex in brain mitochondria. *Trans. Am. Soc. Neurochem.* **16,** 193.

Sheu K.-F. R., Kim Y.-T., Blass, J. P., and Weksler M. E. (1985c) An immunochemical study of the pyruvate dehydrogenase deficit in Alzheimer's disease brain. *Ann. Neurol.* **17,** 444–449.

Siesjö B. K. (1981) Cell damage in the brain: a speculative synthesis. *J. Cereb. Blood Flow Metab.* **1,** 155–185.

Sims N. R. and Blass J. P. (1986) Expression of classical mitochondrial respiratory responses in homogenates of rat forebrain. *J. Neurochem.* **47,** 496–505.

Sims N. R., Finegan J. M., and Blass J. P. (1986) Effects of postdecapitative ischemia on mitochondrial respiration in brain tissue homogenates. *J. Neurochem.* **47,** 506–511.

Srere P. A. (1985) Organization of proteins within the mitochondrion, in *Organized Multienzyme Systems: Catalytic Properties* (Welch G. R., ed.). pp. 1–61, Academic, Orlando.

Sterling G. H. and O'Neill J. J. (1978) Citrate as the precursor of the acetyl moiety of acetylcholine. *J. Neurochem.* **31,** 525–530.

Sun A. Y., Sun G. Y., and Foudin L. L. (1985) Aging, in *Handbook of Neurochemistry* **Vol. 9** (2nd Ed.) (Lajtha A., ed.). pp. 173–202, Plenum, New York.

Tucek S. and Cheng S.-C. (1974) Provenance of the acetyl group of acetylcholine and compartmentation of acetyl-CoA and Krebs cycle intermediates in the brain *in vivo. J. Neurochem.* **22,** 893–914.

Tzagoloff A. (1982) *Mitochondria,* Plenum, New York–London.

Van Dam K. (1973) The mitochondrion as a model of compartmentation, in *Metabolic Compartmentation in the Brain* (Balázs R. and Cremer J. E., eds.). pp. 321–329, MacMillan, London.

Van den Berg C. J. (1973) A model of compartmentation in mouse brain based on glucose and acetate metabolism, in *Metabolic Compartmentation in the Brain* (Balazs R. and Cremer J. E., eds.). pp. 137–166, MacMillan, London.

Van den Berg C. J., Matheson D. F., Ronda G., Reijnierse G. L. A., Blokhuis G. G. D., Kroon M. C., Clarke D. D., and Garfinkel D. (1975) A model of glutamate metabolism in brain: a biochemical analysis of a heterogeneous structure, in *Metabolic Compartmentation and Neurotransmission* (Berl S., Clarke D. D., and Schneider D., eds.). pp. 515–543, Plenum, New York.

Van Kempen G. W. J., Van den Berg C. J., Van der Helm H. J., and Veldstra H. (1965) Intracellular localization of glutamate decarboxylase, gamma-aminobutyrate transaminase and some other enzymes in brain tissue. *J. Neurochem.* **12,** 581–588.

Vyas I., Heikkila R. E., and Nicklas W. J. (1986) Studies on the neurotoxicity of 1-methyl-4-phenyl-1,2,3,6-tetrahydropyridine: inhibition of NAD-linked substrate oxidation by its metabolite, 1-methyl-4-phenylpyridinium. *J. Neurochem.* **46,** 1501–1507.

Wagner K. R. and Myers R. E. (1986) Hyperglycemia preserves brain mitochondrial respiration during anoxia. *J. Neurochem.* **47,** 1620–1626.

Walsh J. M. and Clark J. B. (1976) Studies on the control of 4-aminobutyrate metabolism in 'synaptosomal' and free rat brain mitochondria. *Biochem. J.* **160,** 147–157.

Whittaker V. P. (1969) The synaptosome, in *Handbook of Neurochemistry,* **Vol. 2** (1st Ed.) (Lajtha A., ed.). pp. 327–364, Plenum, New York.

Whittaker, V. P. (1984) The synaptosome, in *Handbook of Neurochemistry,* **Vol. 7** (2nd Ed.) (Lajtha A., ed.). pp. 1–39, Plenum, New York.

Wilkin G. P., Reijnierse G. L. A., Johnson A. L., and Balázs R. (1979) Subcellular fractionation of rat cerebellum: separation of synaptosomal populations and heterogeneity of mitochondria. *Brain Res.* **164,** 153–163.

Note Added in Proof:

Dunkley and coworkers [Dunkley P. R., Jarvie P. E., Heath J. W., Kidd G. J., and Rostas J.A.P. (1986) A rapid method for isolation of synaptosomes on Percoll gradients. *Brain Res.* **372,** 115–129] have recently reported a new method for the rapid isolation of heterogeneous populations of synaptosomes using discontinuous Percoll gradients. This method may be useful for the isolation of heterogeneous populations of synaptic mitochondria (P. R. Dunkley, personal communications).

The Use of Hippocampal Slices for the Study of Energy Metabolism

Tim S. Whittingham

1. Introduction

1.1. Basic Considerations

Mammalian brain slices are used to study a broad range of cellular neural functions, and their use as a neuroscience tool has greatly expanded in the past decade, especially in the realm of electrophysiological determinations. Warburg (1923) initially developed tissue slice techniques to study metabolism in vitro, and in the 1950s, brain slice research, developed by Quastel and Elliott, was still centered on metabolic phenomena. In those studies, McIlwain, Thomas, and others (Elliott, 1955; McIlwain, 1953, 1961; McIlwain and Bachelard, 1971; Thomas, 1956; Thomas, 1957) measured high-energy phosphate and related metabolite concentrations in cortical slices for different incubation parameters. The consistent deficit in ATP, phosphocreatine, and total adenylates from cortical slices compared to *in situ* cortex caused these investigators to try various additives in their artificial cerebrospinal fluid (ACSF) in an attempt to bolster slice metabolite profiles, such that they resembled the *in situ* condition. Additives such as creatine (Thomas, 1956) and adenosine (Thomas, 1957) were found to alleviate partially the loss of the adenylates and phosphocreatine pool. However, the metabolite profile of cortical slices still failed to match that of *in situ* cortex.

The interest in brain slices greatly increased with the electrophysiological findings by Yamamoto (Yamamoto and McIlwain, 1966; Yamamoto and Kurokawa, 1970) in the olfactory cortex and by Skrede and Westgaard (1971) and others (Schwartzkroin, 1975) in the hippocampal slice. The hippocampal slice preparation in particular has become popular because of the laminar distribution of the intact circuitry that is maintained in slices of hippocampus. It has been shown that hippocampal neurons maintain stable resting membrane potentials, input resistances, and synaptic responsiveness over incubation periods of 8 h and more (Schwartzkroin, 1975;

1977). Thus, it appears that at least some portion of the neuronal population in hippocampal brain slices exists in an essentially functional *in situ* state. Yet, like cortical slices, hippocampal slices are quite different metabolically from *in situ* hippocampus, exhibiting lowered adenylate and creatine levels (Lipton and Whittingham, 1984; Whittingham et al., 1984).

A previous publication (Lipton and Whittingham, 1984) pointed out that many of these metabolic differences can result from the physical trauma of slicing the brain. Thus, along the cut edges of the hippocampal slice there will be severely disrupted tissue, extending 50–100 μm into the slice (Bak et al. 1980; Garthwaite, et al., 1979; Misgeld and Frotscher, 1982). In addition, in thicker slices, there is a possible formation of an anoxic core (Fujii et al., 1982). Both of these events would contribute to lowered metabolite levels in the slice. Metabolite concentration is generally expressed in terms of "per mg protein" or "per μg dry weight" basis. These damaged regions would continue to contribute to the weight or protein factor, but fail to contribute optimal metabolite levels. The metabolic difference between brain slices and *in situ* tissue cannot fully be accounted for by tissue damage or the presence of anoxic tissue, and it appears that neural tissue undergoes some basic change during the process of preparing the brain slices (Lipton and Whittingham, 1984).

If brain slices are dissimilar to *in situ* tissue, then why use them to study metabolism of the brain? Brain slice preparations offer the same benefit for studying metabolism as they do for studying electrophysiology (Schwartzkroin, 1981). The medium bathing the tissue is a simplified environment, devoid of hormonal influences, diurnal rhythms, and extrinsic inputs. The environment can be easily, specifically, and rapidly changed to produce the experimental condition. These changes can include variation of extracellular ionic concentrations, glucose and oxygen availability, ambient temperature, and pharmacological manipulation. In addition, because the tissue is already exposed, it can be accessed rapidly for fixation, especially in relation to changes in the electrophysiological characteristics of the tissue. For instance, it is possible to monitor the electrical responsiveness of postsynaptic neurons during anoxia, and to freeze the tissue rapidly at the time when the postsynaptic response begins to decay (Lipton and Whittingham, 1982). This tissue can then be assayed for high-energy phosphate content. Discrete regions of the hippocampal slice can also be dissected out, and in this way, it was determined that ATP levels in the synaptic layer of the dentate gyrus were significantly decreased at the onset of synaptic transmission failure during anoxia, at a time when overall ATP levels in the hippocampal slice were unaffected.

However, it is important to remember that, when working with metabolic responses in the hippocampal slice, this tissue is different from *in situ* hippocampus. Is it valid to extend metabolic findings from brain slices to in vivo events? In general, most metabolic changes in experimental conditions mirror those changes seen in vivo. For instance, during ischemia in vivo, a rapid fall in phosphocreatine occurs, followed by ATP levels, and concomitant increases in ADP and AMP (Ljunggren et al., 1974). These same changes are observed in hippocampal slices when oxygen and glucose are removed from the ACSF (Whittingham et al., 1984). A good temporal relationship exists between the rate of fall of the high-energy phosphates in both conditions, although the magnitude of the decrease is less in the hippocampal slices because of the lowered control values for ATP and phosphocreatine. The rate of loss of the electrical response is also comparable for in vivo ischemia and the in vitro model of ischemia, and for in vitro anoxia (Lipton and Whittingham, 1979).

The remainder of the chapter will describe the range of metabolites and enzyme activities that can be measured by standard enzymatic techniques, and will describe the general theory and give two specific examples of these assays: (i) a fluorometric measurement of the creatine content of a hippocampal slice, and (ii) the luminescent technique utilizing luciferin–luciferase to measure ATP and phosphocreatine content. A comparison of some in vivo and in vitro metabolite values, and tissue preparation and fixation measurements for optimizing the in vitro values are also included.

2. Theoretical Considerations

Two protocols for measuring brain slice metabolites are described in this chapter. The flurometric technique, using coupled enzymatic systems, has been described extensively and extended to the measurement of many different metabolites and enzyme activities, although the luciferin–luciferase technique has only evolved over the past decade. A brief description of the theory behind these two types of procedures is included in this section.

2.1. Coupled Enzymatic Fluorescent Measurements of Metabolite Levels

Fluorometric measurements of metabolite levels are made possible because of the change in absorbance spectrum that occurs when NAD or NADP is reduced. There is a single absorbance peak at around 260 nm for the oxidized forms of NAD and NADP. When these compounds are reduced by the addition of a hydrogen ion at the #4 position of the pyridine ring, an additional absorbance peak is present at 340 nm. Thus, any metabolite that can be linked by an enzyme system to the production/utilization of NADH or NADPH can be measured indirectly on a spectrophotometer. In addition, fluorometric measurements can greatly increase the sensitivity of these coupled enzymatic assays. This is possible because the reduced form of NADPH or NADH can be excited at a wavelength of 340 nm, and these compounds will emit light at 460 nm. The fluoresced light can be measured by the photomultiplier tube in the fluorometer by placing appropriate filters to omit inappropriate wavelengths. These fluorometric measurements can be performed easily using a Farrand ratio fluorometer with a mercury lamp, Corning #5860 primary filter, and Corning secondary filters #4303 and #3387.

To measure a compound such as ATP by this method, the ATP must be linked to the production or utilization of the reduced form of NAD or NADP. This is accomplished by reacting ATP with glucose in the presence of hexokinase, producing ATP and glucose-6 phosphate. There is a 1:1 production of G-6-P for each ATP used in this reaction. The G-6-P is then reacted with NADP in the presence of G-6-P dehydrogenase, producing 6-phosophogluconolactone, NADPH, and one hydrogen ion (Fig. 1). The NADPH can be measured on a spectrophotometer or fluorometer, where increased NADPH will be reflected by an increase in absorption or fluorescence. The stoichiometric production of 1 NADPH/original ATP present allows sample ATP values to be calculated when compared to standards of known ATP concentrations. In addition, it is a simple procedure to measure phosphocreatine in the same reaction tube by the subsequent addition of creatine kinase and ADP. The phosphocreatine reacts with the ATP in the presence of the creatine kinase, producing creatine and ATP. This newly formed ATP is then utilized by the hexokinase and G-6-P de-

$$\text{Phosphocreatine} + \text{ADP} \xrightarrow{\text{Creatine kinase}} \text{Creatine} + \text{ATP}$$

$$\text{ATP} + \text{glucose} \xrightarrow{\text{Hexokinase}} \text{ADP} + \text{glucose} - \text{6-P}$$

$$\text{Glucose-6-P} + \text{NADP}^+ \xrightarrow[\text{dehydrogenase}]{\text{G-6-P}} \text{6-P-gluconolactone} + \text{NADPH} + \text{H}^+$$

Fig. 1.

hydrogenase system, producing a further increment of NADPH fluorescence.

The great flexibility of the enzymatic-fluorometric system comes from coupling different enzyme systems to produce or utilize reduced forms of the pyridines. Details on how to perform many of these assays are described in the book by Lowry and Passonneau (1972). Also, enzyme activities can be measured from tissue extracts by adding substrate for an enzyme to the reaction tube and measuring the rate of evolution of fluorescence. For instance, creatine kinase activity could be measured by the enzyme system shown in Fig. 1, where a reagent containing ADP, glucose, NADP, hexokinase, and glucose-6-phosphate dehydrogenase is combined with a tissue sample. Initial fluorometric readings would determine the rate of production of NADPH in the absence of phosphocreatine. Excess phosphocreatine is then added to the reagent sample mixture, and NADPH will be formed over time as a function of the amount of creatine kinase present.

2.2. Luminescence Measurements of Adenylate Phosphates and Phosphocreatine

The bioluminescence assays for adenylates and phosphocreatine are based on the light produced when ATP is combined with luciferase from the tail of the firefly (Photinus pyralis) by the reaction:

$$\text{ATP} + \text{D-luciferin} + O_2 \xrightarrow[\text{Luciferase}]{Mg^{+2}}$$

$$\text{Oxyluciferin} + \text{AMP} + \text{PP}_i + CO_2 + 0.9\,hv$$

Approximately 1 photon of 562 nm wavelength is emitted per molecule of ATP consumed. The enzyme is highly specific for ATP, nor do other phosphorylated nucleotides, such as GTP, interfere with the assay. The specific activity of this luciferase is 0.1 μeinstein/min, indicating that the velocity of the luciferase indicator reaction is relatively slow (Wulff et al., 1982). The ATP content of samples can be measured because of this relatively slow reaction rate and because the amount of luminescence produced is a linear function of the ATP concentration in the reaction tube (a higher ATP content will produce a greater amount of luminescence). The reaction rate assures that this luminescence will remain relatively stable over a period of several seconds to many minutes, depending on the relative degree of luciferase activity in the indicator reagent (McElroy et al., 1953; Wulff et al., 1982).

Compounds other than ATP can be measured by this method when they can be enzymatically linked to the production or utilization of ATP. For instance, phosphocreatine can be used to phosphorylate ADP, producing ATP in the presence of creatine kinase. Similarly, ADP and AMP can be phosphorylated to form ATP in the presence of pyruvate kinase and myokinase. These procedures will be described later in the chapter. Other methods have been developed to measure cyclic AMP and the activities of creatine kinase and pyruvate kinase, using the luciferin–luciferase method (Ebadi, 1971; Lundin et al., 1976; Wulff et al., 1983).

A number of commercially available photometers have been produced for primary use with bioluminescent methods. However, a fluorometer can be used for these measurements by inactivating the light source of the fluorometer, and removing or adjusting the secondary filters in the fluorometer (Lust et al., 1981). Then the light emission can be measured using the fluorometers photomultiplier tube.

There are several possible sources for error in the bioluminescent assay. Perhaps the greatest problem can result from ATP contamination in enzyme reagent preparations. The inclusion of appropriate blank tubes containing reagent and enzyme, but theoretically devoid of ATP, will allow calculation and correction for the contaminating ATP levels. However, high levels of contamination will significantly decrease the signal-to-noise ratio for samples and standards. If baseline luminescence is too high, the source of the contaminating ATP must be located and removed, either by enzymatic conversion of the ATP to ADP (Lowry and Passonneau, 1972), or by taking a preliminary luminescence reading in the photometer and injecting the samples to measure the change in luminescence. The light emission of the luciferin–

luciferase assay is also dependent on the temperature of the reagent. Increasing the temperature will shift the wavelength of the emitted light from 562 nm towards 615 nm. Consequently, if filters are being used to measure light emission at 562 nm, or if the photomultiplier sensitivity is greater for 562 nm, there can be a decrease in the luminescence if the reagent is warming. It is important to stabilize the temperature of the luciferin–luciferase reagent either at room temperature or in a controlled temperature water bath to prevent a shift in sensitivity of the assay as sequential samples are being read. This is particularly important since the luciferin–luciferase reagent is generally stored at –70°C and thawed just before the assay. Changes in pH and divalent cations such as zinc may also produce wavelength shifts (Wulff, 1983).

Finally, the emitted photons may be quenched by the turbidity of the reagent. Therefore, a constant dilution in each indicator tube must be maintained. In the assays in this chapter, 250 μL of luciferin–luciferase indicator reagent is added to 10 μL of blank, sample, or standard, maintaining a constant dilution. The intensity of the light emission is also influenced by effects of ionic strength of the reagent on the luciferase. This means that the sensitivity of different batches of luciferin–luciferase reagent may vary, depending on the ionic strength. The sensitivity will also vary if the reagent is repeatedly thawed and refrozen, producing a gradual decrease in luminescence/ATP over time. The problems listed here can be fully accounted for by including appropriate standards for each assay. That is, if ATP is being assayed, ATP standards should be used. Similarly, ADP standards should be used for ADP assays, and blank tubes should be used for all assays. Changes in the sensitivity of the luciferin–luciferase reagent will then be reflected in the standard and blank assay tubes, allowing an accurate calculation of ATP levels from sample tubes.

2.3. Comparison of Metabolite Values

As was mentioned previously, hippocampal brain slice metabolic parameters differ from the *in situ* state. ATP, phosphocreatine, total adenylate (ATP + ADP + AMP) and total creatine (phosphocreatine + creatine) are all reduced by roughly 50% in the brain slice, although AMP, lactate, and the PCr/ATP ratio are all increased. The energy charge of brain slices, the proportion of the total adenylate pool that is readily available as ATP (Atkinson, 1968), is roughly equivalent to that seem in vivo (.903 vs .946, respectively). The unchanged energy charge in slices suggests that the adenylates that remain in the slice are maintained in the same relative proportions that they are maintained *in situ*.

A broad range of situations could exist in the slice that would explain the 50% reduction in the total adenylate pool. The two ends of the spectrum would be that 50% of brain slice tissue is totally destroyed and devoid of metabolities, but contributing to the protein content of the slice (which is the basis for expression of metabolite concentration); or that all of the tissue within the slice has been depleted of 50% of its adenylate and total creatine content. Of these two possibilities, it seems likely that the first is closer to being correct. Brain slices exhibit poor histological preservation within the first 50–100 microns of cut surface. Much of this tissue could be completely destroyed, yet contribute to total slice protein. In addition, it is unlikely that energetic demands could be met and relative ATP proportions be maintained in cells that were 50% depleted of total adenylates.

In addition to the metabolite concentrations, other indices of metabolism are affected in brain slices. The rate of cortical brain slice oxygen consumption (Benjamin and Verjee, 1980) is about 40% of *in situ* cortical respiration (Bertman et al., 1979). Aerobic lactate production is greatly increased in brain slices from about 5% of glucose utilization (Hawkins et al., 1971; Norberg and Seisjo, 1975) to 50–60% in brain slices (Rolleston and Newsholme, 1967). Yet, even with this increased lactate production, there is an apparent alkylinization of intracellular pH in brain slices (Kass and Lipton, 1982) to a value of 7.40, compared to estimates of about 7.0 for *in situ* brain (MacMillan and Seisjo, 1972). The shifts in lactate production and acid–base balance in brain slices cannot readily be explained by the presence of dead tissue. Thus, it appears that, in addition to the presence of a substantial proportion of dead tissue, an underlying shift in brain cell metabolism in vitro also exists.

The two components of changes in brain slice metabolic parameters most likely occurred during the preparation of the slices. During this period, the brain is transiently ischemic for a period of several minutes. It is also during this time that edge damage occurs with slicing of the tissue. Following the onset of incubation, high-energy phosphate values are rapidly optimized, reaching steady-state levels within 2 h (Whittingham et al., 1984). Generally, subsequent transient anoxic periods of less than 5 min do not further decrease the adenylate or creatine pools, suggesting that the metabolic shifts occur during the decapitation ischemia

(Lipton and Whittingham, 1982). For a more detailed consideration of these topics, *see* Lipton and Whittingham (1984).

3. Methodology

The following section will detail the equipment and specific procedures necessary for generating metabolic data from brain slices. Although the emphasis of this section will be on the preparation and analysis of samples for ATP, phosphocreatine, and creatine, preparation and incubation techniques for brain slices are also included.

3.1. Brain Slice Preparation and Incubation

The inherent problems with brain slice preparations create a tissue that is, to some degree, unlike that of in vivo brain. Yet, it is intended that the results obtained from brain slices be applied to the in vivo state. Therefore, it is beneficial to take whatever precautions are possible to produce brain slices that most closely resemble *in situ* brain tissue. Proper incubation and preparation techniques were a primary concern in the early days of brain slices, and a resurgence of such studies has taken place in recent years (for review, *see* Alger et al., 1984; Schwartzkroin, 1981; Teyler, 1980; Whittingham et al., 1984).

A typical brain slice experiment would begin by stunning the animal with a blow to the back of the neck or by anesthetizing the animal with ether or barbiturates. Of these three, barbiturate anesthesia may provide some protective effect against the ensuing ischemia (for review, *see* Shapiro, 1985), and could improve slice function or survival during the incubation period. The animal is then decapitated, the cranium removed, and the brain immediately doused with iced artificial cerebrospinal fluid (ACSF). The brain is rapidly removed from the skull and placed in iced ACSF.

One hemisphere is removed from the ACSF and placed on a piece of moistened filter paper on a chilled, hard surface. If larger rodents are used, such as the rat or guinea pig, the hippocampus can be removed and sliced as an individual unit. However, smaller rodents, such as mice and gerbils, have hippocampi that are difficult to remove without causing some tissue damage during removal or tissue slicing. For this reason, mouse and gerbil hippocampal slices are prepared by slicing the entire hemisphere using a Vibratome. The following procedures would be used for the rat hippocampus. The medial portion of the hemisected brain is placed face up on the filter paper. The brainstem and midbrain portion is then reflected upward and anteriorly, exposing the hippocampus. A small cut of the dorsal cortex and the fimbria ventrally loosens the hippocampus so that a metal spatula can be gently slid into the lateral ventricle, and the hippocampus rolled upward and posteriorly. At this point, the entorhinal cortex can be cut, and the hippocampus is completely detached from the brain. The hippocampus is immediately doused with iced ACSF, and positioned so that the dentate gyrus is exposed. A blood vessel is clearly evident curving along the length of the hippocampus and can be used as a reference for the cutting of hippocampal slices. Cuts should be made roughly perpendicular to this vessel.

Several different ways of slicing this isolated hippocampus exist. These techniques range from cutting free-hand with a razor blade or bow cutter (McIlwain, 1979), where the actual slice thickness is not accurately determined, and which takes some practice in order to produce slices of fairly uniform thickness; to motorized slicing units such as tissue choppers and Vibratomes. Although these units slice with fairly uniform thickness, there are indications that the robustness of the cutting strokes can produce more tissue damage than is seen with gentler procedures (Garthewaite et al., 1979). Perhaps the best apparatus to use is a razor blade attached to a slightly weighted arm, with the hippocampus placed on a mechanized stage. The tissue can then be advanced a known distance, and the weighted arm simply dropped from a predetermined height. However, there appears to be no statistically significant difference in metabolite levels measured in slices prepared in either of these three ways (Whittingham, unpublished observation).

Optimal slice thickness has also been studied by a number of different laboratories with respect to oxygen availability, electrical responsiveness, and metabolite concentrations (Bak et al., 1980; Frotscher et al., 1981; Fujii, 1982; Lust et al., 1982). The slice thickness to be used is a trade-off between minimizing the percentage of tissue that is present in the damaged tissue along the cut edges of the slice, and maximizing the percentage of healthy tissue in the center of the slice without producing an anoxic core region. Oxygen availability studies and theoretical considerations would suggest that slices of about 500 μm are maximal before seeing anoxic core effects. However, slices of 750 μm thickness can be used and still have good metabolite values and, in at least one type of chamber (Newman et al., 1986), brain slices of 1000 μm exhibited good histological preservation, even in the central core areas. In practice, most laboratories cut at a thickness of between 400–500 μm (*see* Dingledine, 1984; Kerkut and Wheal, 1981).

The slices are gently rolled off the razor blade and immediately placed into chilled ACSF until all slices have been prepared. The brain slices can then be incubated in one of several ways. If electrophysiological measurements are to be made, then the slices can be placed in an Oslo-style (Li and McIlwain, 1957; Richards and Sercombe, 1970; Schwartzkroin and Anderson, 1975) or Haas interface-type chamber (Haas et al., 1979), or in one of several submersion-type chambers (Alger et al., 1984). In any of these chambers, the brain slices are suspended on a mesh and maintained at a temperature generally in the range of 32–37°C. The ACSF is equilibrated with 95% O_2/5% CO_2 in the submersion-type chambers; in the interface-type chambers, warmed and humidified gas is also passed across the top of the slices. If only metabolic parameters are to be measured in a given experiment, then the slices can be maintained in beakers containing ACSF that is continually gassed with 95% O_2/5% CO_2, and the beakers are incubated in a heated water bath.

Both the electrophysiological chamber and water bath techniques produce comparable metabolite levels in hippocampal slices. In either case, the slices should be allowed to equilibrate for at least 1 h, and preferably 2 h, before obtaining metabolic measurements (Whittingham et al., 1984), or for electrophysiological measurements (Langmoen and Anderson, 1981). At that time, control slices should be removed and processed, as will be described in the next portion of this section, and the environment changed to the experimental paradigm. Slices can generally be incubated in control conditions for at least 8 h without showing metabolic decline (Whittingham et al., 1984). The content of the ACSF bathing the slices varies between each laboratory, mainly in the composition of potassium, calcium, and glucose. In this laboratory, the standard ACSF contains 125 m*M* NaCl, 3 m*M* KCl, 1.4 m*M* KH_2PO_4, 26 m*M* $NaHCO_3$, 2.4 m*M* $CaCl_2$, and 10 m*M* glucose. The adjusted pH, following aeration by 95% O_2/5% CO_2, is in the range of 7.35–7.45. It should be noted that ACSF containing no calcium and elevated magnesium during slice preparation may decrease the ischemic damage to the slice (Kass and Lipton, 1986).

What problems might be arising during the preparation and incubation procedures that can affect optimal metabolite values? During the preparation procedure, it is important to be as gentle as possible with the tissue and also to minimize the amount of preparation time prior to incubation, for during this time, the tissue is essentially ischemic. Prolonged energy stress can produce increased nonfunctional and dead tissue in the brain slices (Kass and Lipton, 1982; West et al., 1986), as can rough handling. Since metabolite levels are generally expressed in a per milligram protein or per dry weight basis, and because damaged tissue will continue to contribute to the milligram protein content and dry weight of the tissue, ischemia and rough handling can produce aberrantly low metabolite values. Steps that can be taken to minimize the ischemic damage include chilling the tissue as rapidly as possible to retard metabolic rate (Reid et al., 1986; Whittingham et al., 1984), quickness of preparation, and possibly barbiturate anesthesia and removal of calcium from the preparation medium. Tissue disruption by physical events can be minimized by handling the brain gently, and by using gentle slicing procedures (Pohle et al., 1986). When incubating slices in water baths, it may also be beneficial not to have the slices moving vigorously within the bath (Pohle et al., 1986).

Poorly prepared or incubated slices will generally exhibit lowered ATP and phosphocreatine levels with elevated ADP, AMP, and creatine content. A good indication of relative survival of tissue in the slice can be obtained by looking at the total creatine content, the sum of phosphocreatine and creatine within the slice (Alger et al., 1984). In control slices, the higher the total creatine content, generally, the better the energy charge and overall metabolite profile of the given slice. Continual oxygen presence must be maintained for the slices to remain healthy. While submersed in a water bath, slices will frequently maintain their high energy phosphate profiles for quite a while; whereas in a pregassed ACSF, interface slices will rapidly degenerate in the absence of continuous aeration (Whittingham, unpublished observation).

3.2. Preparation of Samples for Analysis

When studying labile metabolites such as the adenylates and phosphocreatine, the tissue must be fixed very rapidly to prevent metabolism of these compounds (*see* Lust et al. in this volume). In the case of brain slices, rapid freezing in either liquid nitrogen or chilled Freon usually accomplishes this. If a number of slices are to be fixed simultaneously, it may be possible to remove the slices on their suspensory mesh from the incubation chamber and freeze the entire mesh in liquid nitrogen. However, it is more common to freeze only one or two sections at a time. In this case, the slices can be picked up with forceps, taking care to grab the tissue by the entorhinal cortex portion, and gently slide onto a small metal spatula. The spatula is then rapidly placed into a small container of liquid nitrogen and stirred. The entire slice will be frozen in a matter of several seconds. The spatula and liquid nitrogen are then transported to a –70°C freezer, which contains prechilled storage vials. The spatula can be slid into the vial, and by gentle pressure with a blunt forceps on the edge of the slice, the slice will come loose from the spatula and fall into the vial. The slices can then be stored for an indefinite period of time at –70°C with no adverse effects on high-energy phosphate levels. Tapered 1.5 mL plastic tubes with caps work well as the storge containers. In addition, there are commercially available pestles that fit the tapers of these tubes, allowing the storage vials to be used for extraction also.

When infraregions of the slice are of interest for a given study, those regions must be dissected out. This is difficult to do on frozen slices, since at –70°C the slices will shatter rather than cut. Slices can be dissected at –20 to –30°C; however, this is rather unwieldy, since it necessitates the use of a glove box-type cryostat or –20°C cold room. It is preferable to perform these dissections on lyophilized tissue. To do this, the slices are first placed in any of a number of commercially available tissue holders that allow free equilibration with ambient pressure. The tissue holders are then placed in evacuation chambers that are available from Vertis or Ace Glassware, taking care to keep the slices chilled at all times. A two-stage vacuum pump is connected to a propylene glycol–dry ice trap, which, in turn, is connected to the evacuation chamber. The evaculation chamber is placed in a –40°C freezer. The pressure in the chamber should drop below 10 Torr, and the lyophilization should be performed for a period of at least 12 h at –40°C. At that point, the evacuation chamber is removed from the –40°C freezer, and an additional 1-h period of lyophilization is performed at room temperature.

The resulting slices can be dissected at room temperature, and all characteristic landmarks of the hippocampal slice are evident in the lyophilized slices. The slices can then be roughly dissected by using either a microscalpel or a #11 scalpel blade. The lyophilized slices contain intrinsic fracture planes along cell body layers. For instance, pressure along the granule cell body layer in the dentate gyrus will cause the molecular layer tissue to fracture from the remainder of the slice. Similar fracture planes are present in areas CA1 and CA3, and along the interfaces between major fiber tracts and neighboring tissues such as the perforant path. If more discrete dissection is necessary, the frozen slices should first be sectioned at about 20-μm thickness in a cutting cryostat and the sections lyophilized as described above. The resulting sections can be very discretely dissected, yielding pieces of tissue of about 1 μg dry wt, and metabolic analyses carried out on these pieces of tissue. The procedures for measuring metabolites in very small tissue samples, such as 1μg dry wt, will not be detailed in this chapter, but are described in the book by Lowry and Passonneau (1972).

Once the tissue has been properly fixed and possibly dissected, the metabolites must be extracted from the cells. The following tissue extraction procedure is for whole slice metabolite analysis, and reagent and extraction volumes should be adjusted for smaller tissue samples. Roughly dissected tissue samples simply require an adjustment of the listed volumes. The 1.5 mL tissue capsules containing frozen slices are removed from the –70°C freezer and placed in an alcohol–dry ice bath. One tube at a time is removed from the dry ice bath, and 100 μL of chilled 0.3*N* perchloric acid (PCA) containing 1 m*M* EGTA is added to the capsule. A Kontes pellet pestle is then pressed into the bottom of the tissue capsule, and the capsule is rotated, crushing the slice. This continues for 1–2 min until the slice is completely dispersed in the PCA. The suspension should look cloudy, with little or no particulate matter present. At this time, an additional 400 μL of the PCA–EGTA mixture is used to rinse the pestle, bringing the total extraction volume to 500 μL. The capsule is then capped and placed in an ice bath. This procedure is repeated for each slice, until all capsules are in the ice bath. Theoretically, the PCA immediately permeates the slice as it thaws, inactivating all enzyme activity. The only metabolism of cellular components that will continue will be nonenzyme mediated, and the solutions are kept chilled to minimize these metabolic changes. The chilled capsules are taken to a cold room, placed in a microfuge, and spun for 10 min at 10,000g to pellet the protein components of the extracts. The capsules are once again placed in an ice bath, and 400 μL aliquots of the supernanant containing the acid soluble metabolites are transferred to 10 × 75 mm tubes that contain 40 μL of chilled 3*N* $KHCO_3$. This neturalizes the acid and produces a white precipitate in the bottom of the tube, which can remain in the tube. These tubes are briefly vortexed and stored in a –70°C freezer until metabolite analyses are to be performed. The fixation and extraction procedures are outlined in Fig. 2.

The remaining supernatant is decanted from the protein pellet, and the pellet dissolved in 400 μL of 1*N* NaOH. The pellets are generally left overnight at room temperature in order to dissolve the protein fully. The protein content of each slice is analyzed in order to express metabolite values on a per mg protein basis. The protein content is analyzed according to the method of Lowry et al. (1951) using a 10 μL sample of the protein extract. The 10 μL aliquot is added to 1 mL of Protein Reagent. This reagent is prepared at the time of the assay by adding 1 mL of 0.5% copper sulfate to 50 mL of stock reagent. The stock reagent can be stored at room temperature, and is prepared by adding 20 g of Na_2CO_3, 200 mg of Na tartrate, and 10 mL of 10*N* NaOH to 1 L of distilled water. The protein content of the sample is photometrically determined by comparing the optical absorbance of protein samples at 700 lambda to the absorbance of known BSA standards. Standards range from 2 to 12 μL of 1 mg/mL BSA protein standard. In addition to the sample and standard tubes, blank tubes are included that contain 10 μL of 1*N* NaOH. The tubes are vortexed and allowed to sit for 20 min prior to the addition of 50 μL of Folin Ciocalteu phenol reagent. The tubes are immediately vortexed and allowed to react for an additional 30 min prior to reading them on the photometer. As the sensitivity of this assay is dampened at higher photometric readings, care should be taken to keep the sample readings within the range of the BSA standards. Readings that exceed the standard range should be repeated using smaller sample aliquots. Each set of standard values is then averaged, with the blank readings subtracted from the average. This net value is divided by the μg of protein that had been added to those tubes. The resulting delta/μg protein is averaged for each of the standards within the range. Sample values are calculated by first subtracting the blank and then dividing by the average delta/μg protein. The μg protein total in a slice can then be calculated by multiplying by the appropriate dilution factor. In this case, the dilution factor is 400 μL total extract divided by 10 μL sample in the assay tube, or 40. The protein can be expressed in terms of mg by dividing this value by 1000. At this point, the tissue has been extracted, and any further metabolite analysis can be expressed on a per mg protein basis for the slice. The protein assay procedure and calculations are shown in Fig. 3.

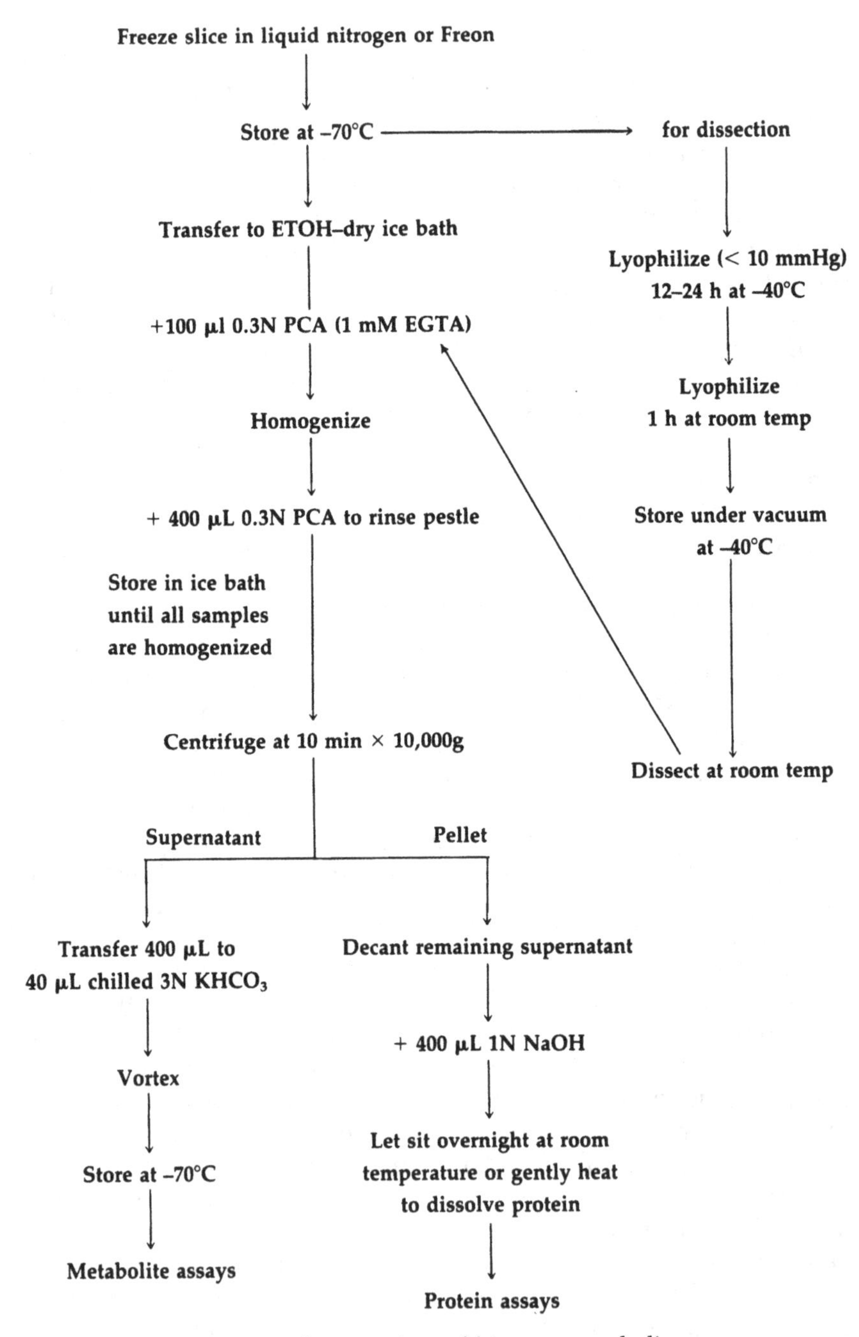

Fig. 2. Fixation and extraction of hippocampal slices.

Procedure

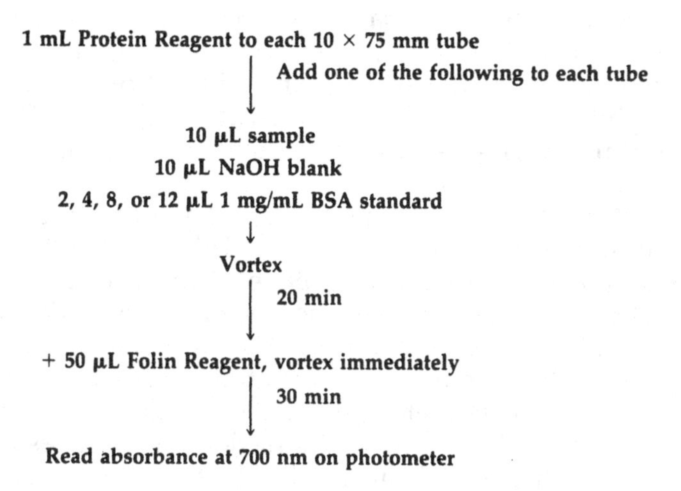

Calculation

$A_{standard} - A_{blank} = A_{net}$

A_{net} / (µg standard) = A_{net} / µg protein

Then to calculate sample protein

$$\left(\frac{(A_{sample} - A_{blank})}{(A_{net}/\mu g\ protein)}\right)(\text{Dilution Factor})\left(\frac{1\ mg}{1000\ \mu g}\right) = \text{mg protein in slice}$$

Dilution Factor = 400 µL NaOH total extract/10 µl assayed = 40

Fig. 3. Protein assay

What are some of the factors that may have occurred during this preparation process and that also may have affected the outcome of the metabolite assays? The tissue may become transiently anoxic or hypoxic during the freezing process when transferring the slice onto the spatula and into the liquid nitrogen. This can result in a rapid fall in phosphocreatine content. Consequently, care should be taken to freeze the slices as rapidly as possible once they are removed from the brain slice chamber or incubation beaker. Stirring the slices in the liquid nitrogen minimizes the gas interface that develops around the slice as it warms the liquid nitrogen, and accelerates the freezing process. Chilled Freon can be used to avoid this gas interface problem. Similarly, warming can occur during any of the transfer steps or while preparing the slices to be lyophilized. In particular, body heat can be rapidly transferred to the tissue holding capsules while handling them, and care should be taken to handle them as little as possible. Often, poorly lyophilized tissue is easily spotted by the absence of a white chalky appearance of the slices, or presence of a sticky interior core of the slice.

More likely, metabolic shifts will occur in frozen slices during the time of the tissue extraction. This can result from excessive warming of the slice prior to penetration of the tissue by the PCA, and also by the incomplete homogenization of the tissue with the pestle. In the latter case, larger chunks of unextracted tissue will be pelleted down in the protein fraction. This tissue will not contribute its complete metabolite content to the supernatant, yet will contribute fully to the protein fraction. This will result in a general decrease in all metabolite levels. The tissue extracts must be maintained in an ice bath or kept frozen at all times. Allowing the extracts to remain at room temperature for any prolonged period of time can result in the loss of high-energy phosphate levels, and consequent increase in creatine, ADP, and AMP. It is also notable that some myokinase activity, in particular, appears to be able to survive the PCA extraction process. Thus, keeping the tissue chilled helps retard any myokinase activity that may be present in the extract. Finally, the incorrect determination of protein content in a slice will result in all calculated metabolite values being different from what they should be, and care should be taken to keep the absorbance values for protein samples within the linear range of the protein assay. Exceeding the linear range will result in an underestimation of the protein content of that slice, and the resulting overestimation of all metabolite concentrations.

3.3. Analysis of ATP, Phosphocreatine, and Creatine from Brain Slices

The photometric determination of ATP and phosphocreatine using luciferin–luciferase luminescence and the fluorometric determination of creatine are described in this section. These serve as two sample assay procedures for brain slice metabolites. The luciferin–luciferase procedure can be adapted to any compound that, by enzymatic methods, can result in the production of ATP, but the flurometric method for creatine can be adapted to a broad range of metabolite determinations where the metabolite of interest can be enzymatically linked to the production or utilization of either NADH or NADPH. Many of these related procedures are described in detail in the book by Lowry and Passonneau (1972).

3.3.1. Luciferin–Luciferase Determination of ATP and Phosphocreatine

This procedure uses the enzyme–coenzyme system found in the tails of fireflies, where increased lumninescence is produced with increased ATP content. Phosphocreatine can be measured by using the high-energy phosphate to phosphorylate ADP, producing a stoichiometric increase in the ATP content of the sample. Similarly, ADP and AMP can be measured using this method by phosphorylating these compounds in the presence of pyruvate kinase and myokinase. The luciferin–luciferase technique has the benefits of being more sensitive than the standard fluorometric assays for adenylates, and of being very rapid to perform. However, this technique is relatively expensive to perform, in that the reagents used are quite costly and the initial cost of specialized equipment to perform the assay is greater than that of the fluorometer.

The luciferin–luciferase indicator reagent is prepared in bulk and stored frozen at –70°C in 50 mL aliquots. The reagent recipe is shown in Table 1. The dithiothreitol and BSA are included to help stabilize the enzymes, and the EGTA is present to complex any calcium. Finally the AP_5A [P^1,P^5-Di(adenosine-5'-)pentaphosphate] is present to inhibit the contaminant myokinase activity found in the luciferase. The indicator reagent is stored in small aliquots, since repeated thawing and refreezing of the reagent will result in a decreased luminescence per ATP aliquot, which would decrease the sensitivity of the assay. Also, the luminescence per ATP available is dependent on the ambient temperature of the reagent. Thus, care should be taken to allow the reagent to reach room temperature prior to the indicator step, assuring that the sensitivity of the assay is not changing during the metabolic readings.

The tissue samples are removed from the –70°C freezer and thawed rapidly in a water bath. Once thawed, the tissue samples are placed in an ice bath and briefly vortexed. ATP and phosphocreatine standards are prepared by taking 5 µL of 1 m*M* stock standard solutions of ATP and phosphocreatine and adding them to separate 200 µL aliquots of neutralized PCA (*N*-PCA). The first ATP standard is then diluted 1:1 four times by transferring 100 µL

Table 1
Luciferin–Luciferase Indicator Reagent (2 L)

Final*	Volume	Stock*	Compound
50 m*M*	200 mL	0.5*M*	Glycyl-Glycine (pH 8)
2 m*M*	2 mL	2 *M*	Dithiothreitol
2 m*M*	40 mL	0.1*M*	EGTA
2 m*M*	4 mL	1 *M*	$MgCl_2$
0.02%	4 mL	10 %	BSA
88 m*M*	1 vial		Luciferin (50 mg)
5 μg/mL	2 mL	5 mg/mL	Luciferase
	4 μL	10 m*M*	AP_5A

*Dilute to 2 L with H_2O.

of the first standard into 100 μL of neutralized PCA. This mixture is vortexed, 100 μL of the second standard is transferred to 100 μL of neutralized PCA, and so forth. This procedure is repeated for the 1 m*M* phosphocreatine standard, producing a total of 5 ATP standards and 5 phosphocreatine standards. Ten microliters of each standard tube are transferred to a corresponding 10 × 75 mm reaction tube, and 10 μL of sample extract are also transferred into corresponding 10 × 75 mm tubes. In addition, a set of blank tubes is prepared that contains 10 μL of neutralized PCA. A 100 μL aliquot of ATP-phosphocreatine reagent is then added to all 10 μL samples, blanks, and standards. The ATP-phosphocreatine reagent is shown in Table 2A.

The 100 μL reagent plus 10 μL of sample, blank, or standard tubes are then briefly vortexed, and 10 μL aliquots are removed from each tube and pipetted into 6 × 50 mm indicator tubes. At this point, sample aliquots will contain all metabolites that were present in the tissue extract. However, only the ATP that was present in the slice at the time of fixation will cause any luminescence in the luciferin–luciferase indicator step. Thus, the ATP content of the tissue extracts can be estimated by simply adding a 250 μL aliquot of the luciferin–luciferase indicator reagent, vortexing immediately, and placing the 6 × 50 mm tube into the photomultiplier tower of a Chem-Glow photometer (Aminco Bowman). The luminescence that is produced as a function of the amount of ATP in the extract can then be directly read from the galvinometer scale on the photometer. Integrator and print out components can also

Table 2
A. ATP-Phosphocreatine Reagent (20 mL)*

Volume used	Stock concentration	Compound
2 mL	1*M*	Imidizole (pH 7.0)
20 μL	1*M*	$MgCl_2$
120 μL	10 m*M*	ADP
4 μL	10 m*M*	AP_5A

*Dilute to 20 mL with H_2O.

B. Reagent for Total Adenylates (20 mL)*

Volume used	Stock concentration	Compound
1 mL	1*M*	Imidizole (pH 7.0)
40 μL	1*M*	$MgCl_2$
1.5 mL	1*M*	KCl
600 μL	0.1*M*	PEP

*Dilute to 20 mL with H_2O.

be purchased for the Chem-Glow unit. If these are present, luminescence can be integrated over a 3-sec period, and the value automatically printed out. This increases the signal-to-noise ratio of the assay and decreases the amount of time necessary to read the indicator tubes.

The ATP standard tubes are used to determine how much luminescence is generated per picomole of ATP in the 6 mm tubes. The blank tubes are used to subtract out any ATP-independent luminescence from the indicator reagents and the ATP-phosphocreatine reagent. This luminescence results from ATP contamination of the ADP added to the ATP-phosphocreatine reagent, and also from any ATP contamination or nonspecific luminescence in the luciferin–luciferase indicator reagent. If the blank readings are substantial in relation to the sample readings, then the blank readings must be reduced. The primary contribution to the blank generally results from ATP contamination of the ADP added to the ATP-phosphocreatine reagent. Consequently, it may be necessary to remove ATP from the ADP solution, and this can be done as described in the book by Lowry and Passonneau (1972).

The phosphocreatine content of the extracts can be measured by adding 10 μL of a creatine kinase solution. This solution is prepared by adding 5 μL of a 50 mg/mL creatine kinase stock to 2 mL of reagent, and then pipetting 10 μL of this into each 100 μL that remain in the 10 × 75 mm reaction tubes. These tubes are vortexed and allowed to react for 30 min. A 10 μL aliquot of the reaction reagent is then removed from each tube and pipeted into a corresponding 6 × 50 mm tube. The presence of the creatine kinase in a reagent containing ADP and sample containing phosphocreatine results in the stoichiometric 1:1 production of ATP for each phosphocreatine present. Incomplete reaction of all phosphocreatine in the sample will automatically be corrected for, because the phosphocreatine standards will similarly not be totally reacted. Once again, 250 μL of the luciferin–luciferase indicator reagent is added to each new indicator tube, vortexed, and placed in the photometer. The resulting luminescence is a function of the sum of all ATP and phosphocreatine that is present in the indicator tube. The ATP-PCr assay procedure and calculations are outlined in Fig. 4.

The ATP content of the tissue is calculated by subtracting the blank luminescence from the luminescence produced by the standard, and dividing by the total picomoles of ATP in each tube. The ATP content for each tissue sample in the 7 × 50 mm tubes is calculated by subtracting the blank from the luminescent values of the sample and dividing by the luminescence/pmol of ATP calculated for the standards. The resulting value must be extended to total ATP content in the whole slice extract by multiplying the ATP content of the indicator tube by a dilution factor. The calculations and estimate for dilution factors in the ATP and phosphocreatine assays are shown in Fig. 4. The final result is in terms of picomoles of ATP in the total slice extract. This value is divided by the mg protein for that slice, yielding a value of pmol ATP/mg protein. The results from the phosphocreatine assay are calculated similarly; however, the final value represents ATP + phosphocreatine, and the value obtained from the ATP assay must be subtracted to obtain the phosphocreatine concentration in the slice.

As was mentioned earlier, the luciferin–luciferase assay can be readily adapted to measuring ADP and AMP, as well as phosphocreatine. The indicator reagent and the procedures outlined for the ATP-phosphocreatine assay are essentially the same in assaying total adenylate content (ATP + ADP + AMP). Adenylate reagent is used in place of the phosphocreatine reagent. This reagent is outlined in Table 2B, and the addition of creatine kinase is replaced by sequential additions of 10 μL pyruvate kinase (200 μL of 10 mg/mL, Boehringer-Mannheim stock in 2 mL reagent) and myokinase (100 μL of 5 mg/mL Boehringer-Mannheim stock in 2 mL reagent). The set of phosphocreatine standards is replaced by two sets of standards: one of ADP and one of AMP.

When performing any of these assays, always use accurately calibrated standards of the particular compound that is being assayed. If for some reason, the assay does not allow full reaction of a given compound, the standards will also be similarly inhibited from reacting, and an accurate estimate of metabolite concentration should still be obtained.

Calibration of an ATP standard solution is performed in the following way: One mL of reagent is added to each cuvet in a photometer. The reagent is composed of 50 m*M* Tris-HCl (pH 8.1), 1 m*M* $MgCl_2$, 0.5 m*M* dithiothreitol, 500 μ*M* NADP, 1 m*M* glucose, and 0.5 μg/mL of glucose-6-phosphate dehydrogenase (yeast). Crystalline ATP from equine muscle (Sigma) is diluted in water to a nominal concentration of 1 m*M*. Fifty μL of the 1 m*M* solution of ATP is added to 1 mL of reagent in the cuvette. Other cuvets will serve as blanks by the addition of 50 μL of water. An initial reading is taken from each cuvet at a wavelength of 340 nm. Each cuvet then receives 2 μg/mL hexokinase (yeast), and subsequent readings are taken until the reaction has run to completion. In the case of ATP, this should take approximately 6 min. The actual concentration of the 1 m*M* ATP standard can be calculated in the following way: The change in optical density of the blank cuvets is subtracted from the change in optical density of the ATP cuvettes. This net value is then divided by 6270, the molar extinction coefficient for NADPH, and multiplied by 1000 to convert to m*M* concentration. Finally, the resulting value is multiplied by a dilution correction factor: the volume of the reagent plus the volume of the sample plus the volume of the enzyme addition, divided by the volume of the sample. The calibrated standard should be kept frozen when not in use, and on ice whenever thawed. In this way, the standard will remain stable for many months.

3.3.2. Fluorometric Determination of Creatine

The fluorometric determination of creatine was selected as an example of a broad range of fluorometric assays for brain metabolite levels. Many of these assays are described in the book by Lowry

Procedure

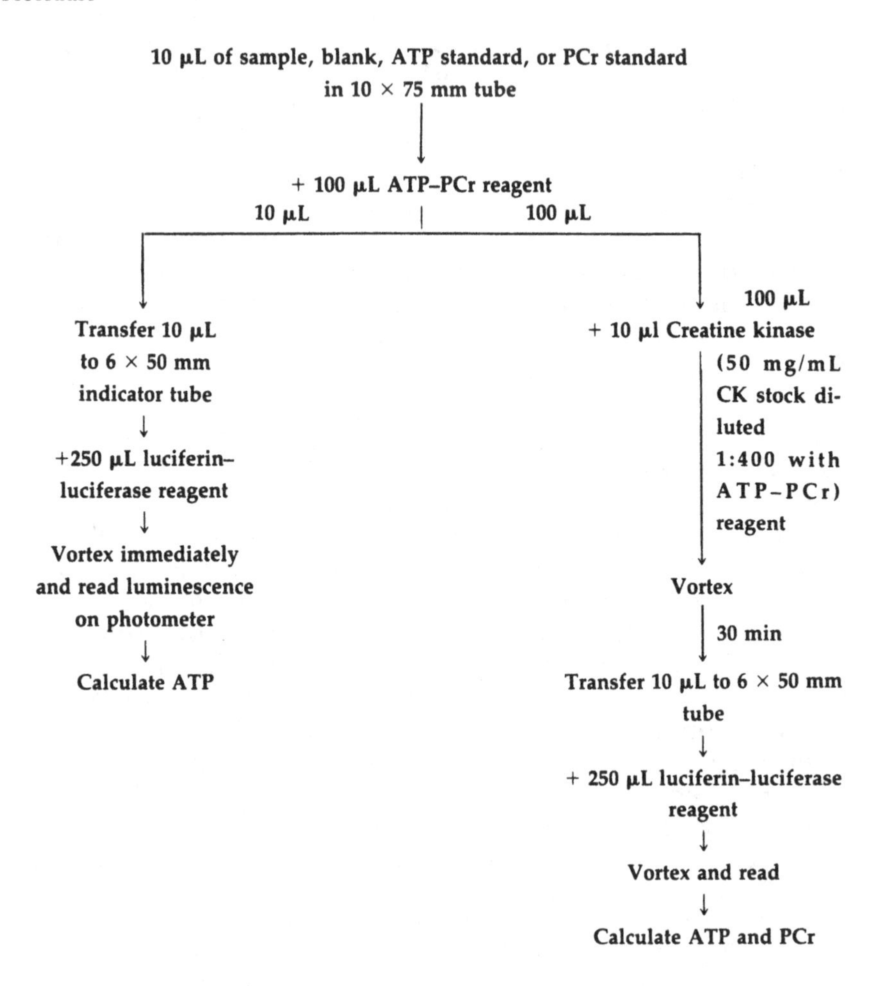

Fig. 4A. ATP and PCr assay.

and Passonneau (1972), and most work by producing NADH or NADPH from their oxidized forms. The creatine assay, however, involves the utilization of existing NADH. Thus, there is a stoichiometric decrease in fluorescence related to the amount of creatine that is present in the assay tube. This means that the sensitivity of the assay is determined by the amount of NADH added to the initial reagent, making this assay slightly more com-

Calculation

Standards

$$\frac{(5\ \mu L)\ (1\ mM\ ATP\ or\ PCr\ standard)}{(0.200\ mL\ N\text{-}PCA + 0.005\ mL\ standard)} = [\mu M]\ of\ high\ standard$$

$$([\mu M]_{high})\left(\frac{(10\ \mu L)}{(6 \times 50\ mm\ tube)}\right)\left(\frac{(10\ \mu L\ standard)}{(110\ \mu L\ in\ 10 \times 75\ mm\ tube)}\right) = \text{Picomole standard in } 10 \times 75 \text{ mm tube}$$

Divide by two 4× to obtain range of standards

$$\frac{(\Delta\ standard - \Delta\ blank)}{(picomoles\ of\ ATP\ in\ tube)} = \text{Luminescence/pmol ATP} = \Delta/\text{pmol}$$

Then to calculate sample ATP

$$\left(\frac{(\Delta\ sample - \Delta\ blank)}{(\Delta/pmol)}\right)(\text{Dilution factor})\left(\frac{1}{(mg\ protein)}\right) = \frac{nMoles\ ATP}{mg\ protein}$$

$$\text{Dilution factor: }\left(\frac{(110\ \mu L\ sample + reagent)}{(10\ \mu L\ indicator\ tube)}\right)\left(\frac{(440\ \mu L\ N\text{-}PCA)}{(10\ \mu L\ to\ assay)}\right)\left(\frac{(500\ \mu L\ PCA)}{(400\ \mu L\ transferred)}\right)$$

$$= 605\left(\frac{(1\ nmol)}{(1000\ pmol)}\right) = 0.605$$

For PCr and ATP, the dilution factors for the standards and sample calculations are adjusted for the addition to 10 µL creatine kinase:

$$\text{Standard in } 6 \times 50 \text{ mm tube} = (\text{picomoles for ATP assay})\left(\frac{(100\ \mu L\ before\ CK)}{(110\ \mu L\ after\ CK)}\right)$$

$$\text{Sample dilution factor} = (\text{DF for ATP})\left(\frac{(110\ \mu L\ after\ CK)}{(100\ \mu L\ before\ CK)}\right) = 0.667$$

Fig. 4B.

Table 3
Reagent for Fluorometric Analysis of Creatine

Volume	Stock*	Compound
10 mL	1*M*	Tris-HCl (pH 8.0)
100 μL	10 m*M*	NADH (in carbonate buffer pH 10)
50 μL	0.1*M*	PEP
400 μL	0.1*M*	ATP
1 mL	1*M*	$MgCl_2$
6 mL	1*M*	KCl
100 μL	10 m*M*	AP_5A
100 μL	10 mg/mL	Pyruvate kinase
40 μL	5 mg/mL	Lactate dehydrogenase

*Dilute to 200 mL with water.

plex than if NADH or NADPH were being produced. Some idea of the relative sensitivity of the luciferin–luciferase assay vs the fluorometric procedures is gained by looking at the sample size needed to measure creatine vs phosphocreatine. Roughly 1 μL of total extract can be easily assayed for phosphocreatine content by the luciferin–luciferase technique; however, 25 μL of extract is necessary to accurately measure creatine by the fluorometric technique, even though the concentrations of creatine and phosphocreatine in brain slices are roughly equal. In addition to the decreased sensitivity and the necessity of setting the sensitivity of the assay by the amount of NADH added, the creatine assay is further complicated by a persistent and drifting blank reading that increases over time. By understanding how to perform a creatine assay, other fluorometric procedures should be relatively simple. One mL of creatine reagent is added to each 10 × 75 mm tube to be used. The recipe for the reagent is presented in Table 3. This reagent should be prepared just prior to use, because of the gradual decrease in NADH fluorescence over time. The NADH stock solution is stored in a –70°C freezer, and even at this temperature, it will gradually decrease in fluorescence over a period of months. This decrease in fluorescence can be countered by heating the NADH stock vial in warm water.

A 25 μL aliquot of tissue extract is added to each 10 × 75 mm tube. In addition, a set of 25 μL neutralized PCA blanks are added to a set of tubes, and creatine standards are added to an additional group. These standards will contain 0.5 or 1 μL of a 1 m*M* calibrated creatine standard. All tubes are then briefly vortexed and allowed to sit for about 10 min, allowing any pyruvate and ADP in the tissue to be reacted. The fluorometer sensitivity is optimized by placing an 0.013 μg/mL quinine standard (equivalent to the fluorescence of a 10 μ*M* NADH solution) in the fluorometer and adjusting the sensitivity to full scale. An initial reading is taken for all tubes. Ten microliters of creatine kinase (50 mg/mL stock diluted 1:2 with reagent) are added to each standard, sample, and blank tube, and vortexed briefly. The tubes are allowed to react for 20–30 min before taking a second reading. The second reading will be lower than the first, and this decrease in fluorescence is the result of both a noncreatine-specific decay in fluorescence and also the amount of creatine present. This change in fluorescence, or delta value, is calculated for each sample, standard, and blank. The difference in delta values measured for blanks at the beginning and the end of the assay is significant if many tubes are being measured. This drifting blank is linear over time. Consequently, whenever performing a creatine assay, blank tubes must be placed at the beginning and end of the sample tubes, and to fluorescence measured at relatively consistent time intervals. In this way, one can increment the blank value for each succeeding sample determination. The creatine assay protocol and calculations are outlined in Fig. 5.

The creatine standards are used in the same way as were the standards in the luciferin–luciferase technique. The delta value for each standard is calculated and the blank subtracted from this value, yielding a net delta value specific for the presence of creatine. This value is divided by the pmoles of creatine present in the 10 × 75 mm reaction tube. This value should be the same for both high and low standards. The sample creatine content can then be calculated by obtaining the change in fluorescence from the first to the second reading, subtracting the blank value, and dividing by the delta/pmol calculated from the standards. This number is multiplied by the dilution factor to estimate the total slice creatine content. In the case of the creatine assay, the dilution factor is: (440/25) (500/400), or 22. These values are derived from the following: Of 500 μL initial PCA extract only 400 μL was transferred. This 400 μL was added to 40 μL of $KHCO_3$, bringing the total volume to 440 μL, of which 25 μL was added to the assay tube. Finally, the total creatine content is divided by the mg protein from that slice, yielding a value of nmoles of creatine/mg protein.

Procedure

25 μL of sample or blank, or
0.5 μL or 1 μL of 1 mM creatine standard
in 10 × 75 mm tube

↓

+ 1 mL Creatine reagent

↓ 15 min

1st Fluorometric reading

↓

+ 10 μL creatine kinase
(50 mg/mL stock diluted 1:2 with reagent)

↓

Vortex

↓ 20–30 min

2nd Fluorometric reading

1st reading – 2nd reading = Δ

Calculation

Standards

(0.5 or 1 μL) (1 mM Cr) = nmol of creatine in assay tube

$$\frac{\Delta \text{ creatine} - \Delta \text{ blank}}{(\text{nmol Cr})} = \Delta/\text{nmol creatine}$$

$$\left(\frac{(\Delta \text{ sample} - \Delta \text{ blank})}{(\Delta/\text{nmol creatine})}\right) (\text{Dilution factor}) \left(\frac{1}{(\text{mg protein})}\right) = \frac{\text{nmol Cr}}{\text{mg protein}}$$

$$\text{Dilution factor} = \left(\frac{(440\ \mu\text{L neutralized extract})}{(25\ \mu\text{L sampled})}\right) \left(\frac{(500\ \mu\text{L PCA extract})}{(400\ \mu\text{L PCA transferred})}\right) = 22.0$$

Fig. 5. Creatine assay.

As was mentioned earlier, the primary problems with this assay lie in adjusting the sensitivity by adding more or less NADH and the persistent problem of a drifting blank. Adjusting the sensitivity or knowing how much NADH to add to the reagent is a matter of becoming familiar with what readings are necessary initially in order to have sufficient NADH present to allow all the creatine in the sample to react, while leaving enough fluorescence for the blank in addition to the creatine. Once this is determined, sufficient fluorescence can be produced each time, even though the NADH stock solution may gradually decrease in fluorescence. By preparing the reagent and taking a 1 mL aliquot and reading the fluorescence on the fluorometer, one can determine whether more NADH should be added. Once this is done, and the fluorescence is adjusted to the appropriate level, the reagent can be added to the samples, blanks, and standards. The second problem, that of the drifting blank, can be accounted for by adjusting the blank values for each subsequent calculation. This is done easily on a programmable calculator by dividing the total change in the blank value by the number of tubes that were read between the two sets of blanks. Because the change in the blank is time-dependent, it is important to read each sample tube sequentially and at timed intervals.

As was previously mentioned, the fluorometric assays can be used to determine the concentrations of many metabolites related to intermediary metabolism in brain tissue including: the adenylates and phosphocreatine, glucose and its metabolites glucose 1-phosphate and glucose 6-phosphate, citrate, DHAP, glycogen, oxaloacetate, lactate, and pyruvate. Also, the sensitivity of the fluorometric procedures can be greatly enhanced, to the femptomolar range, by the enzymatic cycling techniques that are described by Lowry and Passonneau (1972). The benefits of the fluorometric procedures include their flexibility compared to the luciferin–luciferase technique, in that they can be used to measure many different compounds; and the assay reagents and equipment necessary are generally cheaper than that needed for the luciferin–luciferase method. Greater sensitivity can be obtained than is possible with the luciferin—luciferase method by using enzymatic cycling techniques. The fluorometric procedures, however, are generally more time-consuming in that two readings are generally needed for a single metabolite measurement.

4. Summary

Brain slices differ from *in situ* tissue in their metabolic content. Much of the depletion of the creatine and adenylate pools can be explained by a substantial proportion of disrupted tissue in the brain slices. Consequently, the surviving tissue would be expected to behave in a manner similar to the in vivo state, although changes in lactate production and intracellular pH make it necessary to interpret some results with caution. The accessibility of the brain slice and the extracellular medium make in vitro preparations very adaptable to the study of brain metabolism in a range of experimental paradigms. Brain slice incubation techniques allow the tissue to be rapidly accessed and frozen, and allow a broad range of metabolite levels to be measured by the two techniques described in this chapter: The luciferin–luciferase luminescent technique and fluorescent measurements of coupled enzymatic systems.

References

Alger B E., Dhanjal S. S., Dingledine R., Garthwaite J., Henderson G., King G. L., Lipton P., North A., Schwartzkroin P. A., Sears T. A., Segal M., Whittingham T. S., and Williams J. (1984) Brain slice methods, in *Brain Slices,* (Dingledine, R., ed.), Plenum, New York, pp. 381–437.

Atkinson D. E. (1968) The energy charge of the adenylate pool as a regulatory parameter. Interaction with feedback modifiers. *Biochemistry* **7,** 4030–4034.

Bak I. J., Misgeld U., Weiler M., and Morgan E. (1980) The preservation of nerve cells in rat neostriatal slices maintained *in vitro:* A morphological study. *Brain Res.* **197,** 341–353.

Benjamin A. M. and Verjee Z. H. (1980) Control of aerobic glycolysis in the brain *in vitro. Neurochem. Res.* **5,** 921–934.

Bertman L., Dahlgren N., and Siesjo B. K. (1979) Cerebral oxygen consumption and blood flow in hypoxia: influence of sympathoadrenal activation. *Stroke* **10,** 20–30.

Dingledine R. (ed.) (1984) *Brain Slices,* Plenum, New York.

Ebadi M. S. (1972) Firefly luminescence in the assay of cyclic AMP. *Advan. Cyclic Nucleotide Res.* **2,** 89–109.

Elliott K.A.C. (1955) Tissue slice technique, in *Methods in Enzymology,* Vol. I; (Colowick, S. P and Kaplan, N. O., eds.), Academic, New York, pp. 3–19.

Frotscher, M., Misgeld, U. and Nitsch, C. (1981) Ultrastructure of mossy fiber endings in in vitro hippocampal slices. *Exp. Brain Res.* **41,** 247–255.

Fujii T., Baumgartl H., and Lubbers D. W. (1982) Limiting section thickness of guinea pig olfactory cortical slices studied from tissue PO_2 values and electrical activities. *Pfluger Arch.* **393,** 83–87.

Garthwaite J., Woodhams P. L., Collins M. J., and Balazs R. (1979) On the preparation of brain slices: Morphology and cyclic nucleotides. *Brain Res.* **173,** 373–377.

Haas H. L., Schaerer B., and Vosmansky M. (1979) A simple perfusion chamber for the study of nervous tissue slices *in vitro. J. Neurosci. Methods* **1,** 323–325.

Hawkins R. A., Williamson D. H., and Krebs H. A. (1971) Ketone body utilization by adult and suckling rat brain *in vivo. Biochem. J.* **122,** 13–18.

Kass I. S. and Lipton P. (1982) Mechanisms involved in irreversible anoxic damage to the *in vitro* hippocampal slice. *J. Physiol.* **332,** 459–472.

Kass I. S. and Lipton P. (1986) Divalent cations and irreversible transmission damage in anoxic rat hippocampal slices. *Soc. Neurosci. Abstr.* **12,** 693.

Kerkut G. A. and Wheal H. V. (eds.) (1981) *Electrophysiology of Isolated Mammalian CNS Preparations,* Academic, London, pp. 402.

Langmoen I. A. and Anderson P. (1981) The hippocampal slice *in vitro.* A description of the technique and some examples of the opportunities it offers, in *Electrophysiology of Isolated Mammalian CNS Preparations,* (Kerkut G. A. and Wheal H. V. eds.), Academic, London, pp. 51–105.

Li C.-L. and McIlwain H. (1957) Maintenance of resting membrane potentials in slices of mammalian cerebral cortex and other tissues *in vitro. J. Physiol.* **139,** 178–190.

Lipton P. and Whittingham T. S. (1979) The effect of hypoxia on evoked potentials in the *in vitro* hippocampus. *J. Physiol.* **287,** 427–438.

Lipton P. and Whittingham T. S. (1982) Reduced ATP concentration as a basis for synaptic transmission failure during hypoxia in the *in vitro* guinea-pig hippocampus. *J. Physiol.* **325,** 51–65.

Lipton P. and Whittingham T. S. (1984) Energy metabolism and brain slice function, in *Brain Slices* (Dingledine R., ed.), Plenum, New York, pp. 113–153.

Ljunggren B., Schultz H., & Seisjo B. K. (1974) Changes in energy state and acid-base parameters of the rat brain during complete compression ischemia. *Brain Res.* **73,** 277–289.

Lowry O. H. and Passonneau J. V. (1972) *A Flexible System of Enzymatic Analysis,* Academic, New York, pp. 291

Lowry O. H., Rosebrough N. J., Farr A. L., and Randall R. J. (1951) Protein measurement with the Folin phenol reagent. *J. Biol. Chem.* **193,** 265–275.

Lundin A., Rickardsson A., and Thore A. (1976) Continuous monitoring of ATP-converting reactions by purified firefly luciferase. *Anal. Biochem.* **75,** 611–620.

Lust W. D., Feussner G. K., Barbehenn E. K., and Passonneau J. V. (1981)

The enzymatic measurement of adenine nucleotides and *P*-creatine in picomole amounts. *Anal. Biochem.* **11,** 258–266.

Lust W. D., Whittingham T. S., and Passonneau J. V. (1982) Effects of slice thickness and method of preparation on energy metabolism in the *in vitro* hippocampus. *Soc. Neurosci. Abstr.* **8,** 1000.

MacMillan V. and Siesjo B. K. (1972) Intracellular pH of the brain in arterial hypoxemia evaluated with the CO_2 method and from creatine phosphokinase equilibrium. *Scand. J. Clin. Lab. Invest.* **30,** 117–125.

McElroy W. D., Hastings J. W., Coulombre J., and Sonnefeld V. (1953) The mechanism of action of pyrophosphate in firefly luminescence. *Arch. Biochem. Biophys.* **46,** 399–416.

McIlwain H. (1953) The effects of depressants on the metabolism of stimulated cerebral tissues. *Biochem. J.* **53,** 403–412.

McIlwain H. (1961) Techniques in tissue metabolism. 5. Chopping and slicing tissue samples. *Biochem. J.* **27,** 213–218.

McIlwain H. (1975) Preparing neural tissues for metabolic study in isolation, in *Practical Neurochemistry,* (McIlwain H., ed.), Churchill Livingston, London, pp. 105–132.

McIlwain H. and Bachelard H. S. (1971) *Biochemistry and the Central Nervous System,* Churchill Livingston, England.

Misgeld U. and Frotscher M. (1982) Dependence of the viability of neurons in hippocampal slices on oxygen supply. *Brain Res. Bull.* **8,** 95–100.

Newman G., Hospod F., and Wu P. (1986) Brain slice thickness. *Soc. Neurosci. Abstr.* **12,** 693.

Norberg K. and Siesjo B. K. (1975) Cerebral metabolism in hypoxic hypoxia. I. Pattern of activation of glycolysis, a re-evaluation. *Brain Res.* **86,** 31–44.

Pohle W., Reymann K., Jork R., and Malisch R. (1986) The influence of experimental conditions on the morphological preservation of hippocampal slices *in vitro. Biomed. Biochim. Acta* **45,** 1145–1152.

Reid K. H., Schurr A., West C. A., Tseng M., Shields C. B., and Edmonds H. L., Jr. (1986) Sensitivity of synaptic activity in the CA1 region of the hippocampal slice preparation of hypoxia. Effects of temperature and ACSF potassium concentration on probability of recovery. *Soc. Neurosci. Abstr.* **12,** 1526.

Richards C. D. and Sercombe R. (1970) Calcium, magnesium, and the electrical activity of the guinea-pig olfactory cortex *in vitro. J. Physiol.* **211,** 571–584.

Rolleston F. S. and Newsholme E. A. (1967) Control of glycolysis in cerebral cortex slices. *Biochem. J.* **104,** 524–533.

Schwartzkroin P. A. (1975) Characteristics of CA1 neurons recorded intracellularly in the hippocampal *in vitro* slice preparation. *Brain Res.* **85,** 423–436.

Schwartzkroin P. A. (1977) Further characteristics of hippocampal CA1 cells *in vitro. Brain Res.* **128,** 53–68.

Schwartzkroin P. A. (1981) To slice or not to slice, in *Electrophysiology of Isolated Mammalian CNS Preparations* (Kerkut G. A. and Wheal H. V., eds.), Academic, London, pp. 15–50.

Schwartzkroin P. A. and Anderson P. (1975) Glutamic acid sensitivity of dendrites in hippocampal slices *in vitro,* in *Advances in Neurology,* Vol 12 (Kreutzberg G. W., ed.), Raven, New York, pp. 45–51.

Shapiro H. M. (1985) Barbiturates in brain ischaemia. *Br. J. Anaesth.* **57,** 82–95.

Skrede K. K. and Westgaard R. H. (1971) The transverse hippocampal slice: A well defined cortical structure maintained *in vitro. Brain Res.* **35,** 589–593.

Teyler T. J. (1980) Brain slice preparation: Hippocampus. *Brain Res. Bull.* **5,** 391–403.

Thomas J. (1956) The composition of isolated cerebral tissues: Creatine. *Biochem. J.* **64,** 335–339.

Thomas J. (1957) The composition of isolated cerebral tissues: Purines. *Biochem. J.* **66,** 655–658.

Warburg O. (1923) Versuche an uberlebendem Carcinomgenebe (Methoden). *Biochem. Z.* **142,** 317–333.

West C. A., Schurr A., Reid K. H., and Shields C. B. (1986) Protection against hypoxia by high glucose—a study using the *in vitro* hippocampal slice. *Soc. Neurosci. Abstr.* **12,** 1526.

Whittingham T. S., Lust W. D., Christakis D. A., and Passonneau J. V. (1984) Metabolic stability of hippocampal slice preparations during prolonged incubation. *J. Neurochem.* **43,** 689–696.

Whittingham T. S., Lust W. D., and Passonneau J. V. (1984) An *in vitro* model of ischemia: Metabolic and electrical alterations in the hippocampal slice. *J. Neurosci.* **4,** 793–802.

Wulff K. (1983) Luminometry, in *Methods of Enzymatic Analysis,* Vol. I, (Bergmeyer, H. U., ed.), Verlag Chemie, Weinheim, pp. 340–368.

Wulff K., Haar H. P., and Michal G. (1982) Constant light signals in ATP assays with firefly luciferase: A kinetic explanation, in *Luminescent Assays: Perspectives in Endocrinology and Clinical Chemistry* (Serio, M. and Pazzagli M., eds.), Raven, New York, pp. 47–52.

Wulff K., Staehler F., and Gruber W. (1983) Standard assay for total creatine kinase and the MB-isoenzyme in human serum with firefly luciferase, in *Methods of Enzymatic Analysis*, Vol. III, Bergmeyer H. U., ed.), Verlag Chemie, Weinheim, pp. 209–214.

Yamamoto C. and Kurokawa M. (1970) Synaptic potentials recorded in brain slices and their modification by change in the level of tissue ATP. *Exp. Brain Res.* **10,** 159–170.

Yamamoto C. and McIlwain H. (1966) Electrical activities in thin sections from the mammalian brain maintained in chemically defined media *in vitro*. *J. Neurochem.* **13,** 1333–1343.

Measurement of Carbohydrates and Their Derivatives in Neural Tissues

Herman S. Bachelard

1. Introduction

The mammalian brain has very small reserves of carbohydrates, which under normal circumstances are the preferred sources of energy in vivo. Thus, the cerebral cortex contains glucose at 1–1.5 μmol/g fresh weight and a little more of the equivalent amount of glucose stored as glycogen (McIlwain and Bachelard, 1985). On death, very rapid rates of anaerobic metabolism continue postmortem, so within a few seconds, cerebral glucose in the mouse or rat brain falls to undetectable levels; glycogen is then mobilized, and the total stores of readily utilizable carbohydrate are severely depleted in less than 1 min. Concomitantly with these decreases, lactate increases to very high levels of some 3- to 4-fold normal within 1 min (Lowry et al., 1964). Other carbohydrates or their derivatives occur in minute amounts and do not contribute significantly to this process.

In order therefore to study levels of carbohydrates and their derivatives accurately, great care must be exercised to ensure rapid cessation of such endogenous postmortem metabolism. Various methods of rapid fixation have been evolved as described briefly below.

The techniques available for measurement of such intermediates have changed markedly over the past 50 yr. Up to relatively recently, research workers depended upon laborious chemical separation methods, using for example the different solubilities of their barium salts to separate individual sugar phosphates and a variety of slow chromatographic techniques. The separated constituents were then assayed by organic chemical, essentially colorimetric, methods. Of these, certain electrophoretic methods are still usefully employed, and the earlier chromatographic methods have evolved into rapid, nonlaborious gas-liquid chromatographic (glc) and high performance liquid chromatographic (hplc) techniques (McGinnis and Tang, 1980; Geiger et al., 1980). The most commonly used current methods are based on enzymological techniques, often with great sensitivity; nuclear magnetic resonance methods are increasingly being applied (Barker and Walker, 1980; Bachelard et al., 1986).

2. Preparation of Tissue Extracts

It should always be remembered that the chemical composition of the brain may be correlated to the behavioral and physiological state of the animal at death, so precautions should be taken where possible to avoid any unnecessary trauma. The animal should therefore be kept in its habitual environment and, where possible, unrestrained immediately before death. In many experimental situations, this may not be appropriate, and such considerations should be kept in mind when interpreting the results.

In order to prevent the rapid postmortem autolysis referred to above, cerebral metabolism must be stopped as rapidly as possible. Of the techniques described below, some trauma is unavoidable, as will be indicated where appropriate.

2.1. Rapid-Freezing Techniques

The most widely used method of rapid fixation is rapid freezing of the brain in liquid N_2 first described by Kerr some 50 yr ago (Kerr, 1935). This original technique was designed for larger experimental animals (such as cat and dog). The animals were anesthetized, and the brain exposed and frozen by pouring liquid N_2 over the surface. The method has two disadvantages: (1) the freezing is relatively uneven and some postmortem changes in certain metabolites occur; (2) anesthesia causes changes in the levels of many metabolic intermediates. For example, barbiturates and halothane may cause considerable increases in brain glucose, glucose 6-phosphate, creatine phosphate, and glycogen; metabolism is inhibited, and decreases in such intermediates as lactate and glutamate also occur (Strang and Bachelard, 1973; McIlwain and Bachelard, 1985; *see* Table 1).

The method of rapid freezing was subsequently modified for small laboratory animals (rat and mouse) by total immersion in liquid N_2 (Stone, 1938). The frozen brain is then chipped out and placed immediately in liquid N_2. Care is taken to prevent any thawing before enzymes can be denatured by extraction into deproteinizing reagents *(see below)*. The method has proved most

Table 1
Results Obtained from Different Methods of Tissue Fixation[a]

Technique	Glucose	Lactate	Pyruvate	α-Oxyglutarate	Malate	Creatine P
Freezing by immersion	1.3	2.0	0.11	0.13	0.44	3.3
Freezing by immersion[b]	4.3	1.6	0.13			4.3
Funnel freezing[b]		4.1	1.6	0.12	0.37	4.8
Freeze blowing	1.4	1.2	0.10	0.21	0.28	3.8
Microwave	1.4	1.2				3.0
Freezing after decapitation	0.5	5.0				1.0

[a]Results (μmol/g fresh weight) are means of reported values for mouse and rat brain. Data from Bachelard et al. (1974), Guidotti et al. (1974); Bachelard (1975); Medina et al., (1975); Siesjö, (1978); Strang and Bachelard (1971b, 1973); Veech et al. (1973).

[b]With anesthesia.

valuable for the analysis of labile constituents in whole brain, but is not suitable for regional analysis. For this, the frozen brain can be placed in a cryostat and "warmed up" to *ca* –20°C when the brain is softer and can be dissected into major regions (Folbergrova et al., 1969). Freezing in liquid N_2 causes extensive bubbling of the N_2, and in view of the possibility that the gaseous N_2 may insulate the brain, Lowry and Passonneau (1972) froze the animal in Freon-12 precooled with liquid N_2. The freon is more "greasy" and does not "boil off." Therefore, the insulation problems were considered to be circumvented. However, the results of subsequent studies indicated that no practical advantage had been achieved (Ferrendelli et al., 1972). Whatever method of rapid freezing is used, all are infinitely superior to decapitation followed by freezing, where postmortem changes inevitably occur.

One major criticism of rapid-freezing methods is that the temperature gradient through the tissue may not achieve rapid fixation of lower layers, which will then be subject to some postmortem ischaemia and autolysis. Measurements with thermocouples showed that the brains of totally-immersed mice froze within 3 s (Stone, 1938), but young rats required up to 20 s for deeper layers to freeze (Richter and Dawson, 1948). However, it was noted that in these deeper layers the temperature remained close to 37°C before it suddenly dropped to below 0°. This has been argued to be true also for the "funnel-freezing" technique described below, i.e., if normal rates of circulation are preserved, the tissue is likely to remain metabolically viable at 37° until the temperature drops very rapidly (Pontèn et al., 1973). However, in large rats (over 100 g body weight), freezing after total immersion took up to 150 s in deeper regions (Swaab, 1971), by which time much of the glucose and creatine phosphate would have disappeared (*see* Lowry et al., 1964). As shown in Table 1, results obtained for these labile constituents by rapid freezing of mice and young rats are comparable to those obtained by alternative methods, such as microwave fixation and freeze-blowing *(see below).*

Problems of the temperature gradient in larger animals remain. Siesjö and his coworkers therefore revived Kerr's surface freezing method, and modified it to devise the "funnel-freezing" technique for animals such as larger rats or the rabbit. The animal is anesthetized with N_2O, the surface of the brain is exposed, taking care not to break the dura, and a plastic funnel is sewn into place over the exposed brain. Freezing is then achieved by pouring liquid N_2 into the funnel. Results comparable with other techniques were obtained (Table 1), although the anesthesia clearly affected some metabolism as noted above.

The technique has been further adapted for freezing the cerebral cortex of large animals (cat and baboon). The above funnel freezing method was modified by Yang et al. (1983), so that a specially constructed styrofoam box was fitted, thus allowing the whole upper part of the head to be frozen by liquid N_2. Measurements of blood flow, pH, and selected metabolites (ATP, creatine phosphate, lactate) indicated that structures less than 10 mm from the surface retained normal metabolic states (Yang et al. 1983; Obrenovitch et al. 1988).

A further limitation of rapid freezing by total immersion is that the animals may undergo transient convulsions on first contact. Although the convulsions may be very brief, some metabolic constituents may be affected.

2.2. Freeze-Blowing

The above-mentioned uncertainties with respect to the rapidity of the freezing of deeper regions of the brain led Veech et al. (1973) and Veech and Hawkins (1974) to devise the technique of freeze-blowing. The tissue from the supratentorial part of the brain is expelled by air pressure into a chamber precooled with liquid N_2. Large rats can be used, so the method gets over the previous limitations for larger animals. Freezing is extremely rapid, and the results may indicate some improvement over the total immersion method (Table 1), but the method cannot be applied to analysis of different brain regions.

2.3. Microwave Fixation

This method originally involved placing the conscious animal in a microwave oven and subjecting the whole animal to irradiation comparable to that of a domestic microwave oven (Schmidt et al., 1971). This technique gave relatively slow fixation (see Lust et al., 1973), and when used in the author's laboratory, perceptible brief convulsions were observed. However the results for a very labile constituent, cyclic AMP, indicated some improvement over rapid freezing methods. The technique has since been modified to produce more intense irradiation directed more specifically towards the head of the animal (Guidotti et al., 1974; Medina et al., 1975). One advantage of microwave fixation is that, after fixation, dissection for regional analysis is possible.

2.4. Metabolic Inhibitors

Before such techniques for rapid fixation had evolved, some workers had successfully prevented postmortem autolytic changes in larger animals by inhibiting glycolysis with prior injections of iodoacetate. Levels of lactate and of glycogen were comparable to those subsequently obtained by rapid freezing (Kinnersley and Peters, 1930; Chesler and Himwich, 1943).

2.5. Tissue Extraction

For the constituents described in this chapter, the most generally satisfactory method is to prepare a neutralized perchloric acid extract. The frozen tissue is pulverized under liquid N_2 and rapidly homogenized in 0.6*M* perchloric acid (usually 5–10 mL/g tissue). The microwave-fixed samples can be homogenized directly in the perchloric acid. The homogenate is centrifuged (e.g., at 10,000 g for 10 min) at 0°C and the supernatant collected. It is often advisable to wash the pellet by resuspension in a small volume of perchloric acid and recentrifugation. The pooled supernatants at 0° are then neutralized* with the equivalent amount of ice-cold $KHCO_3$ solution, held at 0°C for about 15 min, and then centrifuged. Aliquots of the supernatant are then suitable for direct analysis of most soluble constituents. The perchloric acid pellet can be resuspended for protein determination, and if the tissue dispersion has been gentle, it can also be used for glycogen determination *(see below).*

Other extracting reagents used include trichloracetic acid, picric acid, sulphosalicylic acid, and "Somogyi reagent" ($ZnSO_4$/$Ba(OH)_2$) (*see* Rodnight, 1975 for review).

3. Estimation of Glucose

Numerous methods have been applied in the past. These include separations by various chromatographic techniques followed by colorimetric tests or more directly on crude extracts using enzymes. Earlier, the most commonly used enzyme was glucose oxidase, which is still used in many automated analytical machines. In crude extracts, glucose oxidase may give high results compared with the more specific hexokinase, which is now the

**Note:* For some intermediates such as oxaloacetate, which are more stable in acid, it is best to store the unneutralized perchloric acid extract.

basis of the method of choice. Indeed, this widely used method provides an excellent example of the "coupled enzyme" assay technique (*see* Lowry and Passonneau, 1972). The general principles of such techniques are: choose an enzyme specific for the substrate that produces an intermediate that is itself a substrate for another enzyme that has flavin or nicotinamide nucleotide as the coenzyme. The assay is then directly linked to changes in absorbance at 340 nm in a spectrophotometer, or with vastly increased sensitivity, to estimation in a spectrophotofluorometer.

In the case of glucose, the following coupled reaction is applied:

$$\text{Glucose} + \text{MgATP} \xrightarrow{\text{Hexokinase}} \text{Glucose 6-phosphate} + \text{MgADP} \quad (1)$$

$$\text{Glucose 6-phosphate} + \text{NADP}^+ \xrightarrow[\text{dehydrogenase}]{\text{Glucose 6-phosphate}} \text{6-Phosphogluconate} + \text{NADPH} + \text{H}^+ \quad (2)$$

A standard recipe for the enzymic estimation of glucose in a volume of 1 mL in a 1 cm light-path cuvette is given in Table 2.

3.1. Radioactive Glucose

Measurement of the specific radioactivity of glucose in tissues after metabolic experiments depended originally on extraction and purification. An enzymic method was devised for measurement of the specific activity in crude tissue extracts. The neutralized perchloric acid extract is reacted with hexokinase and glucose 6-phosphate dehydrogenase, as described above, to produce 6-phosphogluconate. Complete conversion of the 6-phosphogluconolactone (the immediate product of the dehydrogenase) to 6-phosphogluconate is ensured by brief heating at 100°C. 6-Phosphogluconate dehydrogenase is then added. After 1 h, the solution is acidified with H_2SO_4, and the $^{14}CO_2$ is expelled into Hyamine 10x hydroxide by a stream of air. The specific activity of the glucose can be directly determined from the NADPH produced by the dehydrogenase *(see above)* and the radioactivity of the $^{14}CO_2$ (Strang and Bachelard, 1971a).

Table 2
Glucose Estimation Using $NADP^+$ and Hexokinase

	Spectrophotometry[a], mL	Spectrophotofluorometry[b], mL
0.1*M* Tris-HCl buffer (pH 7.6),10 m*M*-$MgSO_4$	0.5	0.5
ATP	0.1 of 30 m*M*	0.1 of 10 m*M*
$NADP^+$	0.1 of 4 m*M*	0.1 of 0.4 m*M*
Hexokinase (3 U/mL)	0.1	0.03
G6P dehydrogenase (3 U/mL)	0.1	0.01
Water		0.16
Glucose (extract, standard or water)	0.1	0.1

[a]The extinction is read at 340 nm for 5 min before the glucose or water is added and the change in extinction followed until a plateau is reached (usually within 5 min). The normal range of glucose in the 0.1 mL sample would be 0.1–1 m*M*.

[b]All reagents are added to the cuvette except for the glucose 6-phosphate dehydrogenase. The fluorometer is set at 340 nm (excitant) and 440 (emission). The dehydrogenase is then added and the fluorescence recorded until it is constant (usually within 5 min.). A suitable range of glucose in the 0.1 mL sample is 0.01–0.1 m*M*.

3.2. Nuclear Magnetic Resonance

The ^{13}C nuclide is suitable for nuclear magnetic resonance techniques, but it is very insensitive (0.1% compared to the proton) and occurs with an abundance of only 1.1% in nature. Therefore, endogenous carbohydrates can only be detected and quantified if they occur in high concentration. Thus, in the brain, naturally abundant glucose and glycogen can only be seen with difficulty. However, if the metabolites are metabolically labeled with ^{13}C-precursors, many can be readily studied simultaneously. The α and β anomers of glucose are clearly separated, and such intermediates as glutamate, γ-aminobutyrate, lactate, alanine, and glycogen detected and quantified (Bachelard et al., 1986; Morris et al., 1986).

4. Hexose Phosphates

Hexose phosphates and most of the glycolytic intermediates only occur in minute amounts in the brain. They can be measured by coupled enzymic assays analogous to that described above for glucose (Lowry and Passonneau, 1972). This approach is the most suitable and sensitive if concentrations only are required. However, if specific radioactivities are also desired, separation and purification are usually essential. Some, such as glucose 6-phosphate, can be measured by the enzymic radiometric assay described above for glucose.

4.1. Separation Methods

The most commonly used method is still paper chromatography on acid-washed papers, with EDTA present, to prevent interference by cations. The conditions required for hexose phospates and for intermediates of the pentose phosphate pathway have been reviewed by Wood (1968, 1985).

The precise conditions applied will depend on the objectives, since no single method gives complete separation of all constituents. A good example of a two-dimensional paper chromatographic separation of sugar phosphates is that of Dodd et al. (1971). On acid-washed paper, the first dimension was run on butanol-formic acid-water (70:13.2:16.8 by vol) and the second on phenol-water-0.88 NH_3 (80:20:3 w/v/v). The following radioactive constituents clearly separated: glucose 6-phosphate, mannose and fructose phosphates, glucose, and some amino acids.

One- and two-dimensional chromatography on thin-layer plates of ECTEOLA- or MN-cellulose also gives excellent separations of sugar phosphates. Waring and Ziporin (1964) reported good separations of hexose phosphates and triose phosphates on MN-cellulose plates. The first solvent was aqueous *tert*-amyl alcohol with *p*-toluene sulphonic acid, and the second phase, isobutyric acid and ammonia. The one-dimensional separations of Table 3 were achieved on ECTEOLA-cellulose containing EDTA, run in 95% ethanol containing ammonium tetraborate. Individual constituents were visualized using benzidine-trichloroacetic acid sprays for hexose phosphates and molybdate reagent for phosphate esters (Dietrich et al., 1964).

Table 3
Separation of Sugar Phosphates by T.L.C.

	R_p[a] at pH 9	R_p[a] at pH 10
Glucose 1-P	1.2	1.15
Fructose 6-P	0.68	0.68
Glucose 6-P	0.56	0.39
Fructose 1,6-bis P	0.33	0.20
3 P-glycerate	0.97	0.95
2,3 bis P-glycerate	0.78	0.80

[a] R_p is the ratio of the distance traveled to that of inorganic phosphate (Dietrich et al., 1964).

4.2. Enzymic Methods

These are the most commonly used for estimations in crude extracts such as the neutralized perchloric acid extracts described in section 2, and are analogous to the spectrophotometric and fluorometric methods for glucose shown in Table 2. Table 4 summarizes the assays for the sugar phosphates and related intermediates that can be applied to cerebral extracts.

4.3. Deoxyglucose and Its 6-Phosphate

A rapid and effective method of separating 2-deoxyglucose and 2-deoxyglucose 6-phosphate from neutralized extracts is by passage through an ion exchange column, such as BioRad AG1 × 8. The free sugar passes through and the phosphate ester is rapidly eluted with M-HCl. In the author's laboratory, the separation takes only a few min. The deoxy compounds can be estimated by a sensitive fluorometric method (Blecher, 1961). The sample is boiled for 15 min with 3 mL of 0.01*M* diaminobenzoic acid in 5*M* H_3PO_4, immediately placed on ice in the dark for less than 30 min and the fluorescence recorded (excitant 405, emission 495 nm). The method is suitable for the deoxyglucose or its phosphate over a range of 3–125 nmol/mL.

5. Intermediates of the Pentose Phosphate Pathway

6-Phosphogluconate can be measured very easily by the spectrophotometric or fluorometric methods described above, with 6-phosphogluconate dehydrogenase and $NADP^+$. This intermediate can be measured together with glucose and glucose 6-phosphate (*see* Table 2) by stepwise addition of the appropriate enzymes.

For the other intermediates of the shunt pathway, enzymic methods are available, but may require so many coupling enzymes that such analysis of these metabolites, which occur in very minute amounts in the brain, may be subject to errors resulting from interference by other metabolites present.

Thus, ribulose 5-P and xylulose 5-P can be measured sequentially (Racker, 1984) in the following reaction scheme:

$$\text{Ribulose 5-P} \xrightarrow{\text{epimerase}} \text{Xylulose 5-P} \quad (1)$$

$$\text{Xylulose 5-P} + \text{ribose 5-P} \xrightarrow{\text{transketolase}} \quad (2)$$

$$\text{Sedoheptulose 5-P} + \text{glyceraldehyde 3-P}$$

$$\text{Glyceraldehyde 3-P} + \text{NAD}^+ \xrightarrow{\text{dehydrogenase}} \quad (3)$$

$$\text{3P-glycerate} + \text{NADH} + \text{H}^+$$

Formation of NADH in the final reaction can be measured spectrophotometrically or fluorometrically, as described above (Section 3). Ribose 5-phosphate can be measured by adding xylulose 5-phosphate, transketolase, glyceraldehyde 3-P dehydrogenase, and NAD^+ as part of the above reaction.

Enzymic assay of sedoheptulose 7-phosphate requires the addition of six enzymes (Paoletti, 1984a); the intermediates of the overall reaction include hexose phosphates, which occur at much higher concentrations in the brain, thus causing significant difficulties.

A similar problem occurs with the enzymic assay for erythrose 4-phosphate. Glucose 6-phosphate and fructose 6-phosphate are intermediates in the overall reaction (Paoletti, 1984b).

All in all, it seems preferable for estimation of these intermediates of low content in cerebral preparations to purify by separation methods, and then estimate the individual metabolites by enzymic (as above) or chemical methods. Column and paper chromatographic methods for separating these intermediates have

Table 4
Enzymic Assay of Glycolytic and Related Intermediates[a]

Intermediate	Enzymes required	Coenzyme	Level[b]
Glucose 6-P	Glucose 6-P dehydrogenase	$NADP^+$	0.07
Glucose 1-P[c]	P-glucomutase, glucose 6-P dehydrogenase	$NADP^+$	<0.01
Glucose 1,6-bis P	As in 2 (plus glucose 1-P)	$NADP^+$	0.05
Fructose 6-P[d]	Hexose P isomerase, glucose 6-P dehydrogenase	$NADP^+$	0.02
Fructose 1,6-bis P[e]	Aldolase, triose P isomerase, glycerol P dehydrogenase	NADH	0.12
Dihydroxyacetone P[e]	As in 5	NADH	*ca* 0.2
Glyceraldehyde 3-P[e]	As in 5	NADH	*ca* 0.1
Glycerate 1,3-bis P	Glyceraldehyde 3-P dehydrogenase	NADH	—
Glycerate 2,3-bis P[f]	P-glucomutase, P-glycerate kinase, glyceraldehyde 3P-dehydrogenase	NADH	—
3-P glycerate[g]	P-glycerate kinase, glyceraldehyde 3-P dehydrogenase	NADH	*ca* 0.2
2-P glycerate[h]	Enolase, pyruvate kinase, lactate dehydrogenase	NADH	*ca* 0.2
P-enolpyruvate[h]	Pyruvate kinase, lactate dehydrogenase	NADH	0.8

Pyruvate[h]	Lactate dehydrogenase	NADH	0.12
Lactate[i]	Lactate dehydrogenase	NAD^+	*ca* 2
UDP-glucose	UDP-glucose dehydrogenase	NAD^+	—

[a]Care should be taken to ensure that any contaminating enzyme will not interfere. The catalogs of the manufacturers usually provide such information.

[b]Levels (μmol/g) are means of those reported for rat and mouse brain. Methods and data are from Lowry and Passonneau (1972); Bachelard et al. (1974); Bachelard (1975); Bergmeyer (1984); McIlwain and Bachelard (1985).

[c]One can estimate G6P first with the dehydrogenase (as in 1), and then add the mutase to determine GIP.

[d]One can estimate G6P first (as in 1), and then add the isomerase to determine F6P.

[e]If the enzymes are added stepwise, dihydroxyacetone P, glyceraldehyde 3-P, and fructose 1,6-bis P can be measured sequentially.

[f]Glycolate 2-P must be present to convert the phosphoglucomutase to the phosphatase, which produces glycerate 3-P. Care has to be taken to correct for metabolites present in crude extracts that may be acted upon by the coupling enzymes.

[g]MgATP is required for the kinase.

[h]MgADP is required to drive the kinase. Note that 2P-glycerate, P-enolpyruvate, and pyruvate can all be estimated sequentially by stepwise addition of the enzymes.

[i]The pyruvate produced in this reaction must be trapped or removed, because the equilibrium is in favor of the reverse reaction. Pyruvate can be trapped with a hydrazine buffer (pH 9.6), or removed by adding glutamate and glutamate-pyruvate transaminase to the reaction mixture. *Note:* Lactic acid is often difficult to remove from glassware; acid washing of all glassware is recommended.

recently been comprehensively reviewed by Wood (1985), who also gives details of colorimetric estimations.

6. Intermediates of the Tricarboxylic Acid Cycle

In similar fashion to the methods for glycolytic intermediates (section 4), those of the tricarboxylic acid cycle can be estimated accurately by analogous sensitive enzymic assays coupled to nicotinamide or flavin coenzymes. These are summarized in Table 5.

7. Glycogen

7.1. Extraction

Cerebral glycogen is present in the perchloric acid precipitate (*see* section 2) if the tissue dispersion has been gentle (Bachelard and Strang, 1974). This pellet can be dispersed in 60% (w/v) KOH, (2 ml/g original tissue). Alternatively, the frozen powdered brain tissue can be treated directly with KOH.

The dispersion in KOH is heated at 100°C for some 10 min in a boiling water bath. Ethanol (2 mL of 90%) is then added and thoroughly mixed. Heating is continued at 100°C until the mixture begins to boil. After cooling in ice for 1–2 h, the mixture is centrifuged at 3,000 g for 10 min. The supernatant is discarded, and the pellet is washed twice with 60% ethanol, once with chloroform-methanol (1:4), dried at 100°C, and suspended in 0.1*M* acetate buffer, pH 4.8 (ca 1 mL). Cold water or buffer is preferable, since it was found that nonglycogen carbohydrate was extracted into hot water (Strang and Bachelard, 1971b; Bachelard and Strang, 1974).

7.2. Estimation

The glycogen can be hydrolyzed with acid to form glucose, which can then be measured chemically or enzymically (section 3). However, studies in the author's laboratory showed that glucose from nonglycogen complex carbohydrates was produced, giving artificially high results. Analysis of this carbohydrate showed it to also contain galactose, fucose, and uronic acids (Strang and Bachelard, 1971b), but it was not further characterized.

Table 5
Intermediates of the Tricarboxylic Acid Cycle

Intermediate	Enzymes required	Coenzyme	Level[a]
Citrate[b]	(i) Aconitase, isocitrate dehydrogenase	NAD^+	
	(ii) Citrate lyase, malate dehydrogenase	NADH	0.3
Isocitrate[b]	Isocitrate dehydrogenase	NAD^+	0.01
α-Oxoglutarate[b]	Glutamate dehydrogenase	NADH	0.12
Succinate	(i) Succinate dehydrogenase	FAD	
	(ii) Succinate thiokinase, pyruvate kinase, lactate dehydrogenase[d]	NADH	0.7
Fumarate[e,f]	Fumarase, malate dehydrogenase, glutamate-OAA transaminase	NAD^+	0.07
Malate[f]	Malate dehydrogenase, glutamate-OAA transaminase	NAD^+	0.4
Oxaloacetate	Malate dehydrogenase	NADH	0.005

[a]Levels (μmol/g) are means of those reported for rat or mouse brain. Data and methods from the sources of Table 4.

[b]Isocitrate can be measured together with citrate using stepwise addition of the enzymes of reaction 1 (i).

[c]NH_4^+ is required.

[d]The reaction requires MgATP, coenzyme A, and phosphoenolpyruvate. Note that interference from other intermediates may occur.

[e]Glutamate is required to drive the final reaction, which removes oxaloacetate. This is superior to an alternative method where a hydrazine buffer is used to trap the oxaloacetate, in place of the glutamate plus transaminase.

[f]Malate and fumarate can be estimated together by stepwise addition of enzymes.

The method of choice is to use the glycogen-specific amyloglucosidase (Bachelard and Strang, 1974). The glycogen sample in the above acetate buffer (1 mL) is added to 10 μL of amyloglucosidase (140 U/mL) and allowed to stand at room temperature for 2 h. A reaction mixture is prepared containing 0.70 mL of 0.1*M* Tris buffer (pH 7.6, with 0.5$MgSO_4$), 0.05 mL of 6 m*M*-ATP, 0.05 mL of 10 m*M*-$NADP^+$ and 25 μL hexokinase (50 U/mL). The enzymically treated glycogen, glucose standard, or water blank (0.1 mL) are added, and the fluorescence (340 nm excitant and 440 nm emission wavelengths) is recorded. Glucose 6-phosphate dehydrogenase (75 U/mL; 25 μL) is added, and the change in fluorescence recorded until a plateau is reached (usually within 20 min). Glucose standards of 0.1–1 m*M* are used. Alternatively, a spectrophotometric method can be used, but as noted in section 3, it is about 10 × less sensitive.

7.3. Specific Activity of ^{14}C-labeled Glycogen

The glycogen extract prepared as above in cold buffer is unlikely to be more than 75% pure (Strang and Bachelard 1971b), and the ^{14}C-glucose obtained by acid hydrolysis may, as noted above, be derived from carbohydrate other than glycogen. It is best, therefore, to hydrolyze the glycogen with amyloglucosidase *(see above)*, and the radioactivity of the glucose specifically produced can be measured as described in section 3. The normal levels of cerebral glycogen are of the order of 2 μmol/g fresh tissue measured as described above (Table 1).

8. Complex Carbohydrates—Glycosaminoglycans

Many complex carbohydrates occur in the brain associated with lipids (the glycolipids) or with proteins (glycoproteins and glycosaminoglycans). The carbohydrate constituents may include glucose, fucose, N-acetylneuraminic acid, hexosamines (glucosamine and galactosamine, which may be *N*-acetylated), and hexuronic acids (glucuronic acid and iduronic acid, which may be sulphated in either or both the 4 and 6 positions). Some may also be sulphated at the amino group of the hexosamine. The glycolipids are usually classed among the complex lipids; the principal groups are the gangliosides, cerebrosides, and sulphatides (McIlwain and Bachelard, 1985), but will not be considered here.

The carbohydrate parts of individual classes of glycoprotein and glycosaminoglycans contain some microheterogeneities, which result from variations in chain length, the sequence of the disaccharide constituents, and variations in the extent of sulphation. Little characterization of the carbohydrate moieties of cerebral glycoproteins has been achieved. Much more is known about the glycosaminoglycans, though their precise structures remain unclear.

8.1. Extraction of Glycosaminoglycans

The brain is first defatted by gentle homogenization in 19 vol of chloroform-methanol (2:1 v/v) and centrifugation. The partially defatted pellet is then washed by repeating the procedure with chloroform-methanol (1:2 v/v) and filtration through Whatman GF/F glass-fiber filters under suction. The defatted tissue is then rinsed with acetone and pulverized in a mortar precooled with liquid nitrogen, and finally dried in a vacuum desiccator over P_2O_5. The dried defatted tissue is then extracted into an aqueous environment, and proteins are digested using a combination of papain and pronase (see Schmid et al., 1982). Any undigested protein is precipitated by the addition of trichloroacetic acid to a final 7% and centrifuged. The pellet is washed with a small volume of 5% trichloroacetic acid. To the combined supernatants is added 5 vol of 5% potassium acetate in absolute ethanol, and after standing at 4° for 48 h, the mixture is centrifuged. The pellet can then be dissolved in 0.1*M* sodium acetate buffer, pH 5.5. Treatment with cetylpyridinium salts will precipitate the glycosaminoglycans, leaving the sialic acid-containing constituents in solution (Brunngraber et al., 1969; Di Benedetta et al., 1969).

The precipitated glycosaminoglycans can be separated into broad classes by fractionation into NaCl solutions. Hyaluronic acid (which is nonsulphated) dissolves in 0.4*M*-NaCl, and the sulphated constituents can be extracted with 1.2*M*-NaCl (Singh and Bachhawat, 1965). The main classes can be separated also by anion exchange chromatography (Murata, 1980) or by high-voltage electrophoresis on cellulose acetate plates (Capelletti et al., 1979), and visualized by staining with Alcian blue. Reasonable semiquantitative estimations of constituents on the stained plates can be achieved by densitometry.

Table 6
Glycosaminoglycans of Rat Cerebral Cortex[a]

Constituent	Age of rat (wk)		
	1	40	100
Hyaluronic acid	1.55	0.55	0.70
Chondroitin sulphates	0.85	0.62	0.50
Dermatan sulphate	0.65	0.1	0.1
Keratan sulphate	0	0	0
Heparan sulphates	0.8	0.2	0.19
Heparin	0.24	0.06	0.05

[a]Data, as μg/mg defatted dry tissue, are from Jenkins and Bachelard (1988).

8.2. Enzymic Analysis

The partially purified glycosaminoglycans are sequentially degraded with the following enzymes:

1. Hyaluronidase (from *Streptomyces hyalurolyticus*), which is specific for hyaluronic acid
2. Chondroitinase AC (from *Arthrobacter auresceus*), which acts on hyaluronic acid and attacks the hexosamine-glucuronic acid linkages of the sulphated constituents
3. Chondroitinase ABC (from *Proteus vulgaris*), which attacks the hexosamine-glucuronate and also the hexosamine-iduronate linkages.

These enzymes produce the unsaturated constituent disaccharide units, which can then be separated by high-voltage electrophoresis, visualized by staining, and quantified as above, or can be subjected to a variety of specific colorimetric analysis (Chandreskaan and BeMiller, 1980).

In the author's laboratory, the high-voltage electrophoretic method of Capellatti et al. (1979) gave good yields to give the results of Table 6 (Jenkins and Bachelard, 1988). The sialoaminoglycans, which are not precipitated by the above cetylpyridinium salt treatment, contain, in addition to *N*-acetylneuraminic acid, hexosamine, fucose, and mannose but no hexuronic acids. Constituents can be analzyed by techniques similar to the above (Brunngraber et al., 1969; Di Benedetta et al., 1969).

References

Bachelard, H. S. (1975) Constituents of neural tissues: intermediates in carbohydrate and energy metabolism, in *Practical Neurochemistry* (McIlwain, H., ed.), London: Churchill-Livingstone, pp. 33–59.

Bachelard, H. S., Cox, D. W. G., and Morris, P. G. (1986) NMR as a tool to study brain metabolism, in *Cerebral Metabolism in Aging and Neurological Disorders* (Lechner, H., ed.) Amsterdam: Elsevier Medical.

Bachelard, H. S., Lewis, L. D., Pontèn, U., and Siesjö, B. K. (1974) Mechanisms activating glycolysis in the brain in arterial hypoxia. *J. Neurochem.* **22,** 395–401.

Bachelard, H. S. and Strang, R. H. C. (1974) Determination of glycogen in nervous tissue, in *Research Methods in Neurochemistry* (Marks, N. and Rodnight, R., eds.), vol. 2, Chapter 12, Plenum, New York, pp. 301–318.

Barker, R. and Walker, T. E. (1980). ^{13}C-N.M.R. spectroscopy of isotopically-enriched carbohydrates. *Meths. Carbohydrate Chem.* **8,** 151–165.

Bergmeyer, H. U., ed. (1984) *Methods of Enzymatic Analysis* (3rd. ed.) Vol. 6, Metabolites 1: Carbohydrates, Weinheim: Verlag Chemie.

Blecher, M. (1961). A fluorometric method for the determination of 2-deoxy-*D*-glucose *Anal. Biochem* **2,** 30–38.

Brunngraber, E. G., Brown, B. D., and Aguilar, V. (1969) Isolation and determination of non-diffusible sialofucohexosaminoglycans derived from brain glycoproteins and their anatomical distribution in bovine brain. *J. Neurochem.* **16,** 1059–1070.

Capelletti, R., Del Rosso, M., and Chiorugi, V. P. (1979) A new electrophoretic method for the complete separation of all known animal glycosaminoglycans in a mono-dimensional run. *Anal Biochem.* **99,** 311–315.

Chandreskaan, E. V. and BeMiller, J. N. (1980) Constituent analysis of glycosaminoglycans. *Meths. Carbohydrate Chem.,* **8,** 89–96.

Chesler, A. and Himwich, H. E. (1943) Glycogen content of various parts of the central nervous system of dogs and cats at different ages. *Arch. Biochem.* **2,** 175–181.

Di Benedetta, C., Brunngraber, E. G., Whitney, G., Brown, B. D., and Aro, A. (1969). Compositional patterns of sialofucohexosaminoglycans derived from rat brain glycoproteins. *Arch. Biochem. Biophys.* **131,** 403–413

Dietrich, C. P., Dietrich, S. M. C., and Pontis, H. G. (1964). Separation of sugar phosphates and sugar nucleotides by thin-layer chromatography. *J. Chromatogr.* **15,** 277–278.

Dodd, P. R., Bradford, H. F. and Chain, E. B. (1971). The metabolism of glucose 6-phosphate by mammalian cerebral cortex *in vitro*. *Biochem. J.* **125,** 1027–1038.

Ferrendelli, J. A., Gay, M. H., Sedgwick, W. G., and Chang, M. M. (1972) Quick freezing of the murine CNS: comparison of regional cooling rates and metabolic levels when using liquid nitrogen or Freon-12. *J. Neurochem.* **19,** 979–987.

Folbergrova, J., Passonneau, J. V., Lowry, O. H., and Schulz, D. W. (1969). Glycogen, ammonia and related metabolites in the brain during seizures evoked by methionine sulphoximine. *J. Neurochem.* **16,** 191–203.

Geiger, P. J., Ahn, S., and Bessman, S. P. (1980). Separation and automated analysis of phosphorylated intermediates. *Meths. Carbohydrate Anal.* **8,** 21–32.

Guidotti, A., Cheney, M., Trabucchi, M., Doteuchi, M., Wang, C., and Hawkins, R. A. (1974). Focussed microwave irradiation: a technique to minimize post-mortem changes of cyclic nucleotides, dopa and choline, and to preserve brain morphology. *Neuropharmacology,* **13,** 1115–1122.

Jenkins, H. G. and Bachelard, H. S. (1988). Developmental and age-related changes in rat brain glycosaminoglycans. *J. Neurochem.* **51,** pp. 1634–1640.

Kerr, S. E. (1935). Studies on the phosphorus compounds of brain. 1. Phosphocreatine. *J. Biol. Chem.* **110,** 625–635.

Kinnersley, H. W. and Peters, R. A. (1930). Carbohydrate metabolism in birds; brain localisation of lactic acidosis in avitaminosis B, and its relation to origin of symptoms. *Biochem. J.* **24,** 711–722.

Lowry, O. H. and Passonneau, J. V. (1972) *A Flexible Method of Enzymatic Analysis*. New York: Academic Press.

Lowry, O. H., Passonneau, J. V., Hasselberger, F. X., and Schulz, D. W. (1964) Effect of ischemia on known substrates and cofactors of the glycolytic pathway in brain. *J. Biol. Chem.* **239,** 18–30.

Lust, D. W., Passonneau, J. V., and Veech, R. L. (1973) Cyclic adenosine monophosphate, metabolites and phosphorylase in neural tissue: a comparison of methods of fixation. *Science,* **181,** 280–282.

McGinnis, G. D. and Tang, P. (1980) High performance liquid chromatography. *Meths. Carbohydrate Chem.* **8,** 33–43.

McIlwain, H. and Bachelard, H. S. (1985) *Biochemistry and the Central Nervous System* (5th ed.) London: Churchill-Livingstone.

Medina, M. A., Jones, D. J., Stavinhoa, W. B., and Ross, D. H. (1975) The levels of labile intermediary metabolites in mouse brain following rapid tissue fixation with microwave irradiation. *J. Neurochem.* **24,** 223–227.

Morris, P. G., Bachelard, H. S., Cox, D. W. G., and Cooper, J. C. (1986).

^{13}C Nuclear magnetic resonance studies of glucose metabolism in guinea-pig brain slices. *Biochem. Soc. Trans.* **14,** 1270–1271.

Murata, K. (1980) Enzymic analysis of acidic glycosaminoglycans. *Meths. Carbohydrate Chem.* **8,** 81–88.

Obrenovitch, T. P., Bordi, L., Garofalo, O., Ono, M., Momma, F., Bachelard, H. S., and Symon, L. (1988). In situ freezing of the brain for metabolic studies: evaluation of the "box" method for large experimental animals. *J. Cereb. Blood Flow Metab.* **8,** 742–749.

Paoletti, F. (1984a) *D. Sedoheptulose 7-phosphate,* in Bergmeyer vol. 6, no. 1, pp. 149–157.

Paoletti, F. (1984b) *D-Erythrose 4-phosphate,* in Bergmeyer, vol. 6, no. 1 pp. 490–497.

Pontèn, U., Ratcheson, R. A., Salford, L. G., and Siesjö, B. K. (1973) Optimal freezing conditions for cerebral metabolites in rats. *J. Neurochem.* **21,** 1127–1138.

Racker, E. (1984) *D-Ribulose 5-phosphate,* in Bergmeyer, vol. 6, no. 1 pp. 437–441.

Richter, D. and Dawson, R. M. C. (1948) Brain metabolism in emotional excitement and in sleep. *Am. J. Physiol.* **154,** 73–79.

Rodnight, R. (1975) Obtaining, fixing and extracting neural tissues, in *Practical Neurochemistry* (McIlwain, H., ed.), London, Churchill-Livingstone, pp. 1–16.

Schmidt, M. J., Schmidt, D. E., and Robinson, G. A. (1971) Cyclic AMP in brain areas: microwave irradiation as a means of tissue fixation. *Science,* **173,** 1142–1143.

Schmid, K., Wernli, M., and Ninberg, R. B. (1982) A microtechnique for the rapid determination of the glycosaminoglycans of vascular tissues. *Anal. Biochem.* **121,** 91–96.

Siesjö, B. K. (1978) *Brain Energy Metabolism,* Chichester, John Wiley and Sons.

Singh, M. and Bachhawat, B. K. (1965) The distribution and variation with age of different uronic acid-containing mucopolysaccharides in brain. *J. Neurochem.* **12,** 519–525.

Stone, W. E. (1938) The effects of anaesthetics and of convulsants on the lactic acid content of the brain. *Biochem. J.* **32,** 1908–1918.

Strang, R. H. C. and Bachelard, H. S. (1971a) Rapid enzymic methods for the determination of the specific radioactivities of metabolic intermediates in unpurified tissue extracts. *Anal. Biochem.* **41,** 533–542.

Strang, R. H. C. and Bachelard, H. S. (1971b). Extraction, purification and turnover of rat brain glycogen. *J. Neurochem.* **18,** 1067–1076.

Strang, R. H. C. and Bachelard, H. S. (1973) Rates of cerebral glucose utilization in rats anaesthetized with phenobarbitone. *J. Neurochem.* **20,** 987–996.

Swaab, D. F. (1971) Pitfalls in the use of rapid freezing for stopping brain and spinal cord metabolism in rat and mouse. *J. Neurochem.* **18,** 2085–2092.

Veech, R. L. and Hawkins, R. A. (1974). Brain blowing: a technique for *in vivo* study of brain metabolism. *Research Methods in Neurochemistry,* **2,** 171–182.

Veech, R. L., Harris, R. L., Veloso, D., and Veech, E. H. (1973) Freeze-blowing: a new technique for the study of brain *in vivo*. *J. Neurochem.* **20,** 183–188.

Waring, P. P. and Ziporin, Z. Z. (1964) The separation of hexose phosphates and triose phosphates by thin-layer chromatography. *J. Chromatogr.* **15,** 168–172.

Wood, T. (1968) The detection and identification of intermediates of the pentose phosphate cycle and related compounds. *J. Chromatogr.* **35,** 352–361.

Wood, T. (1985) *The Pentose Phosphate Pathway*. London. Academic Press.

Yang, M. S., Lutz, H., De Witt, D. S., Becker, D. P., and Hayes, R. L. (1983). An improved method for in situ freezing of cat brain for metabolic studies. *J. Neurochem.* **41,** 1393–1397.

The [^{14}C]Deoxyglucose Method for Measurement of Local Cerebral Glucose Utilization

Louis Sokoloff, Charles Kennedy, and Carolyn B. Smith

1. Introduction

The brain is a complex, heterogeneous organ composed of many anatomical and functional components with markedly different levels of functional activity that vary independently with time and function. Other tissues are generally far more homogeneous, with most of their cells functioning similarly and synchronously in response to a common stimulus or regulatory influence. The central nervous system, however, consists of innumerable subunits, each integrated into its own set of functional pathways and networks, and subserving only one or a few of the many activities in which the nervous system participates. Understanding how the nervous system functions requires knowledge not only of the mechanisms of excitation and inhibition, but even more so of their precise localization in the nervous system and the relationships of neural subunits to specific functions.

Tissues that do physical and/or chemical work, such as heart, kidney, and skeletal muscle, exhibit a close relationship between energy metabolism and functional activity. This relationship has been utilized to develop a method that maps alterations in functional activity simultaneously in all components of the central nervous system during physiological, pharmacological, or pathological induced changes of local functional activity (Sokoloff et al., 1977). The method employs radioactive deoxyglucose (DG), an analog of glucose, to trace glucose metabolism in the brain. The procedure is so designed that the concentration of radioactivity in the tissue is more or less proportional to the rate of glucose utilization. The concentrations of radioactivity in the local cerebral tissues are measured by a quantitative autoradiographic technique. The method not only allows quantification of the actual rates of glucose utilization in the individual cerebral tissues, but autoradiographs obtained with it provide pictorial representations of the relative rates of glucose utilization in all the cerebral structures seen in autoradiographs of 20 μm serial sections of the entire brain. It is, therefore, now possible to obtain not an anatomical map, but a functional map of the entire central nervous system in normal and experimental states.

2. Theoretical Basis of Radioactive Deoxyglucose Method

The radioactive deoxyglucose method was developed to measure the local rates of energy metabolism simultaneously in all components of the brain in conscious laboratory animals. It was designed specifically to take advantage of the extraordinary spatial resolution made possible by quantitative autoradiography (Sokoloff et al., 1977). The dependence on autoradiography prescribed the use of radioactive substrates for energy metabolism, the labeled products of which could be assayed in the tissues by the autoradiographic technique. Although oxygen consumption is the most direct measure of energy metabolism, the volatility of oxygen and its metabolic products and the short physical half-life of its radioactive isotopes precluded measurement of oxidative metabolism by the autoradiographic technique. In most circumstances, glucose is essentially the sole substrate for cerebral oxidative metabolism, and its utilization is stoichiometrically related to oxygen consumption. Radioactive glucose is, however, not fully satisfactory because its labeled products are lost too rapidly from the cerebral tissues. The labeled analog of glucose, 2-deoxy-D-[^{14}C]glucose, was, therefore, selected because its biochemical properties make it particularly appropriate to trace glucose metabolism and to measure local cerebral glucose utilization by the autoradiographic technique.

The method was derived by analysis of a model based on the biochemical properties of 2-deoxyglucose in the brain (Fig. 1A) (Sokoloff et al., 1977). 2-Deoxyglucose is transported bidirectionally between blood and brain by the same carrier that transports glucose across the blood–brain barrier. In the cerebral tissues, it is phosphorylated by hexokinase to 2-deoxyglucose-6-phosphate (DG-6-P). Deoxyglucose and glucose are, therefore, competitive substrates for both blood–brain transport and hexokinase-catalyzed phosphorylation. However, unlike glucose-6-phosphate, which is metabolized further eventually to CO_2 and water, DG-6-P cannot be converted to fructose-6-phosphate, and it is also not a substrate for glucose-6-phosphate dehydrogenase. There is relatively little glucose-6-phosphatase activity in the brain and even less deoxyglucose-6-phosphatase activity. Deoxyglucose-6-phosphate can be converted into deoxyglucose-1-phosphate, then into UDP-deoxyglucose, and eventually into glycogen, glycolipids, and glycoproteins, but these reactions are slow and, in mammalian tissues, only a very small fraction of the deoxyglucose-6-phosphate formed proceeds to these products (Nelson et al., 1984). In any case, these compounds are secondary, relatively stable products of deoxyglucose-6-phosphate, and all together represent the products of deoxyglucose phosphorylation. Deoxyglucose-6-phosphate, once formed, remains, therefore, essentially trapped in the cerebral tissues, at least for the duration of the experimental period.

If the interval of time is kept short enough—for example, less than 1 h—to allow the assumption of negligible loss of [^{14}C]DG-6-P from the tissues, then the quantity of [^{14}C]DG-6-P accumulated in any cerebral tissue at any given time following the introduction of [^{14}C]DG into the circulation is equal to the integral of the rate of [^{14}C]DG phosphorylation by hexokinase in that tissue during that interval of time. This integral is, in turn, related to the amount of glucose that has been phosphorylated over the same interval, depending on the time courses of the relative concentrations of [^{14}C]DG and glucose in the precursor pools and the Michaelis–Menten kinetic constants for hexokinase with respect to both [^{14}C]DG and glucose. With cerebral glucose consumption in a steady state, the amount of glucose phosphorylated during the interval of time equals the steady-state flux of glucose through the hexokinase-catalyzed step times the duration of the interval, and the net rate of flux of glucose through this step equals the rate of glucose utilization.

These relationships can be rigorously combined into a model (Fig. 1A) that can be mathematically analyzed to derive an operational equation (Fig. 1B), provided that the following assumptions are made

1. Steady state for glucose (i.e., constant plasma glucose concentration and constant rate of glucose consumption) throughout the experimental period
2. Homogeneous tissue compartment within which the concentrations of [^{14}C]DG and glucose are uniform and exchange directly with the plasma and

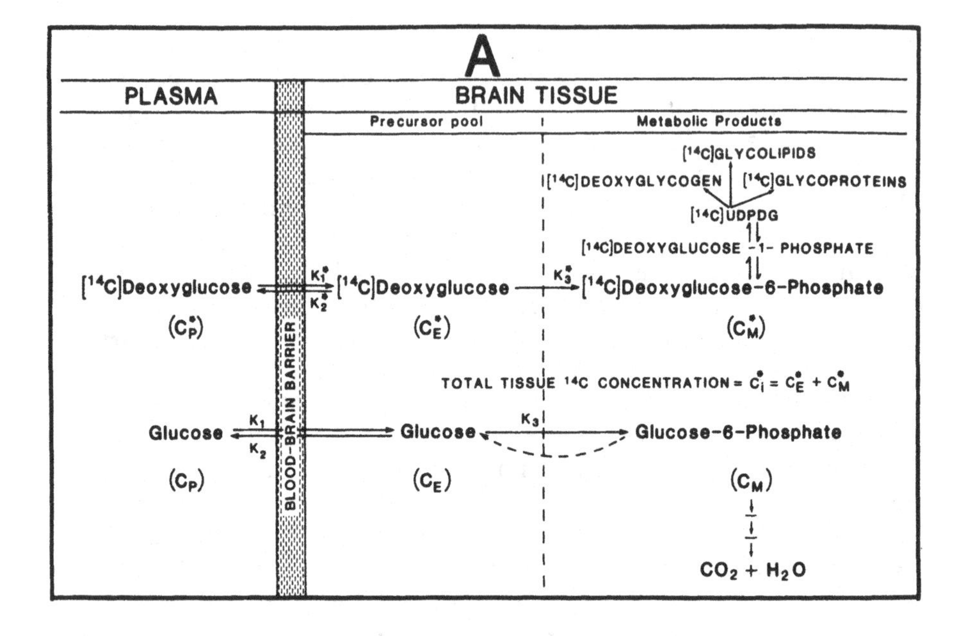

Fig. 1A. Theoretical basis of radioactive deoxyglucose method for measurement of local cerebral glucose utilization (Sokoloff et al., 1977). Diagrammatic representation of the theoretical model. C_i^* represents the total ^{14}C concentration in a single homogeneous tissue of the brain. C_P^* and C_P represent the concentrations of [^{14}C]deoxyglucose and glucose in the arterial plasma, respectively; C_E^* and C_E represent their respective concentrations in the tissue pools that serve as substrates for hexokinase. C_M^* represents the concentration of [^{14}C]deoxyglucose-6-phosphate in the tissue. The constants k_1^*, k_2^*, and k_3^* represent the rate constants for carrier-mediated transport of [^{14}C]deoxyglucose from plasma to tissue, for carrier-mediated transport back from tissue to plasma, and for phosphorylation by hexokinase, respectively; the constants k_1, k_2, and k_3 are the equivalent rate constants for glucose. [^{14}C]Deoxyglucose and glucose share and compete for the carrier that transports both between plasma and tissue and for hexokinase, which phosphorylates them to their respective hexose-6-phosphates. The dashed arrow represents the possibility of glucose-6-phosphate hydrolysis by glucose-6-phosphatase activity, if any.

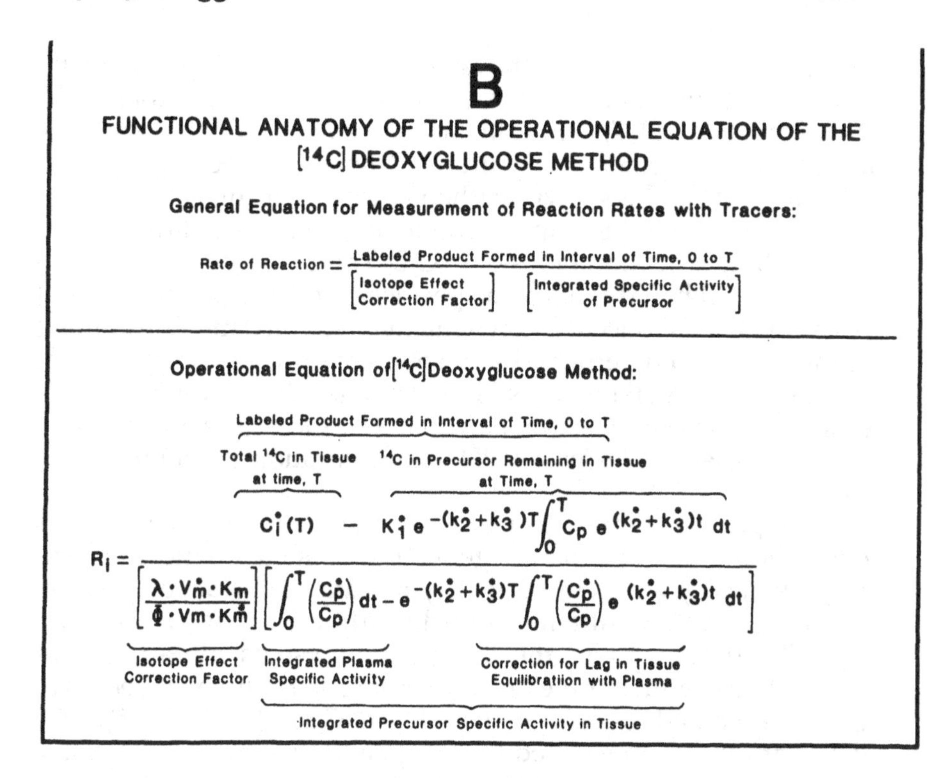

$$R_i = \frac{C_i^*(T) - k_1^* e^{-(k_2^*+k_3^*)T} \int_0^T C_p^* e^{(k_2^*+k_3^*)t}\, dt}{\left[\frac{\lambda \cdot V_m^* \cdot K_m}{\Phi \cdot V_m \cdot K_m^*}\right]\left[\int_0^T \left(\frac{C_p^*}{C_p}\right) dt - e^{-(k_2^*+k_3^*)T} \int_0^T \left(\frac{C_p^*}{C_p}\right) e^{(k_2^*+k_3^*)t}\, dt\right]}$$

Fig. 1B. Operational equation of the radioactive deoxyglucose method and its functional anatomy. T represents the time of termination of the experimental period; λ equals the ratio of the distribution space of deoxyglucose in the tissue to that of glucose; Φ equals the fraction of glucose that, once phosphorylated, continues down the glycolytic pathway; and K_m^* and V_m^* and K_m and V_m represent the familiar Michaelis-Menten kinetic constants of hexokinase for deoxyglucose and glucose, respectively. The other symbols are the same as those defined in A.

3. Tracer concentrations of [^{14}C]DG (i.e., molecular concentrations of free [^{14}C]DG essentially equal to zero).

The operational equation, which defines R_i, the rate of glucose utilization/unit mass of tissue, i, in terms of measurable variables, is presented in Fig. 1B.

The rate constants, k_1^*, k_2^*, and k_3^*, are determined in a separate group of animals by a nonlinear, iterative process that provides the least-squares best fit of an equation that defines the time course of tissue ^{14}C concentration in terms of the time, the history of the plasma concentration, and the rate constants to the experimentally determined time courses of tissue and plasma concentrations of ^{14}C (Sokoloff et al., 1977).The λ, Φ, and the enzyme kinetic constants are grouped together to constitute a single, lumped constant (*see* equation). It can be shown mathematically that this lumped constant is equal to the asymptotic value of the product of the ratio of the cerebral extraction ratios of [^{14}C]DG and glucose and the ratio of the arterial blood to plasma specific activities when the arterial plasma [^{14}C]DG concentration is maintained constant. The lumped constant is also determined in a separate group of animals from arterial and cerebral venous blood samples drawn during a programmed intravenous infusion that produces and maintains a constant arterial plasma [^{14}C]DG concentration (Sokoloff et al., 1977).

Despite its complex appearance, the operational equation is really nothing more than a general statement of the standard relationship by which rates of enzyme-catalyzed reactions are determined from measurements made with radioactive tracers (Fig. 1B). The numerator of the equation represents the amount of radioactive product formed in a given interval of time; it is equal to C_i^*, the combined concentrations of [^{14}C]DG and [^{14}C]DG-6-P in the tissue at time, T, measured by the quantitative autoradiographic technique, less a term that represents the free unmetabolized [^{14}C]DG still remaining in the tissue. The denominator represents the integrated specific activity of the precursor pool times a factor, the lumped constant, which is equivalent to a correction factor for an isotope effect. The term with the exponential factor in the denominator takes into account the lag in the equilibration of the tissue precursor pool with the plasma.

3. Procedure

The operational equation dictates the variables to be measured to determine the local rates of cerebral glucose utilization. The specific procedure employed is designed to evaluate these variables and to minimize potential errors that might occur in the actual application of the method. If the rate constants, k_1^*, k_2^*, and k_3^*, are precisely known, then the equation is generally applicable with any mode of administration of [^{14}C]DG and for a wide range of time intervals. At the present time, the rate constants have been fully determined only in the conscious rat (Sokoloff et al., 1977) (Table

1). Partial determination of the rate constants indicates that they are similar in the monkey (Kennedy et al., 1978), dog (Duffy et al., 1982), and sheep (Abrams et al., 1984). These rate constants can be expected to vary with the condition of the animal, however, and for most accurate results, they should be redetermined for each condition studied. The structure of the operational equation suggests a more practical alternative. All the terms in the equation that contain the rate constants approach zero with increasing time if the [^{14}C]DG is so administered that the plasma [^{14}C]DG concentration also approaches zero. From the values of the rate constants determined in normal animals and the usual time course of the clearance of [^{14}C]DG from the arterial plasma following a single intravenous pulse at zero time, it has been determined that an interval of 30–45 min after a pulse is adequate for these terms to become sufficiently small that considerable latitude in inaccuracies of the rate constants is permissible without appreciable error in the estimates of local glucose consumption (Sokoloff et al., 1977). An additional advantage derived from the use of a single pulse of [^{14}C]DG followed by a relatively long interval before killing the animal for measurement of local tissue ^{14}C concentration is that, by then, most of the free [^{14}C]DG in the tissues has been either converted to [^{14}C]DG-6-P or transported back to the plasma (Fig. 2); the optical densities in the autoradiographs then represent mainly the concentrations of [^{14}C]DG-6-P and, therefore, reflect directly the relative rates of glucose utilization in the various cerebral tissues.

The following steps are taken in the conduct of each individual experiment:

1. Preparation of the animal
2. Administration of [^{14}C] deoxyglucose and timed sampling of arterial blood
3. Analysis of arterial plasma for [^{14}C] deoxyglucose and glucose concentrations
4. Processing of brain tissue
5. Preparation of autoradiographs
6. Densitometric analysis of autoradiographs
7. Calculation of rates of glucose utilization.

3.1. Preparation of Animals

Catheters are inserted into any conveniently located artery and vein. In most of the studies done to date, femoral or iliac

Table 1
Values of Rate Constants in the Normal Conscious Albino Rat[a]

Structure	Rate constants (min^{-1}) ± standard error of estimates			Distribution volume, mL/g	Half-life of precursor pool, min
	k_1^*	k_2^*	k_3^*	$k_1^*/(k_2^* + k_3^*)$	$Log_e 2/(k_2^* + k_3^*)$
		Gray matter			
Visual cortex	0.189 ± 0.048	0.279 ± 0.176	0.063 ± 0.040	0.553	2.03
Auditory cortex	0.226 ± 0.068	0.241 ± 0.198	0.067 ± 0.057	0.734	2.25
Parietal cortex	0.194 ± 0.051	0.257 ± 0.175	0.062 ± 0.045	0.608	2.17
Sensory-motor cortex	0.193 ± 0.037	0.208 ± 0.112	0.049 ± 0.035	0.751	2.70
Thalamus	0.188 ± 0.045	0.218 ± 0.144	0.053 ± 0.043	0.694	2.56
Medial geniculate body	0.219 ± 0.055	0.259 ± 0.164	0.055 ± 0.040	0.697	2.21
Lateral geniculate body	0.172 ± 0.038	0.220 ± 0.134	0.055 ± 0.040	0.625	2.52

Hypothalamus	0.158 ± 0.032	0.226 ± 0.119	0.043 ± 0.032	0.587	2.58
Hippocampus	0.169 ± 0.043	0.260 ± 0.166	0.056 ± 0.040	0.535	2.19
Amygdala	0.149 ± 0.028	0.235 ± 0.109	0.032 ± 0.026	0.558	2.60
Caudate-putamen	0.176 ± 0.041	0.200 ± 0.140	0.061 ± 0.050	0.674	2.66
Superior colliculus	0.198 ± 0.054	0.240 ± 0.166	0.046 ± 0.042	0.692	2.42
Pontine gray matter	0.170 ± 0.040	0.246 ± 0.142	0.037 ± 0.033	0.601	2.45
Cerebellar cortex	0.225 ± 0.066	0.392 ± 0.229	0.059 ± 0.031	0.499	1.54
Cerebellar nucleus	0.207 ± 0.042	0.194 ± 0.111	0.038 ± 0.035	0.892	2.99
Mean ± SEM	0.189 ± 0.012	0.245 ± 0.040	0.052 ± 0.010	0.647 ± 0.073	2.39 ± 0.40
		White matter			
Corpus callosum	0.085 ± 0.015	0.135 ± 0.075	0.019 ± 0.033	0.552	4.50
Genu of corpus callosum	0.076 ± 0.013	0.131 ± 0.075	0.019 ± 0.034	0.507	4.62
Internal capsule	0.077 ± 0.015	0.134 ± 0.085	0.023 ± 0.039	0.490	4.41
Mean ± SEM	0.079 ± 0.008	0.133 ± 0.046	0.020 ± 0.020	0.516 ± 0.171	4.51 ± 0.90

[a]From Sokoloff *et al.*, 1977.

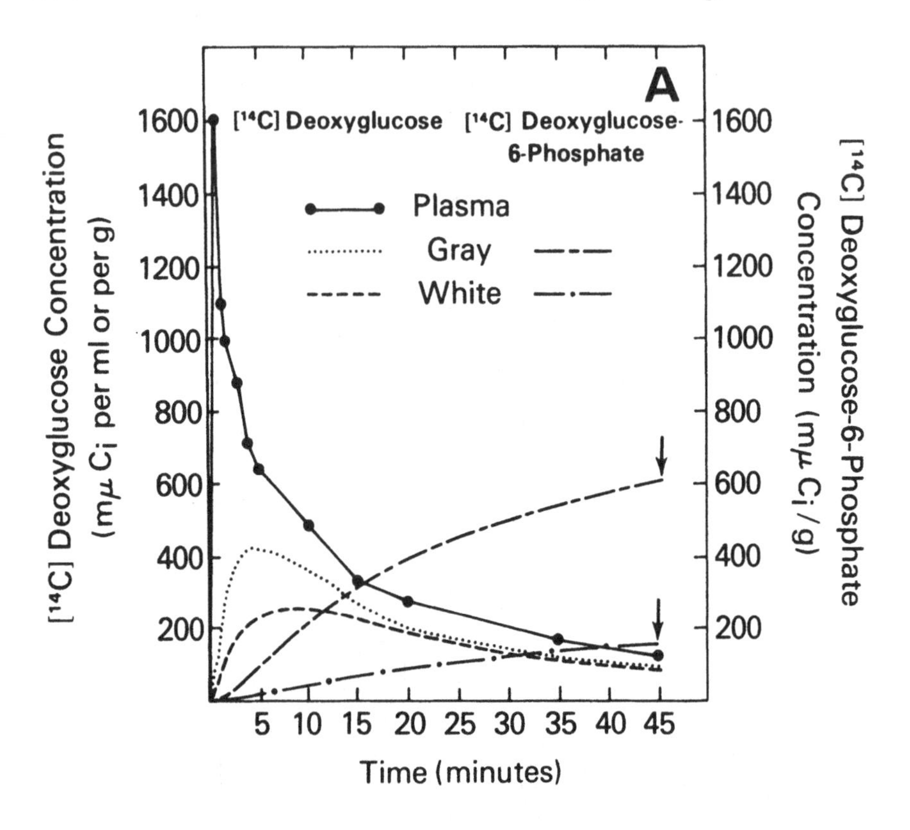

Fig. 2A. Graphical representation of the significant variables in the operational equation (Fig. 1B) used to calculate local cerebral glucose utilization. Time courses of [^{14}C]deoxyglucose-6-phosphate concentrations in average gray and white matter following an intravenous pulse of 50 μCi of [^{14}C]deoxyglucose. The plasma curve is derived from measurements of plasma [^{14}C]deoxyglucose concentration. The tissue concentrations were calculated from the plasma curve and the mean values of k_1^*, k_2^*, and k_3^* for gray and white matter in Table 1 according to the second term in the numerator of the operational equation. The [^{14}C]deoxyglucose-6-phosphate concentration in the tissues was calculated from the integral of the free deoxyglucose concentration in the tissue and k_3^*; the autoradiographic technique measures the total ^{14}C content at the time of killing, [(C^*_1 (T)], the first term in the numerator of the operational equation. Note that at the time of killing the total ^{14}C content represents mainly [^{14}C]deoxyglucose-6-phosphate concentration, especially in gray matter.

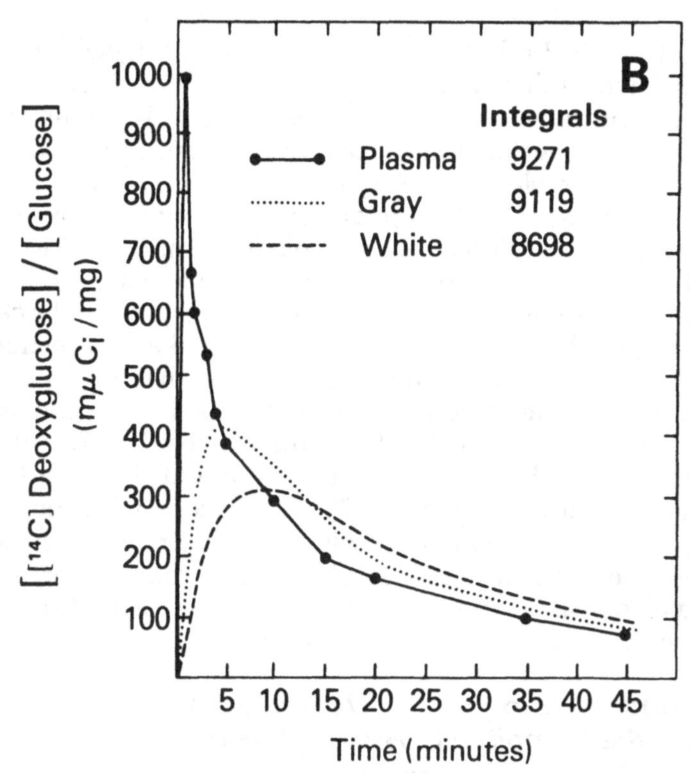

Fig. 2B. Time courses of ratios of [^{14}C]deoxyglucose and glucose concentrations (i.e., specific activities) in plasma and average gray and white matter. The curve for plasma was determined by division of the plasma curve in (A) by the plasma glucose concentrations. The curves for the tissues were calculated by the function in the right pair of brackets in the denominator of the operational equation. The integrals in (B) are the integrals of the specific activities with respect to time and represent the areas under the curves. The integrals under the tissue curves are equivalent to all of the denominator of the operational equation, except for the lumped constant. Note that, by the time of killing, the integrals of the tissue curves approach equality with each other and with that of the plasma curve.

vessels have been chosen, but tail and axillary sites have also been used. A general anesthesia is employed, fluothane being preferred because of the relatively short recovery period. The catheters must have the usual characteristics for them to remain patent for repeated blood sampling. Those made of polyethylene, in the size designated as PE-50 by the supplier, Clay Adams, are entirely satisfactory, except in animals weighing less than 100 g, which require the use of size PE-10. The bubble-free catheters, plugged at one end and filled with dilute heparin solution (100 U/mL) before their insertion, will remain patent for many hours. To minimize the need for extensive flushing of dead space during the sampling period, it is desirable that the arterial catheter be as short as possible; 15 cm is suitable for the rat. During recovery from anesthesia, the animal may be placed in a suitable restraining device. In the case of rats, a loosely fitting, bivalved, plaster cast around the lower trunk is applied with the hind legs taped to a lead brick. For cats, a zippered jacket is satisfactory; for monkeys, a restraining chair. When behavioral studies are carried out that require a freely moving animal, the catheters may be threaded under the skin to exit at the back of the neck. If attention is paid to keeping the concentration of fluothane to a minimum and the time for the surgical procedure is kept to 15–20, min, recovery is prompt and the experiment can be initiated within 2–3 h. Immediately before starting the experiment, it is useful to measure hematocrit, arterial pH and blood gases, and mean arterial blood pressure to establish the presence of a normal physiologic state.

3.2. Administration of [^{14}C]Deoxyglucose and the Sampling of Arterial Blood

To ensure that tracer conditions are maintained, the dose of [^{14}C]deoxyglucose should be such that the animal receives no more than 2.5 μmol of deoxyglucose/kg of body weight. If the specific activity is relatively high (50–60 mCi/mmol), 100–125 μCi/kg can be given. This amount of radioactivity is sufficient to attain a desirable optical density of the autoradiographs in a reasonable period of time, namely 4–6 d of exposure. Economic factors may demand the use of a lower dose, in which case, of course, the exposure must be longer. When smaller doses are employed, it is necessary to be sure that the plastic standards used in making the autoradiographs are sufficiently low in their radioactivity for their calibrated values to cover the range of the concentration of ^{14}C in the brain sections. If the [^{14}C]deoxyglucose is supplied in an ethanol solution, it must be evaporated to dryness and the deoxyglucose then redissolved in physiological saline. A suitable concentration for its intravenous infusion is 100 μCi/mL. The experimental period is initiated by the infusion of [^{14}C]deoxyglucose through the venous catheter over a period of 15 s. With zero time marking the start of the infusion, sampling is begun from the arterial catheter to monitor the entire time course of the [^{14}C]deoxyglucose concentration in the plasma. The first 4–5 samples of about 50 μL each are taken in heparin-treated tubes consecutively over the first 20 s, and thereafter at 30 s, 45 s, and 1 min, 2 min, 3 min, 5 min, 7.5 min, 10 min, 15 min, 25 min, 35 min, and 45 min. The samples taken at intervals should be large enough to permit plasma analyses for both glucose and deoxyglucose concentrations (70–80 μL). Care must be taken to clear the dead space of the catheter prior to each sample. The samples are immediately centrifuged in a small, high-speed centrifuge such as the Beckman Microfuge, and then are kept on ice until pipetted for the analyses. The use of heparinized tubes may be unnecessary if the animal is heparinized just prior to the experiment. Unless attention is paid to limiting the amount of blood removed in the course of sampling and clearing the dead space, shock can readily be induced in small animals. Most rats weighing 300–400 g will tolerate the removal of 2 mL over the 45-min experimental period without there being a significant fall in blood pressure. Blood drawn for the purpose of clearing the dead space (three times the volume of the catheter) may, of course, be returned to the animal. Because of the occasional rat that fails to tolerate even small blood losses, it is well to monitor mean arterial blood pressure at intervals during the sampling periods A fall below 90 mm Hg is reason to eliminate the animal from an experimental series.

Immediately after the last sample has been taken, the animal is killed. This may be by decapitation in the case of small animals or, alternatively, by an intravenous infusion of thiopental, followed immediately by a saturated solution of KCl to stop the heart.

3.3. Analysis of Arterial Plasma for [^{14}C]Deoxyglucose and Glucose Concentrations

The concentration of deoxyglucose in each plasma sample is measured by means of counting its ^{14}C content. Twenty μL of plasma are pipeted into 1 mL of water contained in a counting vial. Ten mL of a suitable phosphor solution are added (e.g., Aquasol, New England Nuclear, Boston, MA). With the aid of internal or external standardization, the dpm are determined and the concentration then expressed in μCi/mL. The plasma glucose concentration is most conveniently assayed in a glucose analyzer, such as that made by Beckman Instrument Co. This is quick and requires only 10 μL of plasma/determination. Also suitable for this purpose is the coupled, glucose-dependent, hexokinase–glucose-6-phosphate dehydrogenase-catalyzed reduction of $NADP^+$, which is available in kit form from Calbiochem (La Jolla, CA). Should it be necessary to conserve blood, glucose concentration need not be measured in every sample. Analyses in four samples, spaced over the total experimental period, are sufficient to establish the existence of the required steady state for plasma glucose levels.

3.4. Processing of Brain Tissue

Some investigators have chosen to perfuse the brain with a fixative immediately after the animal is killed. This serves to improve the quality of histologic sections, which is often desirable to establish the anatomic identity of regions of interest in the autoradiographs. Perfusion fixation is also thought by some to reduce artifacts in the sectioning of the frozen brains, especially those of large animals. For this purpose, a 3.5% solution of formaldehyde made in a 0.05*M* phosphate buffer adjusted to a pH of 7.4 may be employed. The animal is heparinized immediately before killing. With a cannula placed in the left ventricle and the reservoir of the perfusate 50 cm above the heart level, the perfusion is carried out for 10 min. The brain is then removed and frozen as described below. In order to determine whether this procedure alters the distribution of ^{14}C in the brain, we removed small sections from one hemisphere of a monkey that had undergone the deoxyglucose procedure just before carrying out the formaldehyde perfusion. The unperfused and perfused tissues were frozen and mounted side by side when the block was sectioned. The autoradiographs made from nonperfused sections revealed clearer definition of fine cortical markings. Calculation of the concentration average of ^{14}C in the two sides indicate that the perfusion process washed out 15% of the label.

Until a perfusion procedure is devised that is shown to prevent loss or movement of the label, we recommend it be omitted and that the brain be frozen immediately after removal. This is done by immersion in isopentane or Freon XII, chilled to –45° with liquid nitrogen. Many prefer the isopentane because the brain sinks and becomes completely covered. Brain in Freon XII floats and thus tends to freeze unevenly. This results in artifacts from the protrusion of expanding tissue at an unfrozen site on the surface. With constant agitation of the brain in the Freon, this, however, can be avoided. There is a tendency for large brains to become grossly distorted in the immersion process. If the brain is placed in a plastic ladle, previously coated with a film of mineral oil, a uniform external configuration of the tissue can be maintained.

After freezing, small brains are mounted on microtome tissue holders with an embedding matrix supplied by Lipshaw Mfg. Co. (Detroit, MI). Brains of larger animals must be cut into two or three smaller blocks in order for them to be accommodated in the microtome. Cutting the frozen brain into blocks is best done with a small bandsaw, the blade of which is precooled immediately before the cut is made. It is crucial to maintain the brain below –30°C at all times during handling to prevent movement of the label by diffusion. Alternatively, blocking can be carried out before freezing. The head with the dorsal aspect of the skull open can be mounted in a stereotaxic device and the brain cut with a large microtome blade guided by needles inserted into the tissue stereotaxically. Storage for any prolonged period prior to sectioning should be in a freezer maintained at –70°C. Sectioning of the brain is carried out in a refrigerated microtome at a thickness of 20 μm and at a temperature cycle that does not go above –22°C. Even at this temperature, there is some movement of the label in a matter of hours, dictating the need to complete the cutting of one block (or entire small brain as the case may be) at one sitting. Errors resulting from inconsistent thickness of sections, and a number of artifacts can be introduced at this stage unless attention is given to a number of operational details of the microtome, such as secure mounting, knife sharpness, antiroll plate adjustment, and the use of smooth, regularly timed strokes of the knife. The 20 μm sections are picked from the knife surface on glass coverslips to which they adhere by thawing. They are immediately transferred to a hot plate maintained at 60°C on which the momentarily thawed section becomes dry within 5–8 s. The coverslips are then placed on an adhesive-coated paper board, cut to fit in a 10″ × 12″ X-ray cassette with a center gap left for a strip holding 6–10 previously calibrated [^{14}C]methyl methacrylate standards. The thickness of the backing that holds the standards must be adjusted so that contact of all surfaces containing radioactivity with the emulsion will be uniform.

An alternate system of sectioning brain is that which employs an LKB 2250 PMV cryomicrotome (Stockholm, Sweden). This has a number of advantages, especially in processing brains of large animals. Because of its capability of cutting bone, the need for removal of the brain from the calvarium is eliminated. The entire head is sectioned and, thus, the brain's normal relationship to other tissues is preserved. This is of particular value in studies involving the entire visual pathway as it includes retina and optic nerves. It has the additional advantage of eliminating various cutting artifacts, especially those resulting from wrinkling or fragmentation. Sections are remarkably uniform in thickness. Other artifacts, however, may be introduced, the most troublesome of which is that caused by slight shrinkage of the tissue on drying, which results in fine lines giving the appearance of parched mud. Some users of this system have noted considerable loss of resolution of detail in autoradiographs. The reason for this is uncertain, but it is possible that it results from transient warming of brain at the time the head is mounted in the embedding medium or from the prolonged period necessary for completion of sectioning. Further experience with the system will determine whether or not this serious drawback can be overcome.

3.5. Preparation of Autoradiographs

A number of different photographic films are suitable for contact autoradiography with ^{14}C. The single-coated blue sensitive X-ray film made by Eastman Kodak and designated SB-5 is generally satisfactory when developed according to the manufacturer's instructions. If the dose and specific activity of [^{14}C]deoxyglucose are employed as suggested above, satisfactory images having an optical density range between 0.1–1.0 are generated in 4–6 d. Films with finer grain, but requiring a longer period of exposure, are Eastman Kodak's OM-1 and Plus-X.

Quantitative autoradiography requires the simultaneous exposure of standards with brain sections on each film (Fig. 3). [^{14}C]Methyl methacrylate standards are available commercially. It is important to note that their ^{14}C content, which may be expressed in µCi/g of plastic material by the manufacturer, is not that which is to be used in the calculations. Each standard must be assigned a value that is equivalent to the ^{14}C concentration in the brain, a 20-µm thick, dried section of which will produce the same optical density as the plastic when they are exposed together on the same film. If calibrated values for the brain are given by the supplier, it is wise to verify them. For this purpose, any freely diffusible, ^{14}C-containing substance that, after being infused in the animal, reaches equilibrium and becomes uniformly distributed in brain, e.g., [^{14}C]antipyrine or [^{14}C]methylglucose, may be employed. A dose given intravenously to a rat equilibrates in the tissues in 45–60

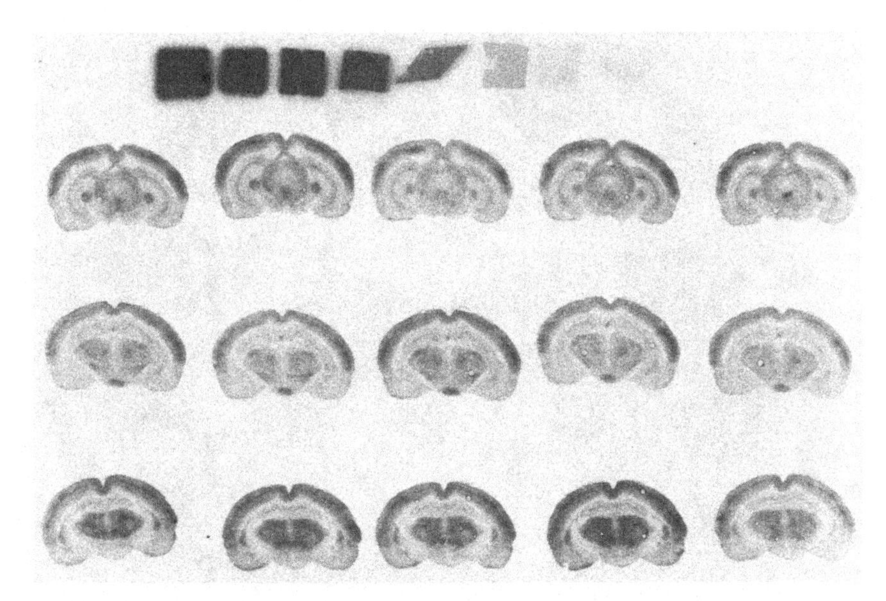

Fig. 3. Autoradiographs of sections of conscious rat brain and calibrated [^{14}C]methyl methacrylate standards used to quantify ^{14}C concentration in tissues. It is difficult to avoid slight variation in section thickness that is evident in a corresponding slight variation in optical density. Where this is present, measurements of optical density must be made in several sections for any given structure.

min. The animal is killed and the brain removed. One hemisphere is immediately weighed, then homogenized in a 1/30 dilution of Triton X-100, and made up to a specified volume. The suspension of the homogenized brain tissue is then assayed for its ^{14}C concentration in a liquid scintillation counter. With a correction for the dilution, the ^{14}C concentration expressed in μCi/unit wet weight of brain is obtained. The other hemisphere is frozen, sectioned at 20 μm, and prepared for autoradiography as described above. Several rat brains are processed in the manner described, each animal having received a different dose of the inert substance containing ^{14}C. These brain sections and the plastic standards are then exposed on film in the same cassette. After development of the film, densitometric measurements are made of the standards and the homogeneously darkened images of the calibrated brain sections *(see below)*. A plot of optical density values of the calibrated brain sections against their measured concentrations of ^{14}C should yield a smooth curve. The optical densities produced by the brain sections must cover a sufficient range to bracket the densities of the

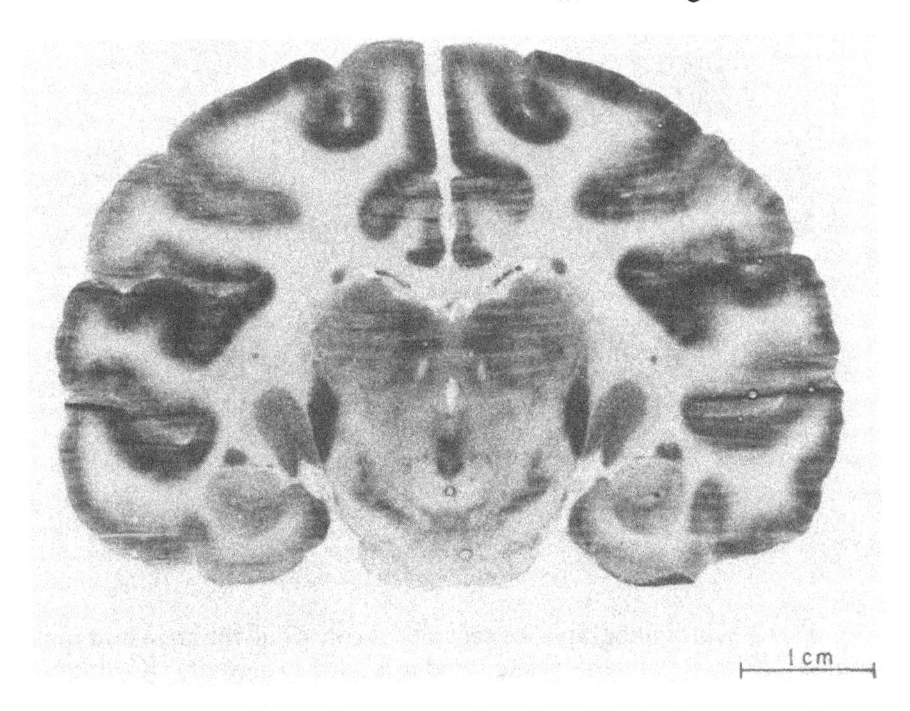

Fig. 4. Autoradiograph of a coronal section of brain from a conscious monkey made with [^{14}C]deoxyglucose prepared as described in the text. The pattern produced by differences in optical density permits recognition of most of the anatomic subdivisions. The right side of the brain is on the right side. Note the right–left symmetry in optical density in all regions, except for regions of the cortex, namely, the superior and inferior parietal lobules. The monkey was engaged in a task involving the left arm and hand throughout the experiment.

plastic standards. From the plot of the ^{14}C concentration of brains and their optical densities, the correct brain-equivalent concentration of ^{14}C can be assigned to the plastic standards.

3.6. Densitometric Analysis of Autoradiographs

The autoradiographs of [^{14}C]DG-labeled brain sections provide a pictorial representation of the relative rates of glucose utilization in various structures of the brain: the darker the region, the higher the rate of glucose utilization. Most of the major subdivisions in gray matter are clearly delineated because of their differences in optical density from that of an adjacent anatomic area (Fig. 4). Verification of the anatomic identity of a region can be made by staining and histologic study of the section from which the autoradiograph is made. With attention paid to the details cited above, the technique can clearly delineate such relatively small structures of the rat brain as the suprachiasmatic nucleus, which measures about 300 μm in coronal section. In the normal rat, Layer IV of the cortex, which measures 100–200 μm thick, is also clearly seen as a dark linear band (Fig. 3).

The determination of the rate of glucose utilization for any given region of the autoradiograph requires an estimate of the concentration of ^{14}C in that region. This is done by a measurement of its optical density. Transmission densitometers of the type widely used in photography are suitable for this purpose if the aperture is 0.2–0.5 mm in diameter. For satisfactory readings to be made in very small structures, a computerized image processing system is necessary *(see below)*. Whatever system is employed for densitometry, it is necessary to make optical density readings for a given structure in several sections. The mean value obtained serves both to reduce errors resulting from variations in section thickness and to give a value that is reasonably representative of the structure in its three dimensions. For larger structures, 12 or more optical density readings should be made. The concentration of ^{14}C in the structure is determined from the plot of optical density vs the concentration of ^{14}C of the calibrated plastic standards. Each film, of course, has its own standard curve.

3.7. Calculation of Rate of Glucose Utilization

The operational equation is given in Fig. 1B. The measured variables are:

1. The entire history of the arterial plasma [^{14}C]deoxyglucose concentration, C_P^*, from zero time to the time of killing, T
2. The steady-state arterial plasma glucose level, C_P, over the same interval and
3. The concentration of ^{14}C in the tissue, $C_i^*(T)$, which was determined densitometrically from the autoradiographs.

The rate constants, k_1^*, k_2^*, and k_3^*, and the lumped constant are not measured in each experiment; the values for these constants are those that have already been determined in our laboratory and are characteristic of the rat (Table 1). Similarly, the lumped constant is characteristic of the species (Table 2). A full discussion of these constants and their possible variation in special situations is given

Table 2
Values of the Lumped Constant in Various Species[a]

Animal	No. of animals	Mean ± S.D.	S.E.M.
Albino rat			
Conscious	15	0.464 ± 0.099[b]	± 0.026
Anesthetized	9	0.512 ± 0.118[b]	± 0.039
Conscious (5% CO_2)	2	0.463 ± 0.122[b]	± 0.086
Combined	26	0.481 ± 0.119	± 0.023
Rhesus Monkey			
Conscious	7	0.344 ± 0.095	± 0.036
Cat			
Anesthetized	6	0.411 ± 0.013	± 0.005
Dog (beagle puppy)			
Conscious	7	0.558 ± 0.082	± 0.031
Sheep			
Fetus	5	0.416 ± 0.031	± 0.014
Newborn	4	0.382 ± 0.024	± 0.012
Mean	9	0.400 ± 0.033	± 0.011
Humans			
Conscious	6	0.568 ± 0.105	± 0.043

[a]The values were obtained as follows: rat (Sokoloff et al., 1977); monkey (Kennedy et al., 1978); cat (M. Miyaoka, J. Magnes, C. Kennedy, M. Shinohara, and L. Sokoloff, unpublished data); dog (Duffy et al., 1982); sheep (Abrams et al., 1984); Human (Reivich et al., 1985).

[b]No statistically significant difference between normal conscious and anesthetized rats ($0.3 < p < 0.4$) and conscious rats breathing 5% CO_2 ($p > 0.9$).

in the section on theoretical and practical considerations *(see below)*. A programmable calculator or personal computer is employed to calculate values for glucose utilization from the operational equation.

3.8. Normal Rates of Local Cerebral Glucose Utilization in Conscious Animals

The rates of local cerebral glucose utilization have been measured in the normal conscious and anesthetized albino rat (Sokoloff et al., 1977) and in the conscious rhesus monkey (Kennedy et al., 1978). The rates in the conscious rat vary widely throughout the brain, with the values in white matter distributed in a narrow low range and the values in gray structures broadly distributed around an average about 3 times greater than that of white matter (Table 3). The highest values are in structures of the auditory system, with the inferior colliculus clearly the most metabolically active structure in the brain. The value for any selected structure is virtually identical to that of the homologous region on the opposite side. The failure to find this symmetry in values in normal conscious animals strongly points to the presence of disease or defective development such as may be found in the relay nuclei of the auditory pathway in rats with otitis media, a common disorder in many rat colonies.

The rates of local cerebral glucose utilization in the conscious monkey exhibit similar heterogeneity to those found in the rat, but they are generally one-third to one-half the values in corresponding structures (Table 3). The differences in rates in the rat and monkey brain are consistent with the different cellular packing densities in the brains of these two species.

4. Theoretical and Practical Considerations

The design of the deoxyglucose method was based on an operational equation, derived by the mathematical analysis of a model of the biochemical behavior of [^{14}C]deoxyglucose and glucose in the brain (Fig. 1). Although the model and its mathematical analysis were as rigorous and comprehensive as reasonably possible, it must be recognized that models almost always represent idealized situations and cannot possibly take into account every single known, let alone unknown, property of a complex biological system. There remained, therefore, the possibility that continued experience with the [^{14}C]deoxyglucose method might uncover weaknesses, limitations, or flaws serious enough to limit its usefulness or even to invalidate it. Several years have now passed since its introduction, and numerous applications of it have been made. The results of this experience generally establish the validity and worth of the method. There still remain, however, some potential problems in specialized situations, and several theoretical and practical issues need further clarification.

The main potential sources of error are the rate constants and the lumped constant. The problem with them is that they are not determined in the same animals and at the same time when local

Table 3
Representative Values for Local Cerebral Glucose Utilization in the Normal Conscious Albino Rat and Monkey (μmoles/100g/min)[a]

Structure	Albino rat[b] (10)	Monkey[c] (7)
	Gray matter	
Visual cortex	107 ± 6	59 ± 2
Auditory cortex	162 ± 5	79 ± 4
Parietal cortex	112 ± 5	47 ± 4
Sensory-motor cortex	120 ± 5	44 ± 3
Thalamus: lateral nucleus	116 ± 5	54 ± 2
Thalamus: ventral nucleus	109 ± 5	43 ± 3
Medial geniculate body	131 ± 5	65 ± 3
Lateral geniculate body	96 ± 5	39 ± 1
Hypothalamus	54 ± 2	25 ± 1
Mamillary body	121 ± 5	57 ± 3
Hippocampus	79 ± 3	39 ± 2
Amygdala	52 ± 2	25 ± 2
Caudate-putamen	110 ± 4	52 ± 3
Nucleus accumbens	82 ± 3	36 ± 2
Globus-pallidus	58 ± 2	26 ± 2
Substantia nigra	58 ± 3	29 ± 2
Vestibular nucleus	128 ± 5	66 ± 3
Cochlear nucleus	113 ± 7	51 ± 3
Superior olivary nucleus	133 ± 7	63 ± 4
Inferior colliculus	197 ± 10	103 ± 6
Superior colliculus	95 ± 5	55 ± 4
Pontine gray matter	62 ± 3	28 ± 1
Cerebellar cortex	57 ± 2	31 ± 2
Cerebellar nuclei	100 ± 4	45 ± 2
	White matter	
Corpus callosum	40 ± 2	11 ± 1
Internal capsule	33 ± 2	13 ± 1
Cerebellar white matter	37 ± 2	12 ± 1
	Weighted average for whole brain	
	68 ± 3	36 ± 1

[a]The values are the means ± standard errors from measurements made in the number of animals indicated in parentheses.
[b]From Sokoloff et al. (1977).
[c]From Kennedy et al. (1978).

cerebral glucose utilization is being measured. They are measured in separate groups of comparable animals and then used subsequently in other animals in which glucose utilization is being measured. The part played by these constants in the method is defined by their role in the operational equation of the method (Fig. 1B).

4.1. Rate Constants

The rate constants, k_1^*, k_2^*, and k_3^*, vary from tissue to tissue, but the variation among gray structures and among white structures is considerably less than the differences between the two types of tissues (Table 1). The rate constants k_2^* and k_3^* appear in the equation only as their sum, and the sum ($k_2^* + k_3^*$) is equal to the rate constant for the turnover of the free [^{14}C]deoxyglucose pool in the tissue. The half-life of the free [^{14}C]deoxyglucose pool can then be calculated by dividing ($k_2^* + k_3^*$) into the natural logarithm of 2 and has been found to average 2.4 min in gray matter and 4.5 min in white matter in the normal conscious rat (Table 1).

The rate constants vary not only from structure to structure but can be expected to vary with the condition. For example, k_1^* and k_2^* are influenced by both blood flow and transport of [^{14}C]deoxyglucose across the blood–brain barrier, and because of the competition for the transport carrier, the glucose concentrations in the plasma and tissue affect the transport of [^{14}C]deoxyglucose and, therefore, also k_1^* and k_2^*. The constant, k_3^*, is related to phosphorylation of [^{14}C]deoxyglucose and will certainly change when glucose utilization is altered. To minimize potential errors resulting from inaccuracies in the values of the rate constants used, it was decided to sacrifice time resolution for accuracy. If the [^{14}C]deoxyglucose is given as an intravenous pulse and sufficient time is allowed for the plasma to be cleared of the tracer, then the influence of the rate constants, and the functions that they represent, on the final result diminishes with increasing time until ultimately it becomes zero. This relationship is implicit in the structure of the operational equation (Fig. 1B); as C_P^* approaches zero, then the terms containing the rate constants also approach zero with increasing time. The significance of this relationship is graphically illustrated in Fig. 5. From typical arterial plasma [^{14}C]deoxyglucose and glucose concentration curves obtained in a normal conscious rat, the portion of the denominator of the operational equation underlined by the heavy bar was computed with

INFLUENCE OF RATE CONSTANTS ON INTEGRATED POOL SPECIFIC ACTIVITY

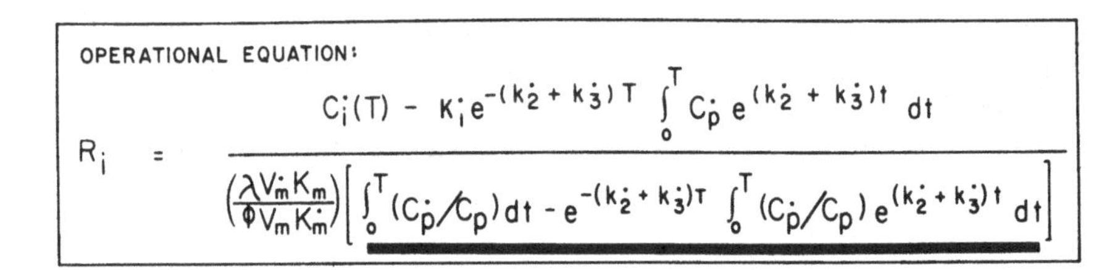

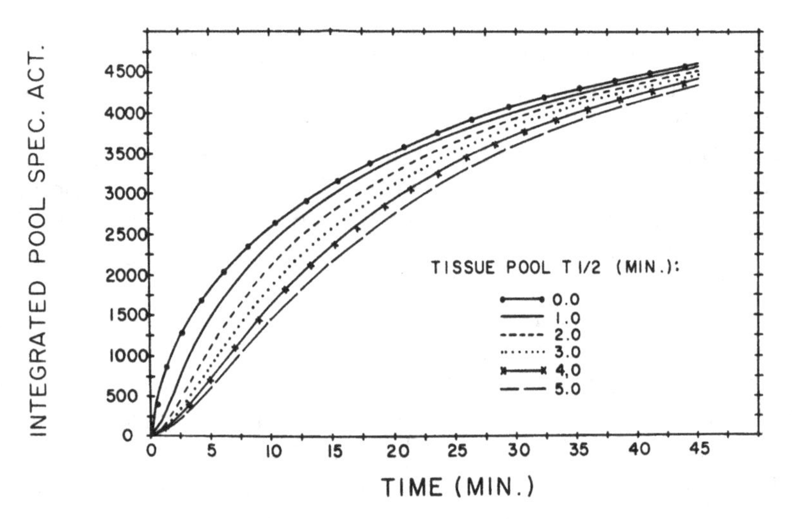

Fig. 5. Influence of time and rate constants, $k_2^* + k_3^*$, on the integrated precursor pool specific activity following a pulse of [^{14}C]deoxyglucose at zero time. The portion of the equation that is underlined corresponds to the integrated pool specific activity; it was computed as a function of time with different values of $k_2^* + k_3^*$, as indicated by their equivalent half-lives, calculated according to $T_{1/2} = 0.693/(k_2^* + k_3^*)$ (*See* text).

a wide range of values for $(k_2^* + k_3^*)$ as a function of time. The values for $(k_2^* + k_3^*)$ are presented as their equivalent half-lives calculated as described above. The values of $k_2^* + k_3^*$ vary from infinite (i.e., $T_{1/2} = 0$ min) to 0.14/min (i.e., $T_{1/2} = 5$ min) and more than cover the range of values to be expected under physiological conditions. The portion of the equation underlined and computed represents the integral of the precursor pool specific activity in the tissue. The curves represent the time course of this function, one each for every value of $k_2^* + k_3^*$ examined. It can be seen that these curves are widely different at early times, but converge with increasing time until at 45 min the differences over the entire range of $k_2^* + k_3^*$ equal only a small fraction of the value of the integral. These curves demonstrate that, at short times, enormous errors can occur if the values of the rate constants are not precisely known, but only negligible errors occur at 45 min, even over a wide range of rate constants of several-fold. In fact, it was precisely for this reason that [^{14}C]deoxyglucose rather than [^{14}C]glucose was selected as the tracer for glucose metabolism. The relationships are similar for glucose. Because the products of [^{14}C]glucose metabolism are so rapidly lost from the tissues, it is necessary to limit the experimental period to short times when enormous errors can occur if the rate constants are not precisely known. [^{14}C]Deoxyglucose permits the prolongation of the experimental period to times when inaccuracies in rate constants have little effect on the final result.

It should be noted, however, that in pathological conditions, such as severe ischemia or hyperglycemia, the rate constants may fall far beyond the range examined in Fig. 5. We have evidence, for example, that this occurs with arterial plasma glucose concentrations above 330 mg%. In such abnormal conditions, it may be necessary to redetermine the rate constants for the particular condition under study.

4.2. Lumped Constant

The lumped constant is composed of six separate constants. One of these, Φ, is a measure of the steady-state hydrolysis of glucose-6-phosphate to free glucose and phosphate. Because in normal brain tissue there is little such phosphohydrolase activity (Hers, 1957), Φ is normally approximately equal to unity. The other components are arranged in three ratios: λ, which is the ratio of distribution spaces in the tissue for deoxyglucose and glucose; V^*_m/V_m; and K_m/K^*_m. Although each individual constant may vary from structure to structure and condition to condition, it is likely that the ratios tend to remain the same under normal conditions. For reasons described in detail previously (Sokoloff et al., 1977), it is reasonable to believe that the lumped constant is the same throughout the brain and is characteristic of the species of animal, but only in normal tissue. Although reasonable, it is not certain, and there are theoretical possibilities that it may not be so. Experience thus far indicates that it is. The greatest experience has been accumulated in the albino rat. In this species, the lumped constant for the brain as a whole has been determined under a variety of conditions (Sokoloff et al., 1977). In the normal conscious rat, local cerebral glucose utilization, determined by the [^{14}C]deoxyglucose method with the single value of the lumped constant for the brain as a whole, correlates almost perfectly ($r = 0.96$) with local cerebral blood flow, measured by the [^{14}C]iodoantipyrine method, an entirely independent method (Sokoloff, 1978). It is generally recognized that local blood flow is adjusted to local metabolic rate, but if the single value of the lumped constant did not apply to the individual structures studied, then errors in local glucose utilization would occur that might be expected to obscure the correlation. Also, the lumped constant has been directly determined in the albino rat in the normal conscious state, under barbiturate anesthesia, and during the inhalation of 5% CO_2; no significant differences were observed (Table 2). The lumped constant varies with the species and has now also been determined in the rhesus monkey (Kennedy et al., 1978), cat (M. Miyaoka, J. Magnes, C. Kennedy, M. Shinohara, and L. Sokoloff, unpublished data), beagle puppy, (Duffy et al., 1982), fetal and newborn sheep (Abrams et al., 1984), and human subjects (Reivich et al., 1985) (Table 2). The values for local rates of glucose utilization determined with these lumped constants in these species are very close to what might be expected from measurement of energy metabolism in the brain as a whole by other methods (Table 3).

Although there is as yet no experimental evidence to indicate that the lumped constants determined in animals with normal brain tissues do not apply to pathological states, such a possibility must be seriously considered. Tissue damage may disrupt the normal cellular compartmentation. There is no assurance that λ, the ratio of the distribution spaces for [^{14}C]deoxyglucose and glucose, is the same in damaged tissue as in normal tissue. This is analogous to the question about the values of the tissue: blood partition coefficients in damaged tissues that are encountered in measurements of regional blood flow with radioactive inert gases. Also in pathological states, there may be release of lysosomal acid hydrolases that may hydrolyze glucose-6-phosphate and thus alter the value of Φ. At the present time, measurements of the lumped constant in pathological states are needed.

4.3. Role of Glucose-6-phosphatase

Glucose-6-Pase activity is known to be very low in the brain (Hers, 1957), and almost all textbooks of biochemistry attest to this fact. Although low, some activity is present, but it appears to have no influence on the deoxyglucose method if the experimental period is limited to the prescribed duration, 45 min. Beyond this time, its effects begin to appear, increasing in magnitude with increasing time. Significant glucose-6-Pase activity would affect the results by hydrolyzing the [^{14}C]DG-6-P and causing loss of product, and its effect would be cumulative with increasing time. If its activity were significant, it would cause the calculated rates of glucose utilization to be too low and to become progressively lower with increasing time. None of this happens in the first 45 min. In groups of rats studied over 20-, 30-, and 45-min periods, the calculated rates of glucose utilization remain constant (Sokoloff, 1982). Also, by determining the average glucose utilization of all the structures in the brain weighted for their relative sizes, it is possible to obtain the weighted average glucose utilization of the brain as a whole and to compare this with the values obtained by the Kety-Schmidt method, the recognized standard method for measuring blood flow and energy metabolism of the brain as a whole by the Fick Principle (Kety and Schmidt, 1948). The deoxyglucose method provides values that are almost exactly those obtained with the Kety-Schmidt method (Table 3). There is, therefore, no detectable influence of glucose-6-Pase during the first 45 min after the pulse of deoxyglucose, because the values for glucose utilization that are obtained are not too low nor do they decrease with time over that interval. After 45 min, however, effects of glucose-6-Pase begin to appear and become progressively greater with increasing time (Sokoloff, 1982). The time course of the effect is compatible with the intracellular distributions of the [^{14}C]DG-6-P and the phosphatase. The [^{14}C]DG-6-P is formed in the cytosol, but the phosphatase is on the inner surface of the cisterns of the endoplasmic reticulum. The [^{14}C]DG-6-P must first be formed in the cytosol and then transported across the endoplasmic reticular membrane by a specific carrier before the phosphatase can act to hydrolyze it. The kinetics of this process, namely, a lag with zero phosphatase activity followed by progressively increasing activity, are exactly those to be expected from the separate compartmentalization of substrate and enzyme and a rate limiting transport of the substrate across the membrane to the enzyme. This compartmentalization, therefore, allows a period of grace before the phosphatase can act on the DG-6-P, and it is a prolonged period of grace because, as Karnovsky and his associates (Fishman and Karnovsky, 1986) have found, the carrier for glucose-6-phosphate and DG-6-P is essentially absent in the brain.

The above and other evidence clearly indicate that phosphatase activity is of no significance to the deoxyglucose method if the experimental period is limited to 45 min. In spite of this evidence, a few reports have appeared alleging that phosphatase activity is a major source of error (Hawkins and Miller, 1978; Huang and Veech, 1982; Sachs et al., 1983). Each of these reports, however, has been shown to be flawed either as a result of the misinterpretation of experimental findings or faulty conduct of biochemical procedures (Nelson et al., 1985; Nelson et al., 1986; Dienel et al., 1986) and, therefore, has no bearing upon this issue.

4.4. Influence of Varying Plasma Glucose Concentration

Because the operational equation of the method (Fig. 1B) was derived on the basis of the assumption that C_P, the arterial plasma glucose concentration, remains constant during the experimental period, the method was applicable only to experiments in which this assumption was satisfied. This restraint has proved to be cumbersome. A new operational equation has, therefore, been derived that does not require this assumption. The equation is as follows:

$$R_i = \frac{C_i^*(\tau) - k_1^* \exp-(k_2^* + k_3^*)\tau \int_0^\tau C_P^* \exp(k_2^* + k_3^*)t\,dt}{\dfrac{\lambda V_m^* K_m}{\Phi V_m K_m^*} \displaystyle\int_0^\tau \left[\frac{(k_2^* + k_3^*) \exp-(k_2^* + k_3^*)T \int_0^T C_P^* \exp(k_2^* + k_3^*)t\,dt}{C_P(0)\exp-(k_2 + k_3)T + (k_2 + k_3)\exp-(k_2 + k_3)T \int_0^T C_P \exp(k_2 + k_3)t\,dt} \right] dT}$$

where $(k_2 + k_3)$ equals the turnover rate constant (i.e., $0.693/T_{1/2}$) of the free glucose pool in the brain; $C_P(0)$ equals the arterial plasma glucose concentration at zero time; τ equals the time of killing; and all other symbols are the same as in the equation in Fig. 1B.

This new equation requires an estimate of the half-life of the free glucose pool in the tissue. A method has been developed to measure the half-life of the free glucose pool in brain tissue that was found to equal 1.2 and 1.8 min in normal conscious and anesthetized rats, respectively, and to vary with the plasma glucose concentration (Savaki et al., 1980). The equation is relatively insensitive to the value of the half-life of the glucose pool in the range in which it usually falls and, therefore, only an approximation of the value of the half-life is necessary. The [^{14}C]deoxyglucose method can now, therefore, be used in the presence of changing arterial plasma glucose concentrations.

4.5. *Animal Behavior During the Experimental Period*

In order for the deoxyglucose method to provide results representative of any given behavioral state, that state should be sustained over the entire 45-min experimental period. Behavior lasting only for a short time, as in the performance of a brief task, is best represented when continuously repeated in a stereotyped manner. Seizures may be difficult to evaluate when the goal is to obtain accurate values for glucose utilization during the ictal episode. Although focal motor seizures, such as those induced by intracortical injections of penicillin, may continue for long periods, generalized attacks are often of only a few minutes in duration. Even with recurrent seizures, there may be intermittent periods of postictal depression. When there is such fluctuation in extremes of behavioral activity, the measured value for glucose utilization reflects some undefined time-weighted average of the effects of all the events during the experimental period. For meaningful quantitative results, it is best to maintain as steady a behavioral state as possible throughout the experimental period.

5. The Use of [^{14}C]Deoxyglucose in Metabolic Mapping and Other Nonquantitative Studies

The use of [^{14}C]deoxyglucose in mapping functional pathways of the brain has been previously described (Kennedy et al., 1975; Kennedy et al., 1976) and has attracted much interest. Even if one foregoes the sampling of blood following the administration of [^{14}C]deoxyglucose and prepares autoradiographs as described, the picture of relative rates of glucose utilization in the autoradiographs may provide useful information. This is especially true in studies that are designed to stimulate or deprive a pathway on one side of the brain, the other side being the control (Kennedy et al., 1975; Kennedy et al., 1976; McCulloch et al., 1980; Collins et al., 1976; Buchner et al., 1979). In view of the side-to-side symmetry in metabolic rates of the brain under normal circumstances, the finding of even small right to left differences in the autoradiographs may be significant. It should be noted that the deoxyglucose method results in labeling of entire functional pathways unattenuated at synaptic junctions. This is illustrated in the demonstration of the ocular dominance columns in the monkey in studies of monocular occlusion. The reduced function of the pathway subserving the occluded eye is clearly evident beyond the terminals of the geniculostriate pathway in Layer IV of the striate cortex and extends to involve not only all cortical layers of the striate cortex (Kennedy et al., 1976), but also the cascade of projections from striate to prestriate and into the temporal lobe (Macko et al., 1982).

Another nonquantitative use of [^{14}C]deoxyglucose is in physiologic alterations and drug-induced behavioral states that result in a redistribution of rates of glucose utilization in various parts of the brain. Autoradiographs in such induced states show regional differences in optical density resulting in a pattern characteristic for a given state (Pulsinelli and Duffy, 1978; Schwartz et al., 1979; Hubel et al., 1978; Kliot and Poletti, 1979; Meibach et al., 1979). A difference in this pattern from that in normal resting animals is apparent by inspection. It is risky to make quantitative interpretations on the basis of comparison of optical density or ^{14}C content in any one region of the brain of one animal with that in the same region of the brain of another animal without employing the fully quantitative method. This is because the ^{14}C concentration for a given brain region is determined not only by its rate of glucose utilization, but also by the entire time courses of the plasma concentrations of [^{14}C]deoxyglucose and glucose. It cannot be assumed that these are the same in different animals simply because each receives the same dose of [^{14}C]deoxyglucose/unit weight. Not only are there normal random physiologic variations from animal to animal that affect these time courses, but there may be systematic alterations that are the result of a given experimental condition. In nonquantitative deoxyglucose studies, the route of administration of [^{14}C]deoxyglucose is largely arbitrary. Subcutaneous or intraperitoneal routes are also suitable. In a study in insect brain, the oral route also proved satisfactory (Buchner et al., 1979).

A number of reports have appeared in which radiolabeled deoxyglucose has been employed in a "semiquantitative" procedure. This consists simply of administering the deoxyglucose, killing the animal, removing and freezing the brain at the prescribed time, and processing the tissue for autoradiography. Optical density readings are made in regions of interest and are reported as ratios to the average optical density readings obtained in some, rather arbitrarily chosen, white matter structures that are assumed to be unaffected in their glucose utilization by the experimental condition. The optical density ratio is presented as an index of the rate of glucose utilization. This approach may occasionally be useful in comparing metabolic rates when it is shown that the white matter used as a reference is unaffected by the experimental condition (Toga and Collins, 1981), but in most reports, this has not been established and, to that extent, such results are of uncertain significance. The rationale often presented to justify this abridged procedure resulting in an index rather than an absolute value for glucose utilization is the need to have the animal unrestrained and "unstressed." However, inasmuch as blood sampling catheters can readily be arranged to permit behavioral studies to be conducted in freely moving animals lacking any evidence of stress (Porrino et al., 1984; Jay et al., 1985), there seems little justification to abandon the original quantitative method.

6. Recent Technological Advances

Several recent technological developments extend the resolution of the [^{14}C]deoxyglucose method in animals and adapt it for use in humans.

6.1. Computerized Image-Processing

The regional localization obtained with the [^{14}C]deoxyglucose method is achieved by the use of quantitative autoradiography. The autoradiographs contain an immense amount of information that cannot be practically recovered by manual densitometry or adequately represented by tabular presentation of the data. A computerized image-processing system has, therefore, been developed to analyze and transform the autoradiographs into color-coded pictorial maps of the rates of local glucose utilization throughout the CNS (Goochee et al., 1980). The autoradiographs are scanned automatically by a computer-controlled scanning microdensitometer. The optical density of each spot on the autoradiograph, from 25–100 μm as selected, is stored in a computer, converted to ^{14}C concentration on the basis of the optical densities of the calibrated ^{14}C plastic standards, and then converted to local rate of glucose utilization by solution of the operational equation. Colors are assigned to narrow ranges of the rates of glucose utilization, and the autoradiographs are then displayed on a monitor in color along with a calibrated color scale for identifying the rate of glucose utilization in each spot of the autoradiograph from its color. The display is similar to that used by Lassen, Ingvar, and associates for regional cortical blood flow (Lassen et al., 1978), but it represents the rates of glucose utilization in all parts of the brain with an increase in spatial resolution such that virtually all macroscopic neuroanatomic subdivisions are delineated.

6.2. Microscopic Resolution

The resolution of the present [^{14}C]deoxyglucose method is at best approximately 100–200 μm (Smith, 1983). The use of deoxyglucose labeled with ^{3}H does not significantly improve this when autoradiographs are made from brain sections of 20 μm in thickness, the limiting factor being the diffusion and migration of the water-soluble labeled compound during the freezing of the brain and the cutting of the brain sections (Smith, 1983). In order to improve significantly the level of resolution, it is necessary to employ procedures for fixation of the label in the tissue so that it cannot migrate during the preparation of tissue sections and their autoradiographs (Des Rosiers and Descarries, 1978). Unfortunately, both conventional fixation with glutaraldehyde (Ornberg et al., 1979) and carbohydrate fixation with periodate, lysine, and paraformaldehyde (Durham et al., 1981) result in large (90%) losses of labeled DG from the tissues. Although chemical fixation of the labeled DG is an ideal approach to this problem, a fixation for sugars that will act rapidly during an in vivo perfusion is not practical. The alternative solution is to remove all water from the tissue either by freeze-drying (Buchner et al., 1979; Duncan et al., 1987) or freeze substitution (Sejnowski et al., 1980; Lancet et al., 1982) or to carry out all procedures including autoradiography on frozen tissues (Hökfelt et al., 1983). These techniques have been used successfully at the electron microscopic level to map odor-induced neuronal activity in the olfactory bulb in rats (Benson et al., 1985) and the effects of visual stimulation on neuronal activity in drosophila (Buchner and Buchner, 1980). The next stage of development will be the implementation of quantitative autoradiographic techniques at the light and electron microscopic level for the determination of cellular and subcellular rates of glucose utilization.

6.3. [^{18}F]Fluorodeoxyglucose Technique

The deoxyglucose method was originally designed for use in animals with quantitative autoradiography and the radioactive isotopes most suitable for film autoradiography, ^{14}C or ^{3}H. Its basic physiological and biochemical principles apply, however, to humans as well, and it is applicable to humans provided the local tissue concentrations of isotope can be measured in the brain. Film autoradiography is a type of emission tomography that, for obvious reasons, cannot be used in humans, but recent developments in computerized tomography have made it possible to determine local concentrations of γ-emitting isotopes in the cerebral tissues. The only possible γ-emitting isotopes that can be incorporated into 2-deoxyglucose are ^{11}C or ^{15}O, but the short half-lives of these isotopes present problems in the synthesis of the compounds. Alternatively, an analog of 2-deoxyglucose with another γ-emitting isotope but with similar biochemical properties could be used. It is a common experience that the substitution of the very small atom, F, in place of a hydrogen at a judicious site in the molecule often does not alter the basic biochemical behavior of metabolic substrates. 2-[^{18}F]Fluoro-2-deoxy-D-glucose has been synthesized, found to retain the biochemical properties of 2-deoxyglucose, and used to measure cerebral glucose utilization in humans by means of single photon emission tomography (Reivich et al., 1979). ^{18}F is actually a positron emitter, and the absorption of positrons in the tissues gives rise to two coincident annihilation γ-rays of equal energy traveling at almost 180° to each other. Positron emission tomography takes advantage of these coincident annihilation γ-rays and, therefore, is inherently capable of better spatial resolution than single photon tomography. The [^{18}F]fluorodeoxyglucose method is, therefore, now generally used with positron emission tomography (Phelps et al., 1979). Positron emission tomography with [^{18}F]fluorodeoxyglucose has been relatively slow with the earlier generations of scanners, and it took up to 2 h to obtain sufficient counts for accurate measurements of local ^{18}F concentrations in all parts of the brain. Although low in the brain, glucose-6-phosphatase activity is not zero, and its effect becomes significant after the first 45 min after the pulse of tracer (Sokoloff, 1982). It was, therefore, necessary to modify the model to include a rate constant for the hydrolysis of the phosphorylated product by glucose-6-phosphatase and to derive a new operational equation that takes this activity into account (Huang et al., 1980; Phelps et al., 1979; Sokoloff, 1982). The [^{18}F]fluorodeoxyglucose technique for the measurement of local cerebral glucose utilization in humans is now operational and in use for studies of the human brain in health and disease in a number of laboratories. It has been used in studies of the visual and auditory systems (Phelps et al., 1981) and of clinical conditions, such as focal epilepsy (Kuhl et al., 1979, 1980), Huntington's disease (Kuhl et al., 1982b), aging (Kuhl et al., 1982a) and dementia (Kuhl et al., 1983; Foster et al., 1983), and cerebral gliomas (DiChiro et al., 1982). MacGregor et al. (1981) and Reivich et al. (1982) have recently succeeded in synthesizing [^{11}C]deoxyglucose and applied it to the measurement of local cerebral glucose utilization in humans with positron emission tomography. Because of the short half-life of ^{11}C, this development should be very useful for sequential measurements in the same subject in a short time period.

7. Summary

The deoxyglucose method provides the means to determine quantitatively the rates of glucose utilization simultaneously in all structural and functional components of the central nervous system, and to display them pictorially superimposed on the anatomical structures in which they occur. Because of the close relationship between local functional activity and energy metabolism, the method makes it possible to identify all structures with increased or decreased functional activity in various physiological, pharmacological, and pathophysiological states. The images provided by the method do resemble histological sections of nervous tissue. The method is, therefore, sometimes misconstrued to be a neuroanatomical method and contrasted with physiological methods, such as electrophysiological recording. This classification obscures the most significant and unique feature of the method. The images are not of structure, but of a dynamic biochemical process, glucose utilization, which is as physiological as electrical activity. In most situations, changes in functional activity result in changes in energy metabolism, and the images can be used to visualize and identify the sites of altered activity. The images are, therefore, analogous to infrared maps; they record quantitatively the rates of a kinetic process and display them pictorially exactly where they exist. The fact that they depict the anatomical structures is fortuitous; it indicates that the rates of glucose utilization are distributed according to structure, and specific functions in the nervous system are associated with specific anatomical structures. The deoxyglucose method represents, therefore, in a real sense, a new type of encephalography, metabolic encephalography. At the very least, it should serve as a valuable supplement to more conventional types, such as electroencephalography. Because, however, it provides a new means to examine another aspect of function simultaneously in all parts of the brain, it is hoped that it and its derivative, the [^{18}F]fluorodeoxyglucose technique, will open new roads to the understanding of how the brain works in health and disease.

References

Abrams R. M., Ito M., Frisinger J. E., Patlak C. S., Pettigrew K. D., and Kennedy C. (1984) Local cerebral glucose utilization in fetal and neonatal sheep. *Am. J. Physiol.* **246,** R608–R618.

Benson T. E., Burd G. D., Greer C. A., Landis D. M. D., and Shepherd G. M. (1985) High-resolution 2-deoxyglucose autoradiography in quick-frozen slabs of neonatal rat olfactory bulb. *Brain Res.* **339,** 67–78.

Buchner E. and Buchner S. (1980) Mapping stimulus-induced nervous activity in small brains by [^{3}H]2-deoxy-d-glucose. *Cell Tissue Res.* **211,** 51–64.

Buchner E., Buchner S., and Hengstenberg R. (1979) 2-Deoxy-D-glucose maps movement-specific nervous activity in the second visual ganglion of *Drosophila, Science* **205,** 687.

Collins R. C., Kennedy C., Sokoloff L., and Plum F. (1976) Metabolic anatomy of focal motor seizures. *Arch. Neurol.* **33,** 536.

Des Rosiers M. H. and Descarries L. (1978) Adaptation de la méthode au désoxyglucose a l'echelle cellulaire: préparation histologique du système nerveux central en vue de la radio-autographie à haute résolution, *C. R. Acad. Sc. Paris,* **287** (Series D), 153.

DiChiro G., DeLaPaz R. L., Brooks R. A., Sokoloff L., Kornblith P. L., Smith B. H., Patronas N. J., Kufta C. V., Kessler R. M., and Wolf, A. P. (1982) Glucose utilization of cerebral gliomas measured by [^{18}F]fluorodeoxyglucose and positron emission tomography. *Neurol.* **32** (12), 1323–1329.

Dienel G., Nelson T., Cruz N., Jay T., and Sokoloff L. (1986) Contaminants in inadequately purified glucose and incomplete recovery of metabolites are responsible for the erroneous conclusion of high glucose-6-phosphatase activity in rat brain. *Soc. Neurosci. Abstr.* **12,** Part 2, p. 1405.

Duffy T. E., Cavazzuti M., Cruz N. F., and Sokoloff L. (1982) Local cerebral glucose metabolism in newborn dogs: effects of hypoxia and halothane anesthesia. *Ann. Neurol.* **11,** 233–246.

Duncan G. E., Stump W. E., and Pilgrim C. (1987) Cerebral metabolic mapping at the cellular level with dry-mount autoradiography of [^{3}H]2-deoxyglucose. *Brain Res.* **401,** 43–49.

Durham D., Woolsey T. A., and Krugher L. (1981) Cellular localization of 2-[^{3}H]deoxyglucose-D-glucose from paraffin-embedded brains. *J. Neurosci.* **1,** 519–526.

Fishman R. S. and Karnovsky M. L. (1986) Apparent absence of a translocase in the cerebral glucose-6-phosphatase system. *J. Neurochem.* **46,** 371–378.

Foster N. L., Chase T. N., Fedio P., Patronas N. J., Brooks R. A., and DiChiro G. (1983) Alzheimer's disease: focal cortical changes shown by positron emission tomography. *Neurol.* **33,** 961–965.

Goochee C., Rasband W., and Sokoloff L. (1980) Computerized densitometry and color coding of [^{14}C]deoxyglucose autoradiographs. *Ann. Neurol.* **7,** 359–370.

Hawkins R. A. and Miller D. L. (1978) Loss of radioactive 2-deoxy-D-glucose-6-phosphate from brains of conscious rats: implications for quantitative autoradiographic determination of regional glucose utilization. *Neurosci.* **3,** 251–258.

Hers H. G. (1957) Le *Métabolisme du Fructose,* Editions Arscia, Bruxelles, p. 102.

Hökfelt T., Smith C. B., Peters A., Norell G., Crane A., Brownstein M., and Sokoloff L. (1983) Improved resolution of the 2-deoxy-D-glucose technique. *Brain Res.* **289,** 311–316.

Huang M.-T. and Veech R. L. (1982) The quantitative determination of the in vivo dephosphorylation of glucose-6-phosphate in rat brain. *J. Biol. Chem.* **257,** 11358–11363.

Huang S.-C., Phelps M. E., Hoffman E. J., Sideris K., Selin C. J., and Kuhl D. E. (1980) Noninvasive determination of local cerebral metabolic rate of glucose in man. *Am. J. Physiol.* **238,** E69–E82.

Hubel D. H., Wiesel T. N., and Stryker M. P. (1978) Anatomical demonstration of orientation columns in Macaque monkey. *J. Comp. Neurol.* **177,** 361.

Jay T. M., Jouvet M., and Des Rosiers M. H. (1985) Local cerebral glucose utilization in the free moving mouse: a comparison during two stages of the activity-rest cycle. *Brain Res.* **342,** 297–306.

Kennedy C., Des Rosiers M. H., Jehle J. W., Reivich M., Sharp F., and Sokoloff L. (1975) Mapping of functional neural pathways by autoradiographic survey of local metabolic rate with [^{14}C]deoxyglucose. *Science* **187,** 850.

Kennedy C., Des Rosiers M. H., Sakurada O., Shinohara M., Reivich M., Jehle J. W., and Sokoloff, L. (1976) Metabolic mapping of the primary visual system of the monkey by means of the autoradiographic [^{14}C]deoxyglucose technique. *Proc. Natl. Acad. Sci. USA* **73,** 4230.

Kennedy C., Sakurada O., Shinohara M., Jehle J., and Sokoloff L. (1978) Local cerebral glucose utilization in the normal conscious Macaque monkey. *Ann. Neurol.* **4,** 293.

Kety S. S. and Schmidt C. F. (1948) The nitrous oxide method for the quantitative determination of cerebral blood flow in man: theory, procedure, and normal values. *J. Clin. Invest.* **27,** 476–483.

Kliot M. and Poletti C. E. (1979) Hippocampal afterdischarges: differential spread of activity shown by the [^{14}C]deoxyglucose technique. *Science* **204,** 641–626.

Kuhl D. E., Engel J., Phelps M. E., and Selin C. (1979) Patterns of local cerebral metabolism and perfusion in partial epilepsy by emission computed tomography of ^{18}F-fluorodeoxyglucose and ^{13}N-ammonia. *Acta Neurol. Scand.* **60** (Suppl. 72), 538–539.

Kuhl D. E., Engel J., Phelps M. E., and Selin C. (1980) Epileptic patterns of local cerebral metabolism and perfusion in humans determined by emission computed tomography of ^{18}F DG and $^{13}NH_3$. *Ann. Neurol.* **8,** 348–360.

Kuhl D. E., Metter E. J., Riege W. H., and Phelps M. E. (1982a) Effects of human aging on patterns of local cerebral glucose utilization determined by the [^{18}F]fluorodeoxyglucose method. *J. Cereb. Blood Metab.* **2,** 163–171.

Kuhl D. E., Phelps M. E., Markham C. H., Metter E. J., Riege W. H., and Winter J. (1982b) Cerebral metabolism and atrophy in Huntington's disease determined by ^{8}FDG and computed tomographic scan. *Ann. Neurol.* **12,** 425–434.

Kuhl D. E., Metter E. J., Riege W. H., Hawkins R. A., Mazziotta, J. C., Phelps M. E., and Kling A. S. (1983) Local cerebral glucose utilization in elderly patients with depression, multiple infarct dementia and Alzheimer's disease. *J. Cereb. Blood Flow Metab.* **3** (Suppl. 1), 494–495.

Lancet D., Greer C. A., Kauer J. S., and Shepherd G. M. (1982) Mapping of odor-related neuronal activity in the olfactory bulb by high-resolution of 2-deoxyglucose autoradiography. *Proc. Natl. Acad. Sci. USA* **79,** 670–674.

Lassen N. A., Ingvar D. H., and Skinhoj E. (1978) Brain function and blood flow. *Sci. Amer.* **239,** 62–71.

McCulloch J., Savaki H. E., McCulloch M. C., and Sokoloff L. (1980) Regina-dependent activation by apomorphine of metabolic activity in the superficial layer of the superior colliculus. *Science* **207,** 313–315.

MacGregor R., Fowler J. S., Wolfe A. P., Shiue C.-Y., Lade R. E., and Wan C.-N. (1981) A synthesis of 2 Deoxy-D-[1-^{11}C]glucose for regional metabolic studies: concise communication. *J. Nucl. Med.* **22,** 800–803.

Macko K. A., Jarvis C. D., Kennedy C., Miyaoka M., Shinohara M., Sokoloff L., and Mishkin M. (1982) Mapping the primate visual system with [2-^{14}C]Deoxyglucose. *Science* **218,** 394–397.

Meibach R. C., Glick S. D., Cox R., and Maayani S. (1979) Localization of phencyclidine-induced changes in brain energy metabolism. *Nature* **282,** 625–626.

Nelson T., Kaufman E. E., and Sokoloff L. (1984) 2-Deoxyglucose incorporation into rat brain glycogen during measurement of local cerebral glucose utilization by the 2-deoxyglucose method. *J. Neurochem.* **43,** 949–956.

Nelson T., Lucignani G., Atlas S., Crane A. M., Dienel G. A., and Sokoloff L. (1985) Reexamination of glucose-6-phosphatase activity in the brain *in vivo:* no evidence for a futile cycle. *Science* **229,** 60–62.

Nelson T., Lucignani G., Goochee J., Crane A. M., and Sokoloff L. (1986) Invalidity of criticisms of the deoxyglucose method based on alleged glucose-6-phosphatase activity in brain. *J. Neurochem.* **46,** 905–919.

Ornberg R. L., Neale E. A., Smith C. B., Yarowsky P., and Bowers L. M. (1979) Radioautographic localization of glucose utilization by neurons in culture. *J. Cell. Biol.* **Abstr. 83,** CN 142A.

Phelps M. E., Huang S. C., Hoffman E. J., Selin C., Sokoloff L., and Kuhl D. E. (1979) Tomographic measurement of local cerebral glucose metabolic rate in humans with (F-18)2-fluro-2-deoxy-D-glucose: validation of method. *Ann. Neurol.* **6,** 371–388.

Phelps M. E., Kuhl D. E., and Mazziotta J. C. (1981) Metabolic mapping of the brain's response to visual stimulation: studies in man. *Science* **211,** 1445–1448.

Porrino L. J., Esposito R. U., Seeger T. F., Crane A. M., Pert A., and Sokoloff L. (1984) Metabolic mapping of the brain during rewarding self-stimulation. *Science* **224,** 306–309.

Pulsinelli W. A. and Duffy T. E. (1978) Local cerebral glucose metabolism during controlled hypoxemia in rats. *Science* **204,** 626–629.

Reivich M., Alavi A., Wolf A., Fowler J., Russell J., Arnett C., MacGregor R. K., Shiue C. Y., Atkins H., Anand A., Dann R., and Greenberg J. H. (1985) Glucose metabolic rate kinetic model parameter determination in humans: the lumped constants and rate constants for [^{18}F]fluorodeoxyglucose and [^{11}C]deoxyglucose. *J. Cereb. Blood Flow Metab.* **5,** 179–192.

Reivich M., Alavi A., Wolf A., Greenberg, J. H., Fowler J., Christman D., MacGregor R., Jones S. C., London J., Shiue C., and Yonekura Y. (1982) Use of 2-deoxy-D-[1-^{11}C]glucose for the determinations of local cerebral glucose metabolism in humans: variation within and between subjects. *J. Cereb. Blood Flow Metab.* **2,** 307–319.

Reivich M., Kuhl D., Wolf A., Greenberg J., Phelps M., Ido T., Casella V., Fowler J., Hoffman E., Alavi A., Som P., and Sokoloff, L. (1979) The [^{18}F]fluorodeoxyglucose method for the measurement of local cerebral glucose utilization in man. *Circ. Res.* **44,** 127–137.

Sacks W., Sacks S., and Fleischer A. (1983) A comparison of the cerebral uptake and metabolism of labeled glucose and deoxyglucose in vivo in rats. *Neurochem. Res.* **8,** 661–685.

Savaki H. E., Davidsen L., Smith C., and Sokoloff L. (1980) Measurement of free glucose turnover in brain. *J. Neurochem.* **35** (2), 495–502.

Schwartz W. J., Smith C. B., Davidsen L., Savaki H., Sokoloff L., Mata M., Fink D. J., and Gainer H. (1979) Metabolic mapping of functional activity in the hypothalamo-neurohypophysial system of the rat. *Science* **205,** 723–725.

Sejnowski T. J., Reingold S. C., Kelley D. B., and Gelperin A. (1980) Localization of [^{3}H]-2-deoxyglucose in single molluscan neurones (1980). *Nature* **287,** 449–451.

Smith C. B. (1983) Localization of activity-associated changes in metabolism of the central nervous system with the deoxyglucose method: prospects for cellular resolution, in *Current Methods in Cellular Neurobiology Vol. I. Anatomical Techniques* (Barker J. L. and McKelvy J. F., eds.), pp. 269–317, John Wiley, New York.

Sokoloff L. (1978) Local cerebral energy metabolism: its relationships to local functional activity and blood flow, in *Cerebral Vascular Smooth Muscle and Its Control,* Ciba Foundation Symposium 56 (new series), pp. 171–197, Elsevier/Excerpta Medica/North-Holland, Amsterdam.

Sokoloff L. (1982) The radioactive deoxyglucose method. Theory, procedure, and applications for the measurement of local glucose utilization in the central nervous system, in *Advances in Neurochemistry*, Vol. 4 (Agranoff B. W. and Aprison, M. H., eds.), pp. 1–82. Plenum Press, New York.

Sokoloff L., Reivich M., Kennedy C., Des Rosiers M. H., Patlak C. S., Pettigrew K. D., Sakurada O., and Shinohara M. (1977) The [^{14}C]deoxyglucose method for the measurement of local cerebral glucose utilization: theory, procedure, and normal values in the conscious and anesthetized albino rat. *J. Neurochem.* **28,** 897–916.

Toga A. W. and Collins R. C. (1981) Metabolic response of optic centers to visual stimuli in the albino rat: anatomical and physiological considerations. *J. Comp. Neurol.* **199,** 443–464.

Determination of Cerebral Glucose Use in Rats Using [^{14}C]Glucose

Richard A. Hawkins and Anke M. Mans

1. Introduction

The postulate that useful information about the activity of cerebral tissue can be deduced from a measure of energy metabolism, as reflected by the cerebral metabolic rate of glucose (CMRglc), is supported by a considerable body of data. Because there are adequate reviews of this subject (e.g., Sokoloff, 1981), it will not be further considered. The purpose of this chapter is to explain how CMRglc may be measured in the laboratory rat with a natural isotope, [6-^{14}C]glucose. The reasons for using [6-^{14}C]glucose are that the experimental period may be as brief as 5 min (enabling the study of short-lived states), the isotope behaves kinetically exactly as glucose and no correction factors are necessary, and label from [6-^{14}C]glucose is almost completely trapped in brain tissue in short experimental periods (i.e., 10 min or less). Whereas [6-^{14}C]glucose measures total CMRglc, the use of [1-^{14}C]glucose instead enables determination of only energy metabolism, without the contribution of the hexose monophosphate shunt, as explained below.

Before discussing the technique in detail, it may be useful to outline the basic procedure and the measurements to be made. The animal model is a rat fitted with arterial and venous catheters, such that [^{14}C]glucose can be introduced into the venous system and samples of arterial blood can be taken at intervals throughout the experimental period. At the end of the experiment, the rat is killed and its brain removed for analysis. The measurements made are: the times of blood sampling, the specific radioactivity of blood glucose at each time point, and the dpm/g in the brain. When states are being studied in which the ratio of brain glucose-to-plasma glucose is expected to be changed, this ratio must be determined in a separate experiment. From these measurements, it is possible to determine CMRglc in the whole brain, dissected brain samples, or by using quantitative autoradiography, at a detailed regional level.

2. Biochemical Theory

An understanding of the biochemical principles underlying the use of [^{14}C]glucose to measure CMRglc is essential. It seems appropriate, therefore, to summarize them here and to describe the fate of each carbon in glucose.

There are four important reactions, each irreversible, that release $^{14}CO_2$ originating from [^{14}C]glucose. These reactions are catalyzed by: pyruvate dehydrogenase, which releases C-3 and C-4; isocitrate dehydrogenase and 2-oxoglutarate dehydrogenase, which release C-2 and C-5 during the second turn of the tricarboxylic acid cycle and C-1 and C-6 during the third and subsequent turns; and 6-phosphogluconate dehydrogenase, which releases C-1 from that small portion of glucose metabolized in the pentose phosphate pathway (about 2–5%). The rate of $^{14}CO_2$ evolution at any moment is determined by the specific radioactivity of the precursor carbon atom destined to form $^{14}CO_2$, and the flux through the decarboxylating reaction. For example, only ^{14}C in the carboxyl group of pyruvate will give rise to $^{14}CO_2$ by the action of pyruvate dehydrogenase; it is the specific radioactivity of the carboxyl group alone that is relevant at this point.

As labeled metabolites pass through the various intermediary metabolite pools, the specific radioactivity decreases by dilution. In addition, there are large pools of amino acids, in transamination equilibrium with oxoacids in the tricarboxylic acid cycle, which sequester ^{14}C (Fig. 1). Because the exchange between oxoacids and amino acids proceeds at a higher velocity than the reactions of the tricarboxylic acid cycle, labeled molecules are rapidly removed and replaced by unlabeled molecules (Balázs, 1969; Hawkins and Mans, 1983; Krebs, 1965; Sacks, 1969). This process lowers the specific radioactivity of intermediary metabolites and thereby retards $^{14}CO_2$ evolution. In other words, ^{14}C is shunted away from the main stream and sequestered in metabolic cul-de-sacs. The fate of each carbon in glucose is illustrated for two different metabolic circumstances: first, where glucose is converted directly to pyruvate, and then through the tricarboxylic acid cycle (Fig. 2); and second, where glucose is metabolized through the pentose phosphate pathway, and then to pyruvate and the tricarboxylic acid cycle (Fig. 3).

Consider a group of [U-^{14}C]glucose molecules on their journey through the glycolytic pathway and the tricarboxylic acid cycle (Fig. 2). Conversion of glucose to pyruvate by glycolysis results in

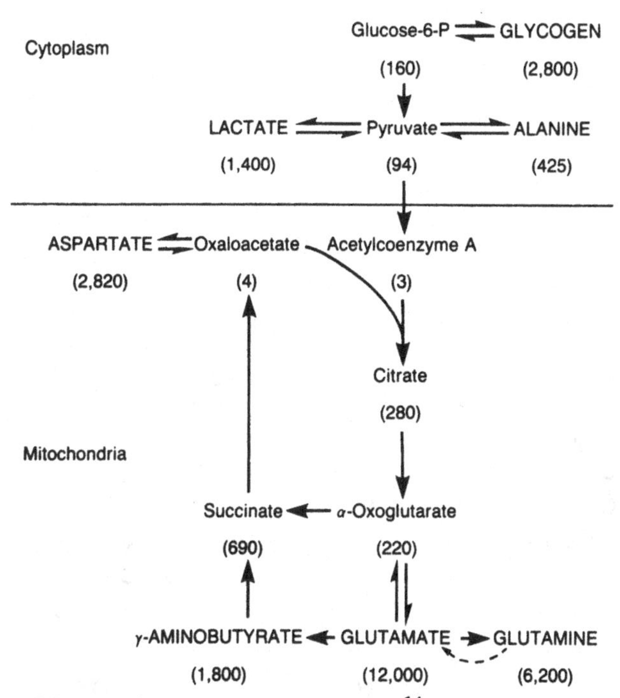

Fig. 1. Major side reactions that divert ^{14}C from oxidation. All reactions indicated as bidirectional are near equilibrium, except glycogen synthesis. The large size of amino acid pools makes them especially effective traps for ^{14}C in brain tissue. Concentrations (in nmol/g) are shown in parentheses.

complete conservation of ^{14}C. Oxidative decarboxylation of pyruvate to acetyl-CoA removes ^{14}C from C-3 and C-4 completely[1], but no further loss of ^{14}C occurs during the first turn of the tricarboxylic acid cycle. Because succinate is symmetrical, ^{14}C is distributed equally to both halves of succinate, fumarate, malate, and oxaloacetate. By the time ^{14}C reaches oxaloacetate, the specific radioactivity is greatly reduced because of exchange with amino acids. It is not until the second turn of the tricarboxylic acid cycle that more ^{14}C is lost; ^{14}C from C-2 and C-5 can be removed by isocitrate dehydrogenase and by 2-oxoglutarate dehydrogenase[2]. (Note that ^{14}C from C-2 and C-5 in 2-oxoglutarate is available for trapping by glutamate during the second turn of the tricarboxylic acid cycle.)

1. If pyruvate is converted directly to oxaloacetate by pyruvate carboxylase, then ^{14}C from C-3 and C-4 can be retained. Pyruvate carboxylase is not, however, believed to be very active in the brain (Cremer, 1980).

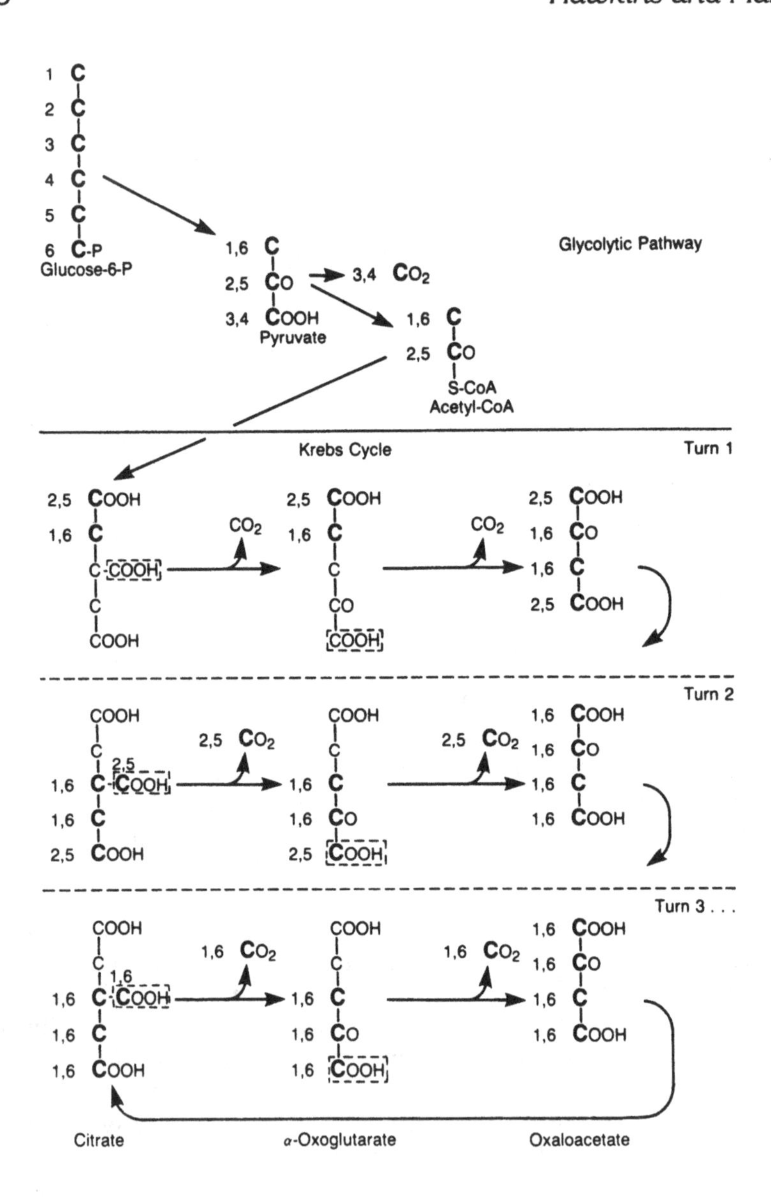

Fig. 2. Fate of carbon in various positions of glucose on metabolism by glycolysis and the tricarboxylic acid cycle. Fate of labeled carbons is shown in selected metabolites at key positions along metabolic pathway. In tricarboxylic acid cycle, boxes indicate the next carbon to be released.

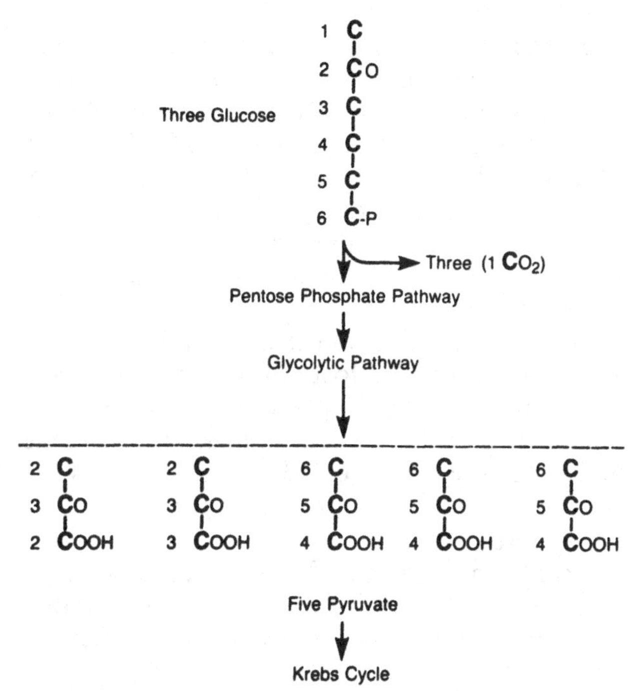

Fig. 3. Fate of carbon in various positions of glucose on metabolism by pentose phosphate and glycolytic pathways. It is assumed here that glucose is metabolized through the pentose phosphate pathway and then through the glycolytic pathway to pyruvate, and that no recycling of intermediary metabolites (e.g., fructose-6-phosphate) occurs.

2. It is possible that net metabolism of some 2-oxoglutarate occurs through the 4-aminobutyrate pathway. Nevertheless, isotope distribution in succinate will be the same whether it is formed from the tricarboxylic acid cycle reactions or the 4-aminobutyrate pathway.

No ^{14}C is lost from C-1 or C-6 positions during the second turn of the cycle, and because label passes again through succinate, every carbon of oxaloacetate becomes labeled. The distribution of label to all carbons of oxaloacetate has the important consequence that only half of the ^{14}C is in a position to be lost during the third turn of the cycle. The remaining ^{14}C passes through the tricarboxylic acid cycle, redistribution of ^{14}C to all possible positions in oxaloacetate occurs again and each subsequent turn of the cycle can remove only half of the remaining label. Because of the above considerations, evolution of $^{14}CO_2$ from [6-^{14}C]glucose is expected to be delayed and then to occur slowly.

Although the pentose phosphate pathway is not known to be especially active in the adult brain, it is interesting to consider the effect it may have on ^{14}C loss. Passage through the 6-phosphogluconate dehydrogenase reaction removes C-1, and in the subsequent reactions, ^{14}C is distributed as shown in Fig. 3 (Landau and Wood, 1983). Pyruvate dehydrogenase removes 33% of C-2 and C-3 and all of C-4. The remaining ^{14}C from C-2 and, of particular note, all the ^{14}C from C-6 are fully retained during at least two turns of the tricarboxylic acid cycle.

To sum up, C-3 and C-4 are the most labile positions. Most of the label from C-3 and C-4 is lost at the pyruvate dehydrogenase reaction, and although some sequestration of label can occur (e.g. in lactate, alanine, glycolytic intermediates, and glycogen), this is not expected to be very effective. Label from C-6 and C-1 is trapped most successfully except, of course, that portion of C-1 lost in the pentose phosphate pathway.

In practical terms, we have found that release of $^{14}CO_2$ from [6-^{14}C]glucose was undetectable (by measuring arteriovenous differences) at 5 min, and less than 2% at 10 min (Hawkins et al., 1985).

3. Mathematical Theory

The following symbols are defined for use in the subsequent treatment:

C_p^* = concentration of [6-^{14}C]glucose in plasma (dpm/mL)
C_p = concentration of glucose in plasma (μmol/mL)
C_b^* = concentration of [6-^{14}C]glucose in blood (dpm/mL)
C_b = concentration of glucose in blood (μmol/mL)
C_e^* = concentration of [6-^{14}C]glucose in brain (dpm/g)
C_e = concentration of glucose in brain (μmol/g)
C_m^* = ^{14}C concentration in brain tissue metabolites (dpm/g)
C_m = concentration of brain intermediary metabolites (expressed as glucose equivalents in μmol/g)
$\Sigma^{14}C$ = total ^{14}C in tissue (dpm/g)
k_1 = rate constant of glucose entry into brain (min^{-1})
k_2 = rate constant of glucose exit from brain (min^{-1})
k_3 = rate constant of glucose phosphorylation (min^{-1})

t = time measured from the moment C_p^* reaches its maximal value (min)
T = final experimental time (min)
T_{max} = maximal velocity of the glucose transport system of the blood–brain barrier (μmol min^{-1} g^{-1})
k_m = affinity constant of the glucose transport system of the blood–brain barrier (m*M*)

It should be noted that, even though k_1, k_2, and k_3 are used as constants in the equations, they take on a range of values depending on factors such as C_p, C_e, T_{max}, and CMRglc. This is taken into account in the calculations.

3.1. Model and Assumptions

A diagram of the three-compartment model of CMRglc is shown in Fig. 4. For the conditions of the experiment, the assumptions are:

1. CMRglc is in steady-state throughout the experimental period. This assumption may not always hold; the longer the experiment, the less likely that it will be true.
2. The concentrations of all nonlabeled intermediary metabolites and substrates are constant throughout the experiment.
3. Brain glucose content is the same in all cerebral structures.
4. Glucose enters and leaves the brain by a carrier mechanism located at the blood–brain barrier that follows saturation kinetics.
5. The transfer of material from one compartment to another is governed by rate constants that are a function of CMRglc (except gluconeogenesis, which is nil in the brain).
6. [6-^{14}C]glucose is the only label introduced into circulation.
7. [6-^{14}C]glucose and its labeled products are present in tracer quantities.
8. The entrance of label into the tissue can only occur by [6-^{14}C]glucose influx.
9. Loss of label from the brain occurs only by the return of [6-^{14}C]glucose to the plasma. Loss of $^{14}CO_2$ is inconsequential.

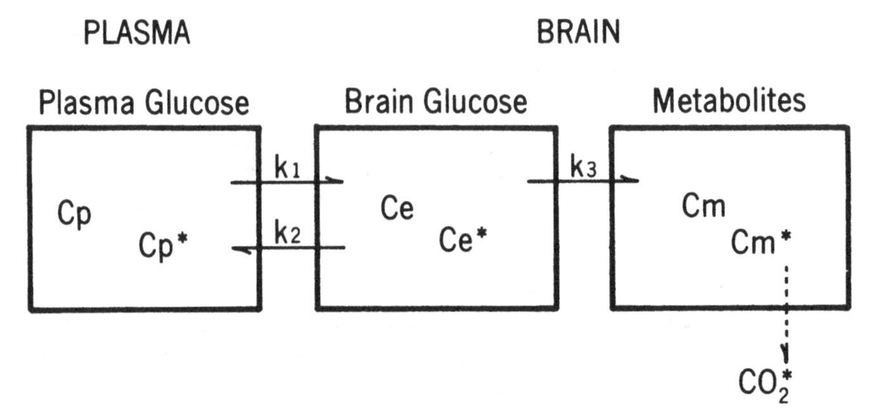

Fig. 4. Compartmental model of cerebral glucose metabolism. After an intravenous injection, [6-^{14}C]glucose moves from the plasma (along with unlabeled glucose) into the brain via a specific carrier-mediated transport system. Normally, brain tissue glucose content is considerably lower than plasma; therefore, plasma and brain [6-^{14}C]glucose specific activities (C_p^*/C_p and C_e^*/C_e) are soon in near-equilibration. In brain tissue, [6-^{14}C]glucose is phosphorylated and enters the metabolite pool. Once within this pool, ^{14}C is incorporated into many intermediary metabolites and, to a lesser extent, structural molecules. The combined quantity of these metabolites is large relative to the amount of ^{14}C glucose taken up, and ^{14}C is effectively conserved at an early stage.

Because the blood–brain barrier of adult mammals is only slightly permeable to lactate, the possibility of loss of label in this form is taken to be nil. This may not be true in immature mammals (Cremer, 1980). Finally, a condition of the experiment is that it is terminated 5–10 min after the injection of [6-^{14}C]glucose.

3.2. *Behavior of Circulating [6-^{14}C]Glucose*

After an intravenous injection of [6-^{14}C]glucose, radioactivity appears in arterial plasma (C_p^*) within 12–16 s. It rises to a peak within a few seconds, and then declines in a multiexponential fashion. The rising phase of C_p^* is short and insignificant. Therefore, C_p^* can be described well by:

$$C_p^* = \sum_{i=1}^{N} A_i \exp{-a_i t} \tag{1}$$

where N is at least 2, and all values of the rate constant (a) are positive. (Increasing N above 2 does not usually improve the fit in short experiments.)

3.3. Behavior of Tissue [6-^{14}C]Glucose

Total tissue ^{14}C is made up primarily of two components: most ^{14}C (after 2 or 3 min) is contained by the metabolite pool C_m^* and a lesser amount is contained by unreacted glucose C_e^*. (Only a very small amount, 1 or 2%, is accounted for by blood contamination.) CMRglc is calculated from C_m^*. However, since it is not possible to differentiate between the radioactivity from C_e^* and C_m^*, C_e^* must be determined and accounted for.

Influx of glucose across the blood–brain barrier can be described by Michaelis–Menten kinetics:

$$\text{Glucose influx} = k_1C_p = C_p\, T_{\max}/(C_p + K_m) \tag{2}$$

It follows that:

$$k_1 = T_{\max}/(C_p + K_m) \tag{3}$$

Glucose entry is also a function of CMRglc (Cremer et al., 1981), and it has been shown that the $T_{\max}$ varies with CMRglc in normal rats according to the relationship (Hawkins et al., 1983).

$$T_{\max} = 3.87(\text{CMRglc})^{0.39} \tag{4}$$

Glucose efflux occurs by the same transport system mediating entry (Pardridge, 1983). Assuming steady-state,

$$dC_e/dt = k_1C_p - (k_2 + k_3)C_e = 0 \tag{5}$$

On rearrangement

$$k_2 = k_1(C_p/C_e) - k_3 \tag{6}$$

It is assumed that glucose and [6-^{14}C]glucose are metabolized in an identical fashion. Therefore, the behavior of C_e^* can be described by

$$dC_e^*/dt = k_1C_p^* - (k_2 + k_3)C_e^* \tag{7}$$

On integration and substitution from Eq. (1)

$$C_e^* = \sum_{i=1}^{N} [k_1A_i/(k_2+k_3-a_i)](e^{-a_it} - \exp - (k_2 + k_3)t) \tag{8}$$

The ^{14}C accumulated in brain metabolites is related to CMRglc by

$$C_m^* = \text{CMRglc} \int_0^T [C_e^*/C_e]dt \qquad (9)$$

Since total tissue radioactivity ($\Sigma^{14}C$) is measured, it is necessary to solve for CMRglc in terms of both C_e^* and C_m^*. Thus:

$$C_m^* = \sum{}^{14}C - C_e^* \qquad (10)$$

and

$$\text{CMRglc} = C_m^* / \int_0^T [C_e^*/ C_e]dt \qquad (11)$$

If [^{14}C]glucose labeled in the 1 position is used, calculation of CRMglc as described underestimates total CMRglc by the flux through the pentose phosphate pathway. The use of [2-^{14}C]glucose measures total glucose use, but requires a correction for [^{14}C]CO_2 loss as described by Lu et al. (1983).

4. Preparation of the Rats

4.1. Acute Preparation

To measure CMRglc, it is necesary to introduce label into circulation and to take frequent samples of arterial blood throughout the experimental period. The simplest model is the acute preparation of a rat with catheters in the femoral artery and vein. The operation can be completed in about 20 min by the following procedure.

After anesthesia is established *(see below)*, the femoral artery and vein are exposed by making a longitudinal incision, 3–4 cm in length, through the skin 1 cm lateral to the abdominal midline. The fascia surrounding the femoral artery and vein is removed, and the nerve is separated. A thread is placed around both vessels distal to the deep femoral branches and tied securely. A cannula is then placed in the femoral vein as follows. Applying light traction, the vein is incised distal to the deep femoral branch and a 19-gage needle is introduced. A blunt catheter (e.g., size 22 g, external diameter 0.65–0.7 mm) is then passed through the needle and positioned in the inferior vena cava at about the level of the renal veins. The needle is then removed and the cannula tied in place. The artery is occluded proximal to the deep femoral branch with a small bulldog clamp or by applying traction on a loose tie. A cut is made halfway through the artery distal to the deep femoral branch.

The beveled end of a cannula (e.g., Bardic cutdown catheter, size 22 g, or P.E. tubing, size 50) is inserted and tied securely, as described above. If the deep femoral artery is not occluded, there is sufficient collateral circulation to perfuse the limb. The two catheters are then tied together, and the wound is closed. After surgery, it is necessary to keep the rat in a restraining cage or to immobilize it in some other way (e.g., by using a plaster cast).

4.2. Anesthesia

Some consideration of the type of anesthetic and the recovery period is required. Inhalational anesthetics are quite satisfactory. A commonly used procedure in our laboratory is to induce anesthesia with 4% halothane in air. Induction takes approximately 2 min, whereupon the rat is transferred to a specially designed mask (Levy et al., 1980), which permits the delivery of anesthetic by inhalation to the rat without contaminating the atmosphere. Maintenance of anesthesia is possible using 1–1.5% halothane carried in nitrous oxide/oxygen (70/30). The elimination of nitrous oxide occurs in minutes, but the elimination of halothane takes much longer. A period of 1–2 h after 30-min surgery is necessary to reduce the halothane concentration to acceptably low levels.

Various injectable anesthetics can also be used, including Brevital (a short-acting barbiturate), etomidate, ketamine, and Althesin. Barbiturates such as thiopental and pentobarbital should be avoided because of their relatively long half-lives.

4.3. Chronic Preparation

The acutely prepared restrained rat may not be an appropriate control for all experiments, because there are data suggesting that CMRglc is elevated by stress. An alternative is to prepare rats with chronically indwelling catheters and to keep them in isolation chambers where they are free to move. Under these circumstances, stress is reduced and CMRglc is lower. A complete description of a chronic, unstressed rat model is given by Bryan et al. (1983).

5. Experimental Procedure

5.1. Preparation of Label

[6-^{14}C]glucose, generally about 5–10 mCi/mmol, is made up in 0.15*M* NaCl (0.9%) to a concentration of 100 mCi/mL. The dose of

Table 1
Blood Glucose and CMRglc in Different Conditions

Condition	Nutritional state	Blood glucose, m*M*	CMRglc, % of normal
Normal	Fed	8	100
Normal	Fasted overnight	6	90
Normal	Fasted 48 h	4	90
Barbiturate anesthetized	Fed	8	50
Bicuculline seizures	Fed	16	200

[6-^{14}C]glucose necessary for an experiment will vary to according to many factors, including rat weight, blood glucose concentration, metabolism, experimental time, and days of film exposure. The dose can be estimated according to the following formula:

$$\text{Dose} = [4200(W)(G)]/[(M)(E)(T)(S)] \tag{12}$$

where W = rat weight in g, G = blood glucose concentration in µmol/mL, M = metabolic rate in percent of normal, E = the film exposure time in days (for DuPont Lo Dose Mammography Film), T = experimental time in min, and S = section thickness in µm. Obviously, in most cases, M and G will be unknown and must be estimated. However, the [6-^{14}C]glucose dose may vary considerably without compromising the method. The critical factor is that the optical densities of the tissue sections must fall within the range of the optical densities of the standards. Table 1 contains some examples of different conditions and the various factors that need to be considered in determining the dose. Thus, in a 5-min experiment using a normal, fed, 300-g rat, where sections are to be cut at 40 µm and images are required in 10 d:

$$\text{Dose} = [(4200)(300)(8)]/[(100)(10)(5)(40)] \simeq 50\ \mu\text{Ci}$$

5.2. Blood Samples

Before beginning the experiment, the rat should be heparinized (about 200 U will last 1.5 h), and a small sample of blood taken to determine the relative glucose concentrations in the blood and plasma. The experiment is initiated by the intravenous injection of [6-^{14}C]glucose over a period of about 10 s followed by a 0.2 mL flush of 0.15 *M* NaCl. The experiment is taken to begin at the end of the injection. Blood samples are then collected frequently throughout the experimental period. Because plasma specific radioactivity drops quickly at early times, it is necessary to take more samples during this period. Samples taken at the following times are adequate for a 5-min experiment: 20–30 s, 1, 2, 3, and 4 min. For 10-min-experiments, samples are taken at 6 and 9 min as well. It is important not to remove too much blood during the experimental period. Rat blood volume is 5–8 mL/100 g, and it is advisable not to remove more than 10% of the total blood volume (Petty 1982). Since only 50 μL of blood is necessary for analysis, the samples should not be much larger than this. The samples of blood are kept on ice until further treatment, as described below. At the end of the experiment, the rat is killed with a cardioplegic dose of barbiturate (150 mg pentobarbital in 1 mL 0.15*M* NaCl). When injected intravenously as a bolus, this dose of barbiturate stops the heart and kills the rat painlessly within about 2 s. The rat is decapitated immediately after.

5.3. Brain Removal and Treatment for Quantitative Autoradiography

Immediately after the experiment, the brain is exposed by removing the bone from the superior and lateral surfaces of the cranial vault with a small Rongeurs. When the bone has been removed, the head is tilted slightly to expose the cranial nerves. All these are carefully severed, including the olfactory tract. It is advisable to touch the brain as little as possible. Considerble care must be given to freezing the brain in order to prevent distortion. If the brain is chilled rapidly, for example by immersion into liquid nitrogen, it will crack. On the other hand, if it is frozen too slowly, ice crystals can grow, thereby distorting the internal anatomy. The following procedure provides optimal freezing. A 50-mL beaker is prepared containing a two-phase mixture: 10 mL of 1-bromobutane (density 1.28) on the bottom and 10 mL of 2-methylbutane (density 0.62 on top). The beaker is placed in a dish of liquid Freon-12, which boils at –29.8°C and precooled for approximately 3 min before introducing the brain. The brain is gently tipped out of the skull and into the beaker containing the two-phase organic mixture. The brain floats gently at the interface and is frozen in about 2 min without apparent distortion. When the brain is hard, it may be removed and placed under Freon-12. If the brain is to be sectioned immediately, it can be transferred to the cryostat. If it is to be sectioned at a later time, the following procedure is recommended. A 50-mL screw-top centrifuge tube is filled with liquid Freon-12. The brain is submerged and held below the fluid level with a small plug of cotton gauze. The lid of the centrifuge tube, with a small hole drilled to allow gas to escape, is then screwed on and the container placed in a freezer at –80°C. When stored in this way, the brain will remain without deterioration for several days.

5.4. Brain Sections

Great care must be taken in the sectioning process, because it is very easy to introduce artifacts at this stage. It is essential that the sections be cut with accuracy and precision. We have found several microtomes to be unsatisfactory because of inadequate precision. Often autoradiographic images would have visibly different densities resulting from different section thicknesses, which varied by as much as ±8 μm. We chose the Slee microtome (Slee International, New York, NY), because it reliably produced sections of the desired thickness.

The sectioning of frozen brains is an art that requires considerable patience and practice before truly excellent sections can be cut reliably. The temperature at which sections are cut can vary from –8 to –20°C. If the temperature is too low, the tissue becomes friable, and develops small horizontal cracks and frayed edges. If on the other hand the tissue is too warm, there is a tendency to form bubbles. Wrinkles can be introduced if there is any impediment to the free movement of the section across the microtome blade. Another artifact occasionally seen is a periodic horizontal deviation in section thickness, giving the image a "washboard" look. This is most often caused by some play in the system caused perhaps by a loose microtome blade or chuck.

Three or four sections are fixed to a glass slide by warming to 50–55°C until dry. (Hotter temperatures will distort the tissue.) The slides are then attached to a thin cardboard mount to restrict movement using double-faced tape, and packed in X-ray cassettes with film.

5.5. Tissue Thickness and Attenuation

There are various factors to be considered when choosing tissue thickness. First, it must be recognized that ^{14}C is attenuated by tissue. The efficiency of β particle penetration from ^{14}C is described by the equation:

$$\text{Efficiency} = 100(e^{-k(\text{tissue thickness})}) \qquad (13)$$

where k has a value of 0.009 μm^{-1} for gray and 0.016 μm^{-1} for white matter, respectively. In sections of 20 μm thickness, the efficiency of β particle release is 83.5% for gray (e.g., cerebral cortex) and 72.6% for white matter. Thus, there is a difference in the efficiency between gray and white matter that must be considered. Although most laboratories cut sections at 20 μm thickness, thicker sections are easier to produce, and because film exposure time can be reduced, so can the dose of radioactivity as described by Eq. (13). Thus, thicker sections can considerably reduce the dose of [6-^{14}C]glucose and the expense of the experiment. However, there is some tradeoff in reduced efficiency and great discrepancy between gray and white matter. We normally cut sections at 40 μm, and this has proved to be quite satisfactory. Attenuation by tissue is taken into account in the calibration of the standads *(see below)*.

5.6. Films and Exposure Time

A variety of films may be used to produce autoradiographs. We prefer Lo Dose Mammography Film, (E. I. DuPont DeNemours and Co., Wilmington, DE) because it produces images of high resolution, fine grain, and low background. The disadvantage of Lo Dose film is that long exposure times are needed; in a typical experiment, exposure times of 10–15 d are necessary. Shorter exposure times (3–5 d) may be used with film such as Kodak SB5 (Eastman Kodak Co., Rochester, New York); the image produced, however, has poorer resolution with larger grain and greater background density. The films can be developed in the laboratory according to the manufacturer's specifications with good results, but automated processes are often more satisfactory. If the experiments are conducted in a medical center environment, there is usually an automatic processor available for X-ray film.

The film is packed against the slides in complete darkness in a standard X-ray cassette, and then wrapped in light-tight paper, a black plastic bag, or aluminum foil. The cassettes are then placed on a shelf and not handled further until it is time for development. It is important not to have any movement of the slide during the incubation period to prevent blurring of the images. To a large extent, movement is prevented by fixing the slides to a piece of cardboard cut to fit just within the X-ray cassettes. Also, the X-ray cassettes are spring-loaded, holding the slides and the film together tightly. If the incubation time is less than 6 or 8 wk, there is no need to refrigerate the film, since loss of the latent image, a temperature-dependent phenomenon, is not a serious problem until after very long periods of time (Rogers, 1967).

The time of image development is to a large degree a matter of judgment. After a brief and insignificant lag period, the production of an autoradiographic image is almost linear over a range of approximately 1 ODU (optical density unit). Thereafter, it curves and flattens when the film is completely saturated at a maximum of 2.5–3 ODU. If the incubation period is too short, the images will have relatively little visible internal detail, and the optical density range may be too narrow to distinguish between different metabolic rates. On the other hand, if the images are incubated for too long, the film becomes saturated and again the OD range becomes too narrow for meaningful determinations. For accurate measurements, the films must be developed before saturation occurs over the most metabolically active structures, keeping in mind that, once radiation from these structures completely develops the film, no further OD changes can take place and, from that point on, the tissue radioactivity, and hence the CMRglc, will be underestimated.

There are a variety of interesting and important facts concerning the properties of X-ray films and the physics of β particles that are beyond the scope of this article. The reader is referred to the excellent text written by Rogers (1967), which contains a great deal of useful information.

Films can be read manually with an inexpensive photographic densitometer. We have found the Tobias densitometer (Tobias, Ivyland, PA) to be satisfactory, after fitting it with an aperture of about 300 μm in diameter. More detailed information can be obtained using a high-speed computer-operated digital microdensitometer (Photoscan Optronics International, Model P-1000, Chelmsford, MA). For comparable tissue areas, both approaches give identical results, and for many experimental circumstances, manual densitometry is entirely satisfactory.

5.7. Standards

It is important to have a set of standards whose OD values (and hence dpm/g values) completely bracket the ODs of the autoradiographic images. The standards we use cover a 30-fold range, whereas the dpm/g expected in a typical experiment cover only a 10-fold range. Thus, there is a considerable margin for biologic or experimental deviation. [^{14}C]polymer standards are available commercially (e.g., Amersham Corp., Arlington Heights, IL; American Radiolabelled Chemicals, Inc., St. Louis, MO). The radioactivity in the standards (dpm/g polymer) can be related to brain tissue autoradiography as described by Reivich et al. (1969). Some companies have calibrated the standards for 20 μm-thick brain tissue sections. These standards can be used for any tissue thickness by taking into account the relative efficiencies and the thickness itself according to the following formula (Hawkins et al., 1979):

$$\text{dpm/g} = (20/x)(e^{-20k}/e^{-xk})(\text{dpm/g at } 20\ \mu\text{m}) \qquad 15$$

where x = tissue thickness in μm and k =0.009 μm^{-1} or 0.016 μm^{-1} for gray and white tissue, respectively. Thus, if for example the calibration for the standard is known at 20 μm, and it is desired to use, say, 30-μm thick tissue sections for a given experiment, then for gray matter:

$$\begin{aligned}\text{dpm/g at } 30\mu\text{m} &= (20/30)(0.835/0.763)(\text{dpm/g at } 20\mu\text{m}) \\ &= 0.73\ (\text{dpm/g at } 20)\end{aligned}$$

5.8. Brain Sampling by Dissection

An alternative procedure to autoradiography that yields results for large brain areas is to dissect the brain and measure the dpm/g after solubilization of the tissue samples.

5.9. Treatment of Blood and Plasma

As soon as possible after the experiment is over, the blood is expressed from the syringes onto a sheet of clean parafilm. Fifty μL are pipetted into 1 mL of 0.5*M* $HClO_4$, mixed, and placed on ice for at least 10 min. After low-speed centrifugation (e.g., 1000 × g) for 5 min, the sample is decanted and neutralized with a solution of 20% KOH in 0.1*M* K_2HCO_3 using a drop of neutral indicator to determine the end point (yellow-green). After allowing the neutralized samples to stand on ice for about 30 min, they are centrifuged and decanted.

Glucose specific radioactivity (dpm/μmol) is determined by measuring the radioactivity (dpm/mL) by scintillation counting and the glucose concentration (μmol/mL). Note that the specific radioactivity is a ratio and is, therefore, independent of sample size

Table 2
Data Necessary for Calculating CMRglc[a]

Data collected in each individual experiment	Comments
C_b and C_p	Measured in a 0.3 mL blood sample taken shortly before injecting [6-^{14}C]glucose
Time course of C_b^*/C_b	Measured in 50 μL blood samples taken during the experiment and used to calculate C_p^* according to $C_p^* = C_p(C_b^*/C_b)[0.3\exp^{-0.38t} + 1.04\exp^{-0.0014t}]$
Brain tissue radioactivity	Total ^{14}C in tissue ($C_e^* + C_m^*$ + blood contamination) determined in brain sections by quantitative autoradiography

[a]Relationships known from other experiments:
$k_1 = 3.87\ (\text{CMRglc})^{0.39}/C_p + 8)$ (Hawkins et al., 1982)
$C_e/C_p = 0.2$ (for normal rats; *see* Table 3)
$C_p^* = C_p(C_b^*/C_b)[0.3\exp^{-0.38t} + 1.04\exp^{-0.0014t}]$ (Lu et al., 1983)

and dilution factor. In our laboratory, we determine the specific radioactivity in the same sample that glucose is assayed spectrophotometrically, which minimizes the effect of pipetting errors, and so on.

In the plasma sample, it is only necessary to measure the glucose concentration. This can be done in an identical fashion to the determination of blood glucose, but accuracy is important, because the brain glucose pool is determined from the plasma glucose concentration and the brain-to-plasma ratio.

6. Collection of Data and Calculation of Rates

The measurements made up to this point are summarized in Table 2. The concentrations of glucose in blood (C_b) and plasma (C_p) were measured in a small blood sample taken shortly before inject-

Table 3
Brain-to-Plasma Glucose Ratios (C_e/C_p)
[mean ± SE (n)]

State	Ratio
Fed, conscious, unrestrained	0.20 ± 0.03 (15)
Fed, conscious, restrained	0.25 ± 0.02 (7)
Fed, conscious, restrained, portacaval shunt	0.26 ± 0.02 (11)
Starved, conscious, unrestrained	0.19 ± 0.01 (8)
Fed, etomidate-anesthetized (2–12 mg/kg)	0.31 ± 0.07 (16)
Fed, ketamine-anesthetized (5–30 mg/kg)	0.23 ± 0.04 (13)
Bicucculine-seizing	0.098 ± 0.020(4)
Thiopental-anesthetized (10 mg/kg)	0.25 ± 0.02 (6)
Thiopental-anesthetized (30 mg/kg)	0.37 ± 0.02 (6)
Thiopental-anesthetized (50 mg/kg)	0.45 ± 0.02 (5)

ing [6-^{14}C]glucose. C_p and the brain-to-plasma ratio of glucose (C_e/C_p) is used to calculate the brain glucose pool with which plasma exchanges. C_b multiplied by the final blood specific activity (the specific radioactivity of blood at the time of death) is used to estimate the dpm/mL of blood. This is needed to correct for the small amount of blood that remains in the brain (approximately 10 µL/g).

The ratio C_e/C_p is 0.2 in normal unstressed rats. However, it is advisable to determine this ratio (Mans et al., 1984) under the particular experimental conditions to be used. Table 3 contains a range of values measured in different states.

Blood specific radioactivity (C_b^*/C_b) at each time point is determined for the experimental period. It is used to calculate the time course of plasma specific radioactivity (C_p^*/C_p), and this is fitted to a multiexponential curve. The total radioactivity in tissue (Σ^{14}C) is determined from the autoradiographic images by comparison of the optical densities with the quantitative standards (or in dissected brain areas by scintillation counting). Σ^{14}C is made up of $C_e^* + C_m^*$ + blood contamination.

The actual form that these measurements take is shown in Table 4, which contains a simulation of a normal experiment. In this example, all the measured values are underlined. Thus, this was a 5-min experiment with relatively normal values of blood

Table 4
Calculation of CMRglc: Example

Experiment: Normal control
Nutritional state: Fed
Experimental time 5 min

Blood values

Blood glucose	6.4 m*M*
Plasma glucose	8.7 m*M*

Sample	Minutes	Blood dpm/μmol(C_b^*/C_b)	Plasma dpm/μmol (C_p^*/C_p)
1	0.5	135,490	174,424
2	1.0	120,328	149,652
3	2.0	105,170	123,826
4	3.0	95,510	108,078
5	4.0	85,510	94,044

Best fitting curve: $C_p^*/C_p = Ae^{-at} + Be^{-bt}$

$A = 64,660$
$a = 1.908$ (These constants are used to generate
$B = 159,800$ the values ^{14}C, C_e^* and C_m^* in Table 4)
$b = 0.1321$

Film values

Standard Number	Standards dpm/g	Standards Optical density
1	0	0
2	54,950	0.07
3	94,360	0.11
4	223,780	0.25
5	671,600	0.66
6	1,178,000	1.09
7	1,233,000	1.14
8	1,725,000	1.53

A = 1,068,162 (These constants are used to relate optical density
B = 1.1119 to dpm/g)

glucose, 6.4 m*M* and plasma glucose, 8.7 m*M*. Blood samples were taken at 0.5, 1, 2, 3, and 4 min, and the corresponding blood specific radioactivities determined.

It is assumed that most glucose exchange occurs between plasma and the brain. Glucose radioactivity is initially in the plasma compartment, but then moves slowly (in rats) into erythrocytes during the experimental period. Therefore, the blood and plasma specific radioactivities are not equal; at early times (e.g., less than 15 min), plasma specific radioactivities are higher than those of blood. Plasma specific radioactivities (C_p^*/C_p) are calculated from blood according to the following formula.

$$C_p^*/C_p = (C_b^*/C_b)(0.3\ e^{-0.38t} + 1.04\ e^{-0.0014t}) \qquad 16$$

The calculated plasma specific radioactivities are shown in the far right column. These values are fit to a multiexponential curve, as described below.

6.1. *Fitting the Plasma Specific Radioactivity Curve*

After an intravenous injection of [6^{14}C]glucose, C_p^* rises to a peak within seconds and then declines in a multiexponential manner, as described by Eq. (1). In practice, only the sum of two exponents is necessary for an accurate description of the exponential decline for the first 5 or 10 min; the addition of more exponents does not usually improve the fit. It is also important to realize that the sum of exponents provides only an empirical fit to the declining plasma specific radioactivity curve in a form that is convenient for calculation of CMRglc; no physiological significance is assigned to the various constants. Programs for fitting exponential declines are available for minicomputers (e.g., International Mathematical and Statistical Libraries, Inc., Houston, Texas) and microcomputer (e.g., McIntosh and McIntosh, 1980), or they can be fit by hand (Riggs, 1963).

Once the plasma specific radioactivity curve is fit and the zero time values for each exponent (*A* and *B*), as well as the rate constants (a and b), are known, it is possible to predict the amount of radioactivity expected in the brain for a given value of CMRglc. The basis of this calculation is that each value of CMRglc corresponds to a unique value of Σ^{14}C. Thus, values for C_e^* and C_m^* are calculated according to Eq. (8) and (11), respectively. A spectrum of values over the range of CMRglc that may be expected in a normal animal is contained in Table 5.

Table 5
Relationship between CMRglc and Tissue ^{14}C for the Example in Table 4

CMRglc, $\mu mol.min^{-1}.100g^{-1}$	C_e^*, dpm/g	C_m^*, dpm/g	$\Sigma^{14}C$,[a] dpm/g
30	115,500	158,100	280,800
40	114,200	215,000	336,400
50	113,200	272,500	392,900
60	112,300	330,300	449,800
70	111,700	388,400	507,300
80	111,100	446,700	565,000
90	110,700	505,200	623,100
100	110,300	563,800	681,300
110	109,900	622,600	739,700
120	109,600	681,400	798,200
130	109,300	740,400	856,900
140	109,100	799,500	915,800
150	108,900	858,600	974,700
160	108,700	917,800	1,033,700
170	108,500	977,100	1,092,800
180	108,300	1,036,400	1,151,900
190	108,200	1,095,800	1,211,200
200	108,000	1,155,200	1,270,400

[a]$\Sigma^{14}C$ includes blood contamination equal to 1% of the dpm/mL at the end of the experiments—in this instance, 7,200 dpm/g. Thus $\Sigma^{14}C = C_e^* + C_m^* + 7200$. See Eqs. 8 and 9 for calculation of C_e^* and C_m^*, respectively.

It can be seen that, if the total amount of radioactivity is known in any particular brain region, it is possible to estimate CMRglc from Table 5 by interpolation. The actual amount of radioactivity is found by measuring the OD of the images and comparing them to the ODs of the standards. This is done in the following way. The ODs of the standards are read and a best fitting line fit according to a power curve ($y = a_x^b$). In general, the power curve gives a very good fit to the data, providing that the highest value is not close to the film's saturation point. A good procedure is to select those standards that bracket the data and eliminate all others. In this way, an adequate fit is almost always obtained. Once the standard curve has been fit, it is possible to determine $\Sigma^{14}C$ for any given OD reading and to look up the corresponding value of CMRglc on

Table 5. Then, a table of OD values vs CMRglc values can be made (Table 6), which is used to find CMRglc for any brain area.

When the procedure is used routinely, it is most convenient to use a computer program for the calculation, enabling the generation of a large table of CMRglc vs OD and accurate interpolation to find the correct values of CMRglc. Examples of CMRglc in brain regions in different states are given in Tables 7–9.

7. Experimental Considerations

7.1. *Experimental Time*

The ability to perform experiments in a short period of time is an important advantage of [6-^{14}C]glucose. A fundamental assumption of any method of measurement of CMRglc is that it is in steady-state in all brain structures throughout the entire experiment. If nonsteady-state conditions exist, it may be difficult to interpret the results. At the present time, there is no clear understanding of the moment-to-moment variation in CMRglc that occurs in normal circumstances. However, there are clearly situations where regional activity will change within seconds, such as portions of the sleep cycle, onset of seizure activity, induction of anesthesia, and so on. To obtain maximum information, it is necessary to study such events or sequences of events separately as single experimental states. Therefore, it is advantageous to reduce the experimental time as much as possible. The procedures and equations described above enable the determination of CMRglc in 5 min in normal, conscious, fed or fasted rats. Under some experimental circumstances, it may be possible to shorten this time. As a rule of thumb, we consider the shortest feasible time to be that at which the amount of ^{14}C accumulated in metabolites exceeds the unreacted [6-^{14}C]glucose (i.e., $C_m^*/C_e^* > 1$). The time for this to occur depends on several factors, most importantly the brain glucose content and the rate of energy metabolism. Figure 5 shows the rate of metabolite accumulation relative to the "background" pool of unmetabolized [6-^{14}C]glucose (C_e^*). The range of metabolic rates likely to be encountered in a typical experiment is from 40–180 μmol min^{-1} 100 g^{-1} (Tables 7–9). Thus, it is predicted that by 5 min the ratio of C_m^*/C_e^* is considerably greater than 1 for all values of CMRglc. If the metabolic rate were doubled, say by seizures, it would be possible to get accurate measures of CMRglc in only 2 or 3

Table 6
Relationship between CMRglc and Optical Density for Example in Table 4[a]

Optical density	CMRglc, $\mu mol.min^{-1}.100g^{-1}$	Optical density	CMRglc, $\mu mol.min^{-1}.100g^{-1}$
0.40	49	0.80	126
0.41	51	0.81	128
0.42	52	0.82	130
0.43	54	0.83	132
0.44	56	0.84	134
0.45	58	0.85	136
0.46	60	0.86	138
0.47	62	0.87	140
0.48	64	0.88	142
0.49	66	0.89	144
0.50	68	0.90	146
0.51	70	0.91	148
0.52	72	0.92	150
0.53	74	0.93	152
0.54	75	0.94	154
0.55	77	0.95	156
0.56	79	0.96	158
0.57	81	0.97	160
0.58	83	0.98	162
0.59	85	0.99	164
0.60	87	1.00	166
0.61	89	1.01	168
0.62	91	1.02	170
0.63	93	1.03	172
0.64	95	1.04	174
0.65	97	1.05	176
0.66	99	1.06	178
0.67	100	1.07	180
0.68	102	1.08	182
0.69	104	1.09	184
0.70	106	1.10	186
0.71	108	1.11	188
0.72	110	1.12	190
0.73	112	1.13	192
0.74	114	1.14	194
0.75	116	1.15	196
0.76	118	1.16	198
0.77	120	1.17	200
0.78	122	1.18	202
0.79	124	1.19	204

[a]The relationship between optical density (OD) and dpm/g is given by dpm/g = A (Optical Density)^{B}. In this instance, dpm/g = 1,068,162 (optical density)$^{1.1119}$. The relationship between dpm/g and CMRglc is given in Table 5.

Table 7
Regional Cerebral Glucose Use in Freely Moving, Unstressed Rats[a]

Gray matter	[1-^{14}C]glucose (9)	[6-^{14}C]glucose (9)
Telencephalon		
Frontal cortex	85 ± 4	94 ± 4
Cingulate gyrus	100 ± 6	105 ± 2
Parietal cortex	97 ± 5	105 ± 2
Pyriform cortex	64 ± 5	66 ± 2
Insular cortex	85 ± 4	90 ± 4
Occipital cortex	93 ± 5	96 ± 3
Caudate nucleus	88 ± 5	93 ± 3
Globus pallidus	49 ± 3	51 ± 1
Amygdala	66 ± 4	68 ± 2
Hippocampus	64 ± 4	66 ± 2
Lateral septal nucleus	58 ± 5	64 ± 3
Medial septal nucleus	63 ± 4	66 ± 4
Diencephalon		
Habenula	100 ± 7	104 ± 5
Hypothalamus	63 ± 5	64 ± 4
Thalamus		
—Anterior nucleus	98 ± 6	103 ± 3
—Ventral nucleus	90 ± 6	92 ± 4
—Medial geniculate	108 ± 8	110 ± 3
—Lateral geniculate	93 ± 5	99 ± 2
Mesencephalon		
Substantia nigra	73 ± 4	73 ± 2
Red nucleus	85 ± 6	92 ± 3
Oculomotor complex	100 ± 7	106 ± 3
Interpeduncular nucleus	102 ± 7	108 ± 3
Reticular formation	66 ± 4	71 ± 2
Superior colliculus	105 ± 7	103 ± 4
Inferior colliculus	175 ± 10	176 ± 5
Metencephalon		
Pons	77 ± 5	84 ± 2
Cerebellar gray		
—Molecular	78 ± 4	81 ± 2
—Granular	87 ± 4	97 ± 3
—Vermis	95 ± 6	105 ± 3
Dentate nucleus	103 ± 8	105 ± 2

Table 7 *(continued)*

Gray matter	[1-^{14}C]glucose (9)	[6-^{14}C]glucose (9)
Myelencephalon		
Vestibular nucleus	126 ± 7	129 ± 3
Cochlear nucleus	124 ± 7	126 ± 3
Superior olive	127 ± 13	144 ± 5
Inferior olive	101 ± 12	96 ± 4
Reticular formation (brain-stem)	73 ± 5	76 ± 2

White matter	[1-^{14}C]glucose (9)	[6-^{14}C]glucose (9)
Corpus callosum	53 ± 5	57 ± 3
Internal capsule	49 ± 4	49 ± 3
Cerebellar white	49 ± 4	53 ± 4

[a]Glucose use was measured using either [1-^{14}C]glucose or [1-^{16}]glucose in 5-min experiments in conscious, freely moving rats kept in isolation chambers to reduce environmental stress. Rates are reported as mean ± SE (μmol min^{-1} 100 g^{-1}) with the number of rats in parentheses. (Data from Hawkins et al., 1985.)

Table 8
Regional Cerebral Glucose Use during Etomidate Anesthesia[a]

Gray matter	Control (10)	Etomidate (5)	% Difference
Telencephalon			
Frontal cortex	87 ± 3.7	57 ± 3.5	–34
Cingulate gyrus	98 ± 5.3	66 ± 1.9	–33
Parietal cortex	96 ± 4.2	64 ± 1.2	–33
Pyriform cortex	65 ± 4.2	40 ± 2.3	–38
Insular cortex	84 ± 4.0	56 ± 2.9	–33
Occipital cortex	94 ± 4.7	54 ± 0.7	–43
Caudate nucleus	88 ± 4.5	57 ± 2.3	–35
Globus pallidus	50 ± 3.0	35 ± 1.6	–30
Amygdala	65 ± 4.0	47 ± 2.6	–28
Hippocampus	64 ± 3.8	47 ± 1.8	–27
Lateral septal nucleus	57 ± 4.7	29 ± 5.4	–49

Table 8 *(continued)*

Gray matter	Control (10)	Etomidate (5)	% Difference
Diencephalon			
Habenula	98 ± 6.1	74 ± 2.3	–24
Hypothalamus	61 ± 4.5	47 ± 2.0	–23
Thalamus			
—Anterior nucleus	99 ± 5.2	65 ± 3.2	–34
—Ventral nucleus	90 ± 4.9	70 ± 4.1	–22
—Medial geniculate	108 ± 7.3	70 ± 3.1	–35
—Lateral geniculate	94 ± 4.5	61 ± 3.0	–35
Substantia nigra	73 ± 4.0	52 ± 1.9	–29
Red nucleus	84 ± 5.0	63 ± 3.1	–25
Occulomotor complex	99 ± 6.2	79 ± 3.1	–20
Interpeduncular nucleus	101 ± 6.1	91 ± 7.3	–10
Reticular formation	73 ± 4.1	53 ± 2.7	–27
Superior colliculus	105 ± 6.1	81 ± 1.6	–23
Inferior colliculus	170 ± 11.3	141 ± 2.1	–17
Metencephalon			
Pons	78 ± 4.5	48 ± 2.6	–38
Cerebellar gray			
—Molecular	78 ± 3.8	66 ± 3.3	–15
—Granular	88 ± 3.9	73 ± 5.3	–17
—Vermis	95 ± 5.2	70 ± 2.2	–26
Dentate nucleus	102 ± 7.6	81 ± 7.5	–21
Myelencephalon			
Vestibular nucleus	123 ± 6.8	119 ± 3.4	– 7
Cochlear nucleus	124 ± 7.6	115 ± 4.0	– 7
Superior olive	125 ± 11.9	141 ± 2.9	+13
Inferior olive	99 ± 10.7	79 ± 3.7	–20

White matter	Control (10)	Etomidate (5)	% Difference
Corpus callosum	52 ± 4.2	45 ± 1.8	–13
Internal capsule	49 ± 3.6	36 ± 2.3	–26
Cerebellar white	48 ± 4.1	45 ± 2.3	– 6

[a]Glucose use was measured with [1-^{14}C]glucose in 5-min experiments in rats kept in isolation chambers. Etomidate (6 mg/kg) was given over 2 min, and glucose use was determined immediately afterward. Rates are reported as mean ± SE (μmol.min^{-1} 100 g^{-1}) with the number of rats in parentheses. (Data from Davis et al., 1986.)

Table 9
Regional Cerebral Glucose Use after Portacaval Shunting[a]

Gray matter	Sham (9)	Shunt (8)	% Difference
Telencephalon			
Frontal cortex	102 ± 3	80 ± 4	–22
Cingulate gyrus	110 ± 5	84 ± 5	–24
Parietal cortex	119 ± 7	92 ± 6	–23
Pyriform cortex	82 ± 5	62 ± 3	–24
Insular cortex	101 ± 2	83 ± 5	–18
Occipital cortex	98 ± 5	77 ± 4	–21
Caudate nucleus	93 ± 4	75 ± 5	–19
Globus pallidus	50 ± 2	39 ± 3	–22
Amygdala	73 ± 2	55 ± 3	–25
Hippocampus	76 ± 3	52 ± 3	–32
Lateral septal nucleus	70 ± 7	50 ± 4	–29
Medial septal nucleus	84 ± 5	57 ± 4	–32
Diencephalon			
Habenula	111 ± 4	98 ± 6	–20
Hypothalamus	79 ± 3	54 ± 3	–32
Thalamus			
—Anterior nucleus	101 ± 6	78 ± 5	–23
—Ventral nucleus	97 ± 3	71 ± 3	–27
—Medial geniculate	114 ± 3	82 ± 4	–28
—Lateral geniculate	98 ± 4	76 ± 5	–22
Mesencephalon			
Substantia nigra	75 ± 3	51 ± 4	–32
Red nucleus	91 ± 2	65 ± 4	–29
Oculomotor complex	109 ± 4	76 ± 4	–30
Interpeduncular nucleus	125 ± 9	77 ± 8	–38
Reticular formation	68 ± 3	54 ± 4	–21
Superior colliculus	113 ± 6	89 ± 5	–21
Inferior colliculus	169 ± 8	121 ± 8	–28
Metencephalon			
Pons	87 ± 3	62 ± 4	–29
Cerebellar gray			
—Molecular	87 ± 4	64 ± 4	–26
—Granular			
—Vermis	108 ± 6	82 ± 5	–24
Dentate nucleus	110 ± 4	88 ± 5	–20

Table 9 *(continued)*

Gray matter	Sham (9)	Shunt (8)	% Difference
Myelencephalon			
Vestibular nucleus	130 ± 4	103 ± 5	–21
Cochlear nucleus	117 ± 4	90 ± 6	–23
Superior olive	132 ± 8	100 ± 6	–24
Inferior olive	93 ± 5	75 ± 3	–19
Reticular formation	77 ± 3	61 ± 3	–21
White matter	**Sham (9)**	**Shunt (8)**	**% Difference**
Corpus callosum	55 ± 3	41 ± 2	–25
Internal capsule	47 ± 3	34 ± 2	–27
Cerebellar white	47 ± 3	37 ± 4	–21

[a]Glucose use was measured using [6-^{14}C]glucose in 8-min experiments in restrained conscious rats 5–6 wk after portacaval shunting or sham operation. Results are given in μmol.min^{-1}.100g^{-1} as the mean ± SE with the numbers of rats in parentheses. (Data from Mans et al., 1986.)

min. On the other hand, there are circumstances where it is advantageous to lengthen the experimental time. For instance, in barbiturate anesthesia the rate of metabolism is depressed, whereas the brain glucose content rises (i.e., increased C_e/C_p ratio). Under these circumstances, it will take longer for the ratio of C_m^*/C_e^* to rise to a satisfactory level.

7.2. Accuracy of the Kinetic Constants

Another conceivable source of miscalculation is an uncertainty in the value of the influx constant k_1. (*See* equation 6 for the relationship between k_1, k_2, and k_3.) The determination of CMRglc is more sensitive to k_1 at lower values of CMRglu and more sensitive at 5 min than at 10 min (Fig. 6). Although at neither time is this a particularly serious problem in normal rats, the possibility of changes in k_1 should be kept in mind. Decreased glucose carrier activity in hyperglycemia (Gjedde and Crone, 1981; McCall et al., 1982) and increased activity in hypoglycemia have been reported (Christensen et al., 1981), although others have reported no

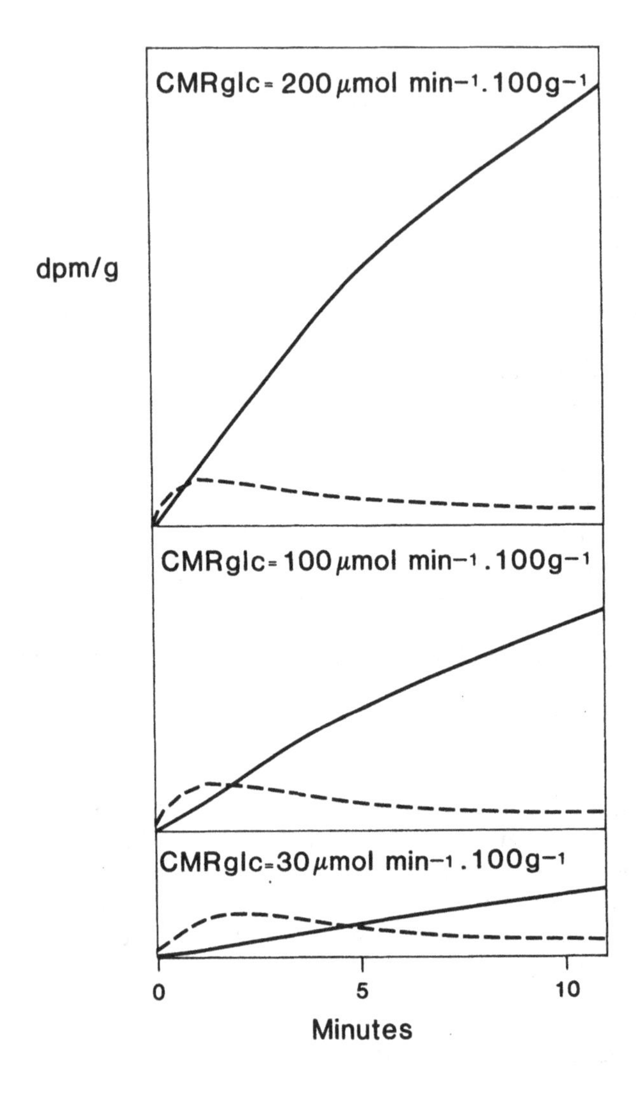

Fig. 5. Relationship between brain [^{14}C]metabolites and background ^{14}C resulting from blood and brain [6-^{14}C]glucose. The curves showing ^{14}C in metabolites (solid line) are shown for various values of CMRglc calculated from equations 8 and 9. The background ^{14}C (^{14}C-glucose, dotted line) is less than ^{14}C in metabolites at 5 min for all CMRglc values greater than 30 μmol.min^{-1}.100g^{-1}.

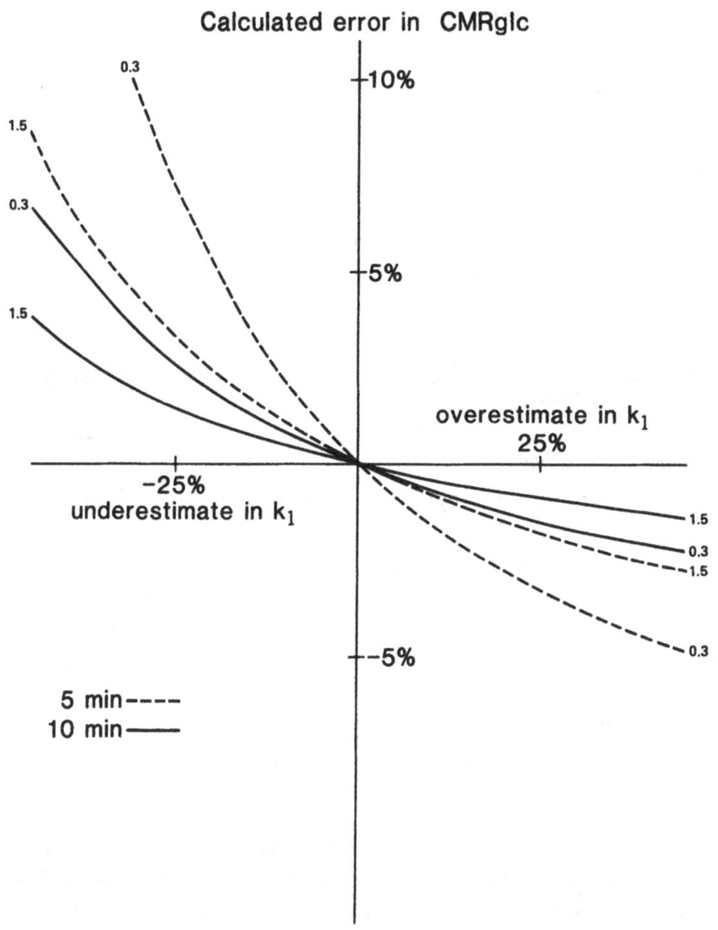

Fig. 6. The effect of overestimating or underestimating the value of k_1 on calculated CMRglc. Vertical axis shows the percent error in CMRglc that would result from incorrect estimates of k_1 at various values of CMRglc.

change (Gjedde and Crone, 1975; Crane et al., 1985; Pardridge and Oldendorf, 1975).

As with k_1, the potential error introduced resulting from an uncertainty in the brain glucose content, which is calculated from the C_e/C_p ratio, is greater at shorter times and in those structures having a lower rate of metabolism (Fig. 7). Brain glucose content is a function of plasma glucose, CMRglc and glucose carrier activity. Figure 8 shows the predicted relationship between brain and plasma glucose concentrations at different values of CMRglc, taking

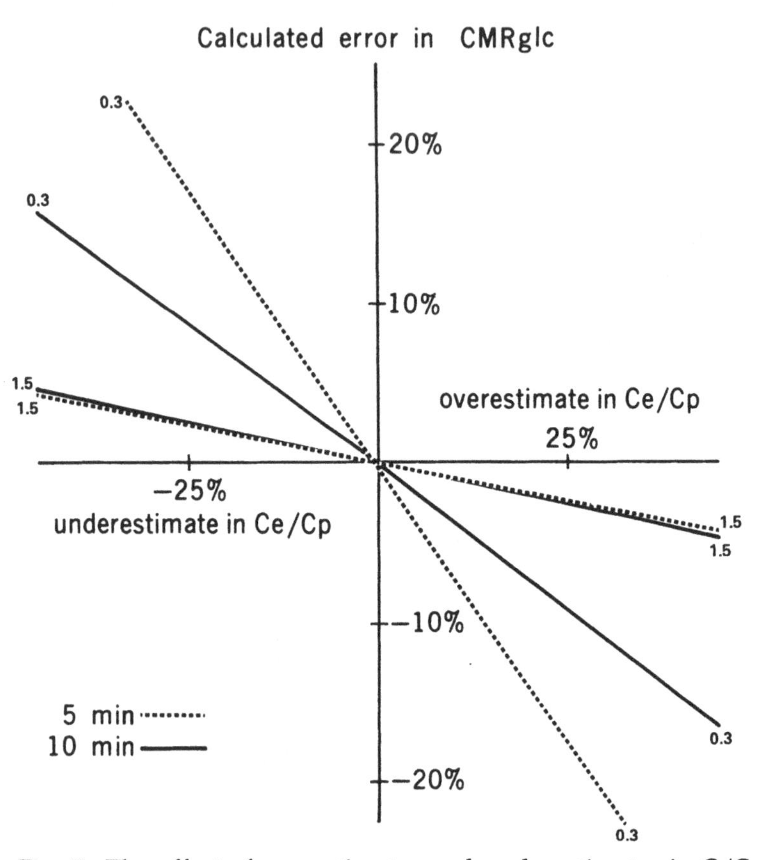

Fig. 7. The effect of overestimates and underestimates in C_e/C_p on calculated CMRglc. Vertical axis shows the percent error in CMRglc that would result from incorrect estimates in C_e/C_p at various values of CMRglc at 5 and 10 min.

transport carrier activity to be constant. Thus, it can be seen that, with several factors interacting, it is possible that the C_e/C_p ratio will change depending on the experimental circumstances. (*See* Table 3 for some experimentally determined values.) Generally, experimental conditions that cause an increase in C_e are associated with a decreased rate of metabolism (e.g., barbiturate anesthesia) or increased plasma glucose (e.g., diabetes). Under these circumstances, it may be possible to choose a longer experimental time consistent with steady-state conditions to minimize the possibility of aberrant results. In any event, it is advisable to determine the C_e/C_p ratio for new experimental circumstances.

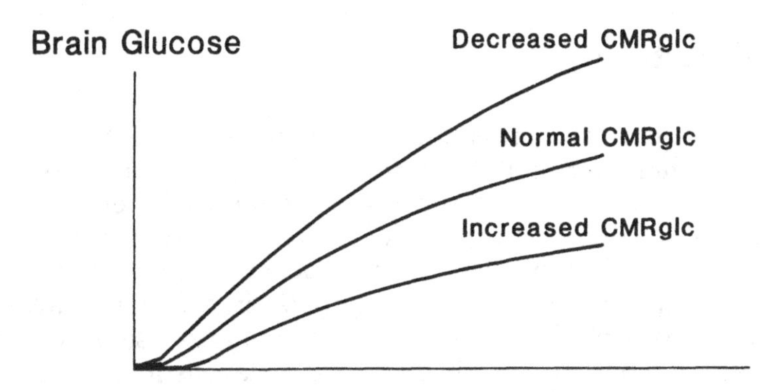

Fig. 8. Relationship between brain glucose content and plasma glucose concentration.

7.3. Preliminary Glucose Use Measurement

In some experiments, it may be useful to obtain a preliminary estimate of whether brain glucose use is altered in the brain as a whole. A simple and less costly method for this measurement could, for example, be used in screening experiments before investing the time and expense required for the full autoradiographic procedure. We have devised a simplified technique based on the method using [6-^{14}C]glucose described above to provide an index of brain glucose use.

The suggested procedure is as follows. Control and experimental rats are given a single injection of 10 μCi of [6-^{14}C]glucose into a tail vein. After 5–10 min, they are decapitated. Blood is sampled from the neck. The glucose concentration and ^{14}C-radioactivity is measured in neutralized perchloric acid extracts of the brain and blood samples. The index of glucose use is calculated as:

$$\text{Glucose use index} = \frac{\text{dpm/g brain tissue}}{SPA \times T}$$

where *SPA* is the blood specific activity and *T* is the duration of the experiment. With this method, it is assumed that the total dpm/g in brain correlates with $C_m{}^*$, and *SPA* with $_0\int^T (C_e{}^*/C_e]]dt$, as defined above. We have found that this index correctly reflected changes in whole brain glucose use, as measured by the full quantitative procedure, in a variety of different states.

8. Caveat

A word of caution is in order relating to studies of CMRglc, because there are limitations regardless of the tracer used. The basic premise is that CMRglc is stoichiometrically related to ATP production. Since almost all energy-requiring reactions rely on ATP, a direct link between cellular activity and CMRglc can be established. An exact relationship, however, may not always exist. Measurement of glucose utilization is primarily useful where animals do not have a severely disturbed metabolism, injured tissue, or interruption of the blood–brain barrier. There are several circumstances where misinterpretation of results could occur.

8.1. Injury

Warburg noted that aerobic glycolysis, i.e., lactate production in the presence of oxygen, was a characteristic feature of injured tissues (malignant tumor tissue is an exception in that it is uninjured tissue that manifests aerobic glycolysis). In fact, he considered aerobic glycolysis to be indicative of dying tissue. The occurrence of increased aerobic glycolysis alters the stoichiometric relationship between CMRglc, oxygen consumption, and ATP production. Without knowledge of the relative proportions, direct determination of a functional relationship is impossible.

8.2. Anaerobic Glycolysis

Under most circumstances, the brain uses glucose as the predominant respiratory fuel, and oxidation of glucose can be considered to be complete. However, there are many circumstances where increased uptake of glucose may not represent oxidative metabolism, but anaerobic glycolysis. These conditions include: injury of ischemia, abnormally large increases in the rate of metabolism, and lowered blood pressure or decreased oxygen tension. Under these circumstances, the rate of glycolysis is generally increased to a greater extent than oxygen consumption. Therefore, the rate of energy metabolism cannot be accurately described by measurement of CMRglc alone.

8.3. Breakdown of the Blood–Brain Barrier

Loss of blood–brain barrier integrity may increase the permeability to glucose, thereby raising the concentration of background [^{14}C]glucose. If the breakdown occurs only in specific areas, the assumption that glucose is distributed equally throughout the whole brain will not be valid. A higher brain-to-blood glucose ratio would suggest an increased metabolic rate, where this was, in fact, not so.

8.4. Compartmentation

To date, virtually all the methods used to measure CMRglc are based on a 2- or 3-compartment analysis. It would be difficult to convince an anatomist that the brain consisted only of two comparments; yet on the other hand, with respect to glucose, the brain does behave as if it consisted primarily of blood separated from a homogeneous group of cells by the blood–brain barrier (Lund-Anderson, 1980). With current experimental limitations, a more detailed analysis is not feasible, because accurate data available for analyzing more compartments are obtained only with great difficulty, and the complexity of systems with more compartments rises almost exponentially.

8.5. Steady-State

A fundamental assumption of the methods is that metabolism in all areas is in a steady-state during the experimetal period. Interpretation becomes difficult if nonsteady-state conditions exist. Presently, there is no clear understanding of the moment-to-moment variation in CMRglc, but it is not difficult to imagine situations where regional activity will change within seconds. Because of this, it is advantageous to shorten the experimental time as much as possible, thus reducing potential errors. In this regard, the use of specifically labeled [^{14}C]glucose has a decided advantage over the more slowly metabolized glucose analogs (Hawkins et al., 1985).

8.6. Autoradiographs

It is often claimed that autoradiographs provide a pictorial description of the metabolic and presumably functional relationships. A variety of photographs have been shown, and conclusions drawn regarding an increased or decreased rate of glucose metabolism without objective quantitation of glucose consumption. Many authors have presented results where the rates are not given in proper chemical units (e.g., μmol min^{-1} $100g^{-1}$). Results not supported by calculated rates of metabolism must be interpreted with considerable caution. The appearance of contrast between different photographs may the result of different development times or different blood isotope concentration.

Acknowledgments

This work was supported in part by grants awarded by the National Institute of Neurological Communicative Disorders and Stroke: NS07366, NS16389, NS16737. Portions of this article were taken from: Bryan et al., 1983; Davis et al., 1986; Hawkins, 1980; Hawkins and Mans, 1983; Hawkins et al., 1974; 1979; 1982; 1983; 1985; Lu et al., 1983; Mans et al., 1984; 1986.

References

Balázs R. (1969) Carbohydrate metabolism, in *Handbook of Neurochemistry* (Latjtha A., ed.), Plenum, New York, Volume 3, pp. 1–36.

Bryan R. M. Jr, Hawkins R. A., Mans A. M., Davis D. W., and Page R. B. (1983) Cerebral glucose utilization in awake, unstressed rats. *Am. J. Physiol.* **244,** C270–275.

Christensen T. G., Diemel N. H., Laursen H., and Gjedde A. (1981) Starvation accelerates blood–brain glucose transfer. *Acta Physiol. Scand.* **112,** 221–223.

Crane P. D., Pardridge W. M., Braun L. D., and Oldendorf W. H. (1985) Two-day starvation does not alter the kinetics of blood–brain barrier transport and phosphorylation of glucose in rat brain. *J. Cereb. Blood Flow and Metab.* **5,** 40–46.

Cremer J. E. (1980) Measurement of brain substrate utilization in adult and infant rats using various ^{14}C-labeled precursors, in *Cerebral Metabolism and Neural Function* (Passonneau J. F., Hawkins R. A., Lust W. D., and Welch F. A., eds). Williams & Wilkins, Baltimore, pp. 300–308.

Cremer J. E., Ray D. E., Sarna G. S., and Cunningham V. J. (1981) A study of the kinetic behavior of glucose based on simultaneous estimates of influx and phosphorylation in brain regions of rats in different physiological states. *Brain Res.* **221,** 331–342.

Davis D. W., Mans A. M., Biebuyck J. F., and Hawkins R. A. (1986) Regional brain glucose utilization during etomidate anesthesia. *Anesthesiology* **64,** 751–757.

Gjedde A. and Crone C. (1975) Induction processes in blood–brain barrier transfer of ketone bodies during starvation. *Am. J. Physiol.* **229,** 1165–1169.

Gjedde A. and Crone C. (1981) Blood brain glucose transfer: repression in chronic hyperglycemia. *Science* **214,** 256–257.

Hawkins R. A. and Mans A. M. (1983) Intermediary metabolism of carbohydrates and other fuels, in *Handbook of Neurochemistry,* (Lajtha A, ed.). Plenum, New York. Volume 3, pp. 259–294.

Hawkins R. A. (1980) Glucose utilization determined in vivo with [2-^{14}C]glucose, in *Cerebral Metabolism and Neural Function* (Passonneau J. V., Hawkins R. A., Lust W. D., and Welsh R. A., eds.). Williams and Wilkins, Baltimore, Maryland. pp. 367–381.

Hawkins R. A., Miller A. L., Cremer J. E., and Veech R. L. (1974) Measurement of the rate of glucose utilization by rat brain in vivo. *J. Neurochem.* **23,** 917–923.

Hawkins R., Hass W. K., and Ransohoff J. (1979) Measurement of regional brain glucose utilization in vivo using [2-^{14}C]glucose. *Stroke* **10,** 690–703.

Hawkins R. A., Mans A. M., and Biebuyck J. F. (1982) Amino acid supply to individual cerebral structures in awake and anesthetized rats. *Am. J. Physiol.* **242,** E1–E11.

Hawkins R. A., Mans A. M., Davis D. W., Hibbard L. S., and Lu D. M. (1983) Glucose availability to individual cerebral structures is correlated to glucose metabolism. *J. Neurochem.* **40,** 1013–1018.

Hawkins R. A., Mans A. M., Davis D. W., Vina J. R., and Hibbard L. S. (1985) Cerebral glucose use measured with [^{14}C]glucose labeled in the 1, 2, or 6 position. *Am. J. Physiol.* **248,** C170–C176.

Krebs H. A. (1965) Metabolic interrelations in animal tissues, in *Proceedings of The Robert A. Welsh Foundation Conferences on Chemical Research VIII.* Selected Topics in Modern Biochemistry (Milligan W. O., ed). Houston, Texas. pp. 101–129.

Landau B. R. and Wood H. G. (1983) The pentose cycle in animal tissues: evidence for the classical and against the L-type pathway. *Trends Biochem. Sci.* **8,** 292–296.

Levy D. W., Zwies A., and Duffy T. E. (1980) A mask for delivery of inhalation gases to small laboratory animals. *Lab. Anim. Sci.* **30,** 868–870.

Lu D. M., David D. W., Mans A. M., and Hawkins R. A. (1983) Regional cerebral glucose utilization measured with [^{14}C]glucose in brief experiments. *Am. J. Physiol.* **245,** C428–C438.

Lund-Anderson H. (1980) Transport of glucose across nerve and glia cell membranes in vitro and in vivo, in *Cerebral Metabolism and Neural Function* (Passonneau J. V., Hawkins R. A., Lust W. D., and Welsh F. A., eds.). Williams & Wilkins, Baltimore. pp. 120–126.

Mans A. M., Biebuyck J. F., Davis D. W., and Hawkins R. A. (1984) Portacaval anastomosis: brain and plasma amino acid metabolite abnormalities and the effect of nutritional therapy. *J. Neurochem.* **43,** 697–705.

Mans A. M., Davis D. W., Biebuyck J. F., and Hawkins R. A. (1986) Failure of glucose and branched-chain amino acids to normalize brain glucose use in portacaval shunted rats. *J. Neurochem.* **47,** 1434–1443.

McCall A. L., Millington W. R., and Wurtman R. J. (1982) Metabolic fuel and amino acid transport into the brain in experimental diabetes mellitus. *Proc. Nat. Acad. Sci. USA* **79,** 5406–5410.

McIntosh J.E.A., and McIntosh R. P. (1980) *Mathematical Modeling in Computers in Endocrinology.* Springer-Verlag, New York.

Pardridge W. M. (1983) Brain metabolism: a perspective from the blood–brain barrier. *Physiol. Rev.* **63,** 1481–1535.

Pardridge W. M. and Oldendorf W. H. (1975) Kinetic analysis of blood–brain barrier transport of amino acids. *Biochim. Biophys. Acta* **401,** 128–136.

Petty C. (1982) *Research Techniques in the Rat,* Charles C. Thomas, Springfield, Illinois. pp. 66–70.

Reivich M., Jehle J., Sokoloff L., and Kety S. S. (1969) Meaurement of regional cerebral blood flow with antipyrine-^{14}C in awake cats. *J. Appl. Physiol.* **27,** 296–300.

Riggs D. S. (1963) *The Mathematical Approach to Physiological Problems.* The Williams & Wilkins Company, Baltimore, Maryland.

Rogers A. W. (1967) *Techniques of Autoradiography,* Elsevier, New York.

Sacks W. (1969) Cerebral metabolism in vivo, in *Handbook of Neurochemistry,* (Lajtha A., ed). Plenum, New York. Volume 3, pp. 301–321.

Sokoloff L. (1981) The radioactive deoxyglucose method: theory, procedure, and applications for the measurement of local glucose utilization in the central nervous system, in *Advances in Neurochemistry,* (Agranoff B. W. and Aprison M. W., eds.). Plenum, New York. pp. 1–82.

Veech R. L., Harris R. L., and Veloso D. (1973) Freeze-blowing: a new technique for the study of brain *in vivo.* *J. Neurochem.* **20,** 183–188.

Glycolytic, Tricarboxylic Acid Cycle and Related Enzymes in Brain

John B. Clark and James C. K. Lai

1. Introduction

1.1. Literature Survey and the Scope of This Review

The present review focuses on the methodological aspects of glycolytic, tricarboxylic acid (TCA) cycle, and related enzymes, and on a discussion of how the methodologies can be usefully exploited in different experimental settings. Because of this primary focus and space limitation, the authors are only able to cite a small number of the references obtained in the authors' computer search (1972–1986). Several reviews have adequately covered different aspects of the literature prior to 1970 [*see* Balázs (1970), Bachelard (1970), Baxter (1970), van den Berg (1970), and Cheng (1971)]. For more recent discussions of papers on topics related to the present review, the excellent contributions by Sokoloff (1983), Hawkins and Mans (1983), Sacks (1983), Pevzner (1983), Kvamme (1983a,b), Thurston and Hauhart (1985), and Miller (1985) should be consulted.

1.2. The Importance of Glucose Utilization in Brain Energy Metabolism and Production

It is generally accepted that the normal adult brain is almost totally dependent on aerobic glucose oxidation for energy [*see* Sokoloff (1983) and Hawkins and Mans (1983) for discussions]. Thus, efficient aerobic oxidation of glucose is vital for providing the brain with ATP. This, in turn, implies that flux through the glycolytic pathway must be well coordinated with the flux through the tricarboxylic acid (TCA) cycle. Moreover, through the coordination of the glycolytic and TCA cycle fluxes, glucose-derived carbon skeleton can be channeled to the syntheses of amino acids (e.g., glutamate, aspartate, GABA) that are closely associated with the TCA cycle.

1.3. Regulatory Aspects of Brain Glycolysis and Citric Acid Cycle

1.3.1. Regulation of Glucose Transport

Although the hexose transporter at the blood–brain barrier may regulate the blood–brain transfer of glucose, the relatively high K_m for glucose of the hexose transporter renders it unlikely that this transporter is saturated with glucose under normal physiological conditions. Consequently, blood-borne glucose is continually transported across the blood–brain barrier into the brain extracellular space [*see* Bachelard (1983) and Hawkins (1985) for discussions]. The glucose in the extracellular space then enters neural cells via the high affinity (low K_m) hexose transporter located at the plasma membranes of neurons and glia.

1.3.2. Regulation of Glycolysis

It is usually assumed that the regulation of glycolysis in brain is probably similar to that in several peripheral tissues [*see* Balázs (1970), Bachelard (1970), and Hawkins and Mans (1983) for discussions]. [For a proper discussion on the various factors that play key roles in metabolic regulation in mammalian tissues, consult Newsholme and Leech (1983).]

1.3.2.1. RATE-LIMITING STEPS. As in nonneural tissues, several rate-limiting and regulated steps have been identified in the brain; these include hexokinase (HK), phosphofructokinase (PFK), and pyruvate kinase (PK) [*see* Newsholme and Start (1973), Newsholme and Leech (1983), Hawkins and Mans (1983), and references cited therein]. Among the glycolytic enzymes in the brain, the activities of hexokinase (HK) and phosphofructokinase (PFK) are particularly low (*see* later Sections of this review). Also low are the activities of aldolase and enolase [*see* Hawkins and Mans (1983) and references cited therein].

1.3.2.2. HEXOKINASE (HK). Being the first enzyme in glycolysis, hexokinase (HK) regulates the flux of glucose-C into the glycolytic pathway. An additional regulatory potential lies in the fact that a major portion (>70%) of brain hexokinase (HK) is bound to mitochondria, leaving only a minor portion of it in the cytosol to interact with other glycolytic enzymes [*see* Land et al. (1977), Wilson (1980, 1983), and Lai and Blass (1985) for discussions].

1.3.2.3. PHOSPHOFRUCTOKINASE (PFK). This enzyme is regulated by substrate inhibition at high concentrations of ATP. In the presence of high ATP concentrations, the affinity of fructose-6-phosphate for the enzyme is decreased, and citrate acts as an allosteric inhibitor. The inhibitory effects of ATP and citrate are antagonized by fructose-1,6-bisphosphate (a product of the PFK-catalyzed reaction), cAMP, AMP, ADP, inorganic phosphate, fructose-6-phosphate, and ammonium and potassium ions. Consequently, TCA cycle intermediates such as citrate (and therefore the flux through the TCA cycle) can regulate glycolytic flux through the effects of these intermediates on PFK [*see* Hawkins and Mans (1983) and Hers and Hue (1983) for discussions].

Two other control mechanisms that may regulate PFK activity have been demonstrated in hepatic and several other tissues. (1) There is a phosphorylation–dephosphorylation mechanism; for example, the cAMP-dependent phosphorylation of hepatic PFK is mediated by glucagon. However, whether or not the cAMP-mediated phosphorylation of PFK leads to alteration(s) of the kinetic properties of the enzyme is still controversial (Hers and Hue, 1983). (2) Fructose-2,6-bisphosphate (a positive effector) has been shown to activate hepatic PFK without directly involving the phosphorylation–dephosphorylation mechanism [*see* Hers and Hue (1983) for a discussion]. However, although likely to occur, whether these two kinds of mechanisms operate in the brain remains to be fully elucidated (Hawkins and Mans, 1983).

1.3.2.4. Aldolase and Enolase. The activities of these two enzymes in the brain are also quite low and may be rate-limiting [*see* Hawkins and Mans (1983); *also see* later Sections]. Moreover, neuron-specific (i.e., the so-called "14-3-2" neuron-specific protein) and glia-specific enolases have been identified and characterized [*see* Zomzely-Neurath (1983) for a discussion and references]. However, the metabolic significance of these two forms of enolase in the brain remains to be clarified.

1.3.2.5. Pyruvate Kinase (PK). In the same order as that of brain lactate dehydrogenase (LDH), the activity of PK in the brain is quite high. Thus, PK may not be rate-limiting.

Brain PK is a regulated enzyme [*see* Bachelard (1970), Hawkins and Mans (1983), and references cited therein for discussions]. For example, ATP inhibits brain PK competitively with respect to ADP, but not phosphoenolpyruvate. On the other hand, fructose 1,6-bisphosphate does not affect brain PK, although its activates hepatic PK [*see* Hawkins and Mans (1983) for a discussion].

Similar to PFK, PK is also regulated by phosphorylation–dephosphorylation, a mechanism shown to occur in liver [*see* Hers and Hue (1983) for a discussion]. Nonetheless, the presence of this mechanism in brain remains to be elucidated.

1.3.2.6. SUBCELLULAR COMPARTMENTATION OF GLYCOLYTIC ENZYMES. It is generally accepted that glycolytic enzymes in mammalian tissues are primarily—if not exclusively—localized in the cytosol [*see* Beattie et al. (1964) for a discussion]. Nonetheless, evidence is accumulating that several of the key, rate-limiting and regulated enzymes in this pathway can also be found in subcellular compartment(s) other than the cytosol [*see* Section 4.2. below and Lai and Clark (1989) for discussions]. The metabolic importance of the subcellular compartmentation of brain glycolytic enzymes obviously needs to be properly evaluated. Nonetheless, it may not be unreasonable to speculate that such a bi- (or multi-) modal distribution of glycolytic enzymes will lead to a decrease in the cytosolic pool of these enzymes and, consequently, a decrease in glycolytic flux in the cytosol.

1.3.2.7. OTHER METABOLIC FACTORS THAT REGULATE GLYCOLYSIS. These include the cytoplasmic redox and phosphate potentials, and levels of key intermediary metabolites. However, detailed discussions [*see* Newsholme and Leech (1983), Hawkins and Mans (1983), and references cited therein] of these important topics is beyond the scope of this review.

1.4. Regulatory Aspects of the Citric Acid Cycle in Brain

It is generally assumed that the control of brain TCA cycle activity is very similar to that in several peripheral tissues, including liver, heart, and skeletal muscle [*see* Hansford (1980), Hawkins and Mans (1983), and Miller (1985) for discussions]. In this respect, the key regulatory features often emphasized consist of

1. the role and activity of the respiratory chain electron transport and oxidative phosphorylation in response to changes in the phosphate and redox potentials [*see* Hansford (1980) for a detailed discussion]
2. key, rate-limiting and regulated enzymes and
3. roles of metabolites (and their transport) and cations [*see* Hansford (1980), Lai and Cooper (1986), and references cited therein for discussions].

Since the present review focuses on the enzymatic aspects, only the second type of regulatory feature will be discussed *(see below)*.

1.4.1. Regulation of the Entry into and the Flux Through the TCA Cycle

Once glycolytically derived pyruvate has entered the mitochondrion via the monocarboxylate translocase, the activity of the pyruvate dehydrogenase complex (PDHC) regulates the entry of carbon into the TCA cycle in the aerobic oxidation of glucose in the brain *(see below)*. Alternatively, carbon from glycolytically derived pyruvate can also enter the TCA cycle via the action of pyruvate carboxylase [*see* Hawkins and Mans (1983) and references cited therein for a discussion]. However, in the normal adult brain, pyruvate carboxylase probably plays only a minor role in this respect.

In brain mitochondria, the fluxes through the TCA cycle are probably controlled at two regulated and/or rate-limiting enzymes—isocitrate dehydrogenase (ICDH) and 2-oxoglutarate dehydrogenase complex (KGDHC) [*see* Lai and Cooper (1986) and Lai and Clark (1989) for discussions]. Consistent with this conclusion are the observations that, in various populations of synaptic and nonsynaptic mitochondria, the activities of NAD-linked isocitrate dehydrogenase (NAD-ICDH) and 2-oxoglutarate dehydrogenase complex (KGDHC) are the lowest among all the TCA cycle enzymes studied (Lai et al., 1977; Lai and Clark, 1979; Clark and Nicklas, 1984; Lai and Cooper, 1986).

1.4.2. Regulated and/or Rate-Limiting Enzymes

1.4.2.1. Pyruvate Dehydrogenase Complex (PDHC). Like the complexes from peripheral tissues, brain PDHC is also regulated by two kinds of mechanisms: (1) product inhibition (e.g., by NADH and acetyl-CoA), and (2) a phosphorylation (inactivation)–dephosphorylation (activation) cycle. However, detailed discussions of these regulatory mechanisms are beyond the scope of the present review [*see* Booth and Clark (1978a), Ksiezak-Reding et al. (1982), Sheu et al. (1983, 1984), Lai and Sheu (1985, 1987), Hinman et al. (1986), Butterworth (1989), and references cited therein for discussions].

1.4.2.2. Citrate Synthase (Cit Syn). Although the activities of this enzyme in synaptic and nonsynaptic brain mitochondria are significantly higher than corresponding activities of PDHC and NAD-ICDH (Lai et al., 1975 and 1977; Lai and Clark, 1979), there is evidence from studies using enzymes from peripheral tissues that the activities of Cit Syn may be regulated by the levels of the substrates (i.e., oxaloacetate and acetyl-CoA) and by the levels of putative effectors such as ATP, ADP, and AMP [*see* Hansford (1980) for a full discussion]. However, these regulatory properties of the brain enzyme remain to be fully elucidated.

1.4.2.3. NAD-Linked Isocitrate Dehydrogenase (NAD-ICDH). This enzyme catalyzes a nonequilibrium reaction. It has been shown in peripheral tissues that NAD-ICDH activity increases in response to decreases in NADH/NAD and ATP/ADP ratios, suggesting that this enzyme may modulate the flux through the TCA cycle in accordance with the mitochondrial energy status and the metabolic demands of the tissues [*see* Hansford (1980) and references cited therein]. Various preparations of NAD-ICDH from several peripheral tissues are under allosteric regulation by a number of metabolites including NADH, NADPH, ADP, and ATP [*see* Hansford (1980) and references cited therein].

Some of the kinetic and regulatory properties of the brain enzyme have been elucidated. Ogasawara et al. (1973) were the first to report that the plot of the activities of a partially purified rat brain NAD-ICDH vs various isocitrate concentrations is sigmoidal, suggesting that rat brain NAD-ICDH is an allosteric protein showing the characteristics of positive cooperativity. At about the same time, Lai and Clark (Lai, 1975; Lai and Clark, 1978) also made very similar observations while studying the kinetic properties of NAD-ICDH in extracts of synaptic and nonsynaptic rat brain mitochondria. More recently, Willson and Tipton (1979, 1980) have confirmed these observations using a more purified preparation of ox brain NAD-ICDH.

Similar to the findings of Lai and Clark (Lai, 1975; Lai and Clark, 1978) using extracts of synaptic and nonsynaptic rat brain mitochondria, Willson and Tipton (1980) have also found that the ox brain enzyme shows normal hyperbolic kinetics with respect to NAD. Moreover, Willson and Tipton (1979 and 1980) have noted that ADP, citrate, and tricarballyate are allosteric activators of the ox brain enzyme, whereas both NADH and NADPH are competitive inhibitors of this enzyme with respect to NAD.

1.4.2.4. 2-Oxoglutarate Dehydrogenase Complex (KGDHC). 2-Oxoglutarate dehydrogenase (EC 1.2.4.2), a component of the KGDHC, catalyzes a nonequilibrium reaction in which 2-oxoglutarate is decarboxylated. Again, evidence from studies on the peripheral tissue KGDHC indicates that the enzyme complex plays a major role in the control of flux through the segment of the TCA cycle from 2-oxoglutarate to malate, and is "down" regulated by increases in the intramitochondrial ATP/ADP and succinyl-CoA/CoA ratios [*see* Hansford (1980) and references cited therein]. However, the regulatory properties of brain KGDHC remain to be fully elucidated (Lai and Cooper, 1986).

1.5. The Concept of Metabolic Compartmentation

It is well established that the metabolism of TCA cycle and related intermediates in the brain is compartmented [*see* Balázs and Cremer (1973), Berl et al. (1975), and articles therein]. This concept implicates a "small" pool of TCA cycle intermediates that are in equilibrium with a "small," but rapidly turning over pool of glutamate that is channeled towards the synthesis of glutamine and a "large" pool of TCA cycle intermediates that are in equilibrium with "large," but slowly turning over pools of glutamate and glutamine [*also see* Lai and Clark (1989) for additional discussions].

1.6. Relations between the Compartmentation of Brain Citric Acid Cycle Metabolism and Neurotransmitter Metabolism

Early tracer studies using radiolabeled glucose and pyruvate have demonstrated the incorporatioon of labeled C (derived from both labeled glucose and pyruvate) into neurotransmitters such as acetylcholine, glutamate, aspartate, and GABA. These observations suggest that the metabolism of these neurotransmitters is closely associated with or linked to the metabolism of the TCA cycle [*see* Clark et al. (1982) and Gibson and Blass (1983) for a full discussion]. Since the metabolic compartmentation concept (*see* the previous Section) predicts the existence of multiple pools of TCA cycle and related intermediates, the question arises as to the origin(s) of the pools of TCA cycle intermediates that are primarily (if not exclusively) responsible for the syntheses of neurotransmitters such as acetylcholine, glutamate, aspartate, and GABA. In considering the question of the origin(s) of such neurotransmitter precursor pools, a conceptual distinction has been made that there are the so-called "neurotransmitter precursor" and "general metabolic" (i.e., nonneurotransmitter precursor) pools of TCA cycle intermediates. However, in practice, the actual sizes of such "pools" have yet to be determined.

2. Assays of the Enzymes of the Glycolytic, Citric Acid Cycle and Related Pathways

One general comment needs to be made in the outset. In the original designs of these assay procedures, it has not been taken into account that crude tissue extracts or subcellular fractions with varying degrees of purity are used as the source(s) of enzymatic activities. One must, therefore, ascertain that complete—or even partial—substrate barrier(s) (e.g., intact mitochondria or other membranous particles) is/are eliminated. This consideration implies that some form of chemical (e.g., treatment with a detergent) or physical (e.g., sonication, freezing and thawing) membrane-disruptive technique has to be implemented in the assay procedures.

All of the enzymatic assays described below are usually carried out at 25°C using a double-beam recording spectrophotometer. One unit (U) of enzymatic activity is defined as 1 μmol of substrate utilized or product formed/min.

2.1. Glycolytic Enzymes

2.1.1. Hexokinase (HK; EC 2.7.1.1)

Principle. HK catalyzes the phosphorylation of D-glucose to D-glucose-6-phosphate. The assay method involves coupling the production of glucose-6-phosphate to the reduction of $NADP^+$ in the presence of excess exogenously added glucose-6-phosphate dehydrogenase (G6PDH) (Lai et al. 1980; Lai and Blass, 1984a).

Assay Mixture. This contains: 0.1*M* MOPS [3-(*N*-morpholino)-propanesulfonic acid]-Tris, pH 7.5, 0.1% (v/v) Triton X-100, 8 m*M* $MgCl_2$, 0.4 m*M* $NADP^+$, 5 m*M* D-glucose, 5 U yeast glucose-6-phosphate dehydrogenase (G6PDH), 1 m*M* ATP, and a small volume of tissue extract (Lai et al., 1980; Lai and Blass, 1984a).

Procedure. The assay mixture is usually made up to a final volume of 1 mL. The reaction is initiated with the addition of ATP. The linear rate of $NADP^+$ reduction is measured after an initial lag of about 1 min by continuously monitoring the increase in absorbance at 340 nm.

2.1.2. Phosphofructokinase (PFK; 6-Phosphofructokinase, EC 2.7.1.11)

Principle. PFK catalyzes the phosphorylation of D-fructose-6-phosphate to D-fructose-1,6-bisphosphate. The assay depends on coupling the production of fructose-1,6-bisphosphate to the oxidation of NADH in the presence of excess aldolase, glycerol-3-phosphate dehydrogenase, and triosephosphate isomerase (*see* Section 2.1.2.1. for a detailed discussion).

Reagents. The assay mixture contains: 0.1*M* MOPS [or HEPES (*N*-2-hydroxyethylpiperazine-*N*'-2-ethanesulfonic acid)]-Tris pH 7.4, 5 m*M* $MgCl_2$, 50 m*M* KCl, 1 m*M* DL-dithiothreitol (DTT), 0.4 U aldolase, glycerol-3-phosphate dehydrogenase/triose phosphate isomerase mixture (0.9 U/2.6 U), 0.2 m*M* NADH, 4 m*M* fructose-6-phosphate, 0.1% (v/v) Triton X-100, a small volume of tissue extract, and 1 m*M* ATP (Leong et al., 1981; Lai and Blass, 1984a).

Procedure. The final volume of the reaction mixture is 1 mL. The reaction is initiated with the addition of ATP, and the oxidation of NADH is determined after an initial lag of 1–2 min by continuously monitoring the decrease in absorbance at 340 nm (Lai and Blass, 1984a).

2.1.3. Other Glycolytic Enzymes

The assay methods for the other glycolytic enzymes not covered by this review have been detailed elsewhere (Wood, 1982a,b).

2.1.3.1. ALDOLASE (AL; FRUCTOSE-BISPHOSPHATE ALDOLASE, EC 4.1.2.13)

Principle. Aldolase catalyzes the conversion of fructose-1,6-bisphosphate to dihydroxyacetone phosphate and glyceraldehyde-3-phosphate. In the presence of added triosephosphate isomerase, one of the reaction products (namely glyceraldehyde-3-phosphate) is converted to dihydroxyacetone phosphate, which is then converted to glycerophosphate by the added α-glycerophosphate dehydrogenase with the concomitant oxidation of NADH. Thus, the assay relies on the determination of the rate of NADH oxidation in the presence of excess exogenously added triosephosphate isomerase and α-glycerophosphate dehydrogenase. As determined by this method, the aldolase activity is equal to half the rate of NADH oxidation.

Reagents. The assay mixture (final volume = 1 mL) contains: 100 m*M* glycylglycine–KOH, pH 7.5, 0.2 m*M* NADH, α-glycerophosphate dehydrogenase–triosephosphate isomerase mixture (1 U/6.7 U), 0.1% (v/v) Triton X-100, a small volume of tissue extract, and 5 m*M* fructose-1,6-bisphosphate (Leong et al., 1981).

Procedure. The reaction is initiated with the addition of 5 m*M* fructose-1,6-bisphosphate and the rate of NADH oxidation is determined by continuously monitoring the decrease in absorbance at 340 nm.

2.1.3.2. PYRUVATE KINASE (PK; EC 2.7.1.40)

Principle. The pyruvate formed by PK is coupled to the reaction catalyzed by lactate dehydrogenase (LDH) and the concomitant oxidation of NADH is measured (Lai and Blass, 1984a).

Reagents. The reaction mixture contains: 50 m*M* MOPS-Tris, pH 7.2, 8 m*M* $MgCl_2$, 0.2 m*M* NADH, 1 m*M* phosphoenolpyruvate (trimonocyclohexylammonium salt), 75 m*M* KCl, 14 U lactate dehydrogenase, 0.1% (v/v) Triton X-100, 1 m*M* ADP (Tris salt), and a small volume of tissue extract (Leong et al., 1981; Lai and Blass, 1984a).

Procedure. The reaction is started with the addition of ADP, and the reduction of NADH is continuously monitored at 340 nm by determining the decrease in absorbance at 340 nm (Lai and Blass, 1984a).

2.1.3.3. LACTATE DEHYDROGENASE (LDH; EC 1.1.1.27)

Principle. LDH catalyzes the conversion of pyruvate to lactate with the concomitant oxidation of NADH. The rate of NADH oxidation is measured.

Reagents. The reaction mixture (final volume = 1 mL) contains: 0.1*M* potassium phosphate buffer (or 50 m*M* MOPS-Tris) pH 7.4, 0.25 m*M* NADH, 0.5% (v/v) Triton X-100, 1.65 m*M* pyruvate (Na or Tris salt), and a small volume of tissue extract (Clark and Nicklas, 1970; Lai and Blass, 1984a).

Procedure. The reaction is initiated with the addition of pyruvate and the initial rate of NADH oxidation monitored at 340 nm by measuring the decrease in absorbance (Clark and Nicklas, 1970; Lai and Blass, 1984a).

2.2. Citric Acid Cycle Enzymes

2.2.1. Pyruvate Dehydrogenase Complex

For details regarding the assay methodology and other aspects of this enzyme complex, consult the chapters by Butterworth and Lai and Clark in this volume.

2.2.2. Isocitrate Dehydrogenases (NAD- and NADP-Linked)

2.2.2.1. NAD-Linked Isocitrate Dehydrogenase (NAD-ICDH; EC 1.1.1.41)

Principle. The enzyme catalyzes the reduction of NAD^+ concomitant with the conversion of isocitrate to 2-oxoglutarate. The assay is based on the measurement of the rate of isocitrate-dependent reduction of NAD^+ (Lai and Clark, 1976, 1978).

Reagents. The assay mixture contains: 35 m*M* Tris-HCl, pH 7.2, 0.7 m*M* $MnCl_2$, 0.7 m*M* NAD^+, 0.71 m*M* ADP, 0.1% (v/v) Triton X-100, 60 m*M* DL-isocitrate, and a small volume of tissue extract (Lai and Clark, 1976 and 1978).

Procedure. The rate of NAD^+ reduction is determined by continuously monitoring the increase in absorbance at 340 nm after the addition of isocitrate to start the reaction.

2.2.2.2. NADP-Linked Isocitrate Dehydrogenase (NADP-ICDH; EC 1.1.1.42)

Principle. This is essentially the same as that for the NAD-linked enzyme, except that $NADP^+$ reduction is coupled to the conversion of isocitrate to 2-oxoglutarate.

Reagents. The reaction mixture is the same as that for the NAD-linked ICDH assay, except that 0.2 m*M* $NADP^+$ substitutes for NAD^+ (Lai and Clark, 1976, 1978).

Procedure. This is the same as that for the NAD-ICDH assay (Lai and Clark, 1976).

2.2.3. 2-Oxoglutarate Dehydrogenase Complex (KGDHC; Oxoglutarate Dehydrogenase, EC 1.2.4.2; Dihydrolipoamide Succinyltransferase, EC 2.3.1.61; and Dihydrolipoamide Reductase (NAD), EC 1.6.4.3)

Principle. The assay involves measuring the reduction of NAD^+ that is concomitant with the conversion of 2-oxoglutarate to succinyl-CoA in the presence of CoA (Lai and Cooper, 1986).

Reagents. The assay mixture contains: 0.2 m*M* TPP, 2 m*M* NAD^+, 0.2 m*M* CoA, 1 m*M* $MgCl_2$, 0.3 m*M* DL-dithiothreitol, 0.1% (v/v) Triton X-100, 130 m*M* HEPES-Tris, pH 7.4, 10 m*M* 2-oxoglutarate (Na salt), and a small volume of tissue extract (Lai and Cooper, 1986).

Procedure. The reaction is started with the addition of either CoA or 2-oxoglutarate. The initial rate of NADH production is measured by monitoring the increase in absorbance at 340 nm (Lai and Cooper, 1986).

2.2.4. Citrate Synthase [Cit Syn; Citrate (si)-Synthase, EC 4.1.3.7]

Principle. The activity of Cit Syn is assayed by measuring the rate of formation of free CoA from acetyl-CoA in the presence of oxaloacetate and DTNB [5,5'-dithio-bis(2-nitrobenzoic acid)] (Clark and Land, 1974).

Reagents. The assay mixture contains: 0.1*M* Tris-HCl, pH 8.0, 0.1 m*M* oxaloacetate (K salt), 0.1 m*M* DTNB, 0.08% (v/v) Triton X-100, 0.05 m*M* acetyl-CoA, and a small volume of tissue extract (Clark and Land, 1974).

Procedure. The reaction is initiated with the addition of oxaloacetate, and the initial velocity of increase in absorbance at 412 nm is measured (Clark and Land, 1974).

2.2.5. Aconitase (Aconitate Hydratase, EC 4.2.1.3)

Principle. The formation of isocitrate from citrate by aconitase is coupled to the reduction of $NADP^+$ concomitant with the conversion of isocitrate to 2-oxoglutarate in the presence of excess added NADP-ICDH [*see* Patel (1974a) for details].

Reagents. The assay mixture is the same as that for NADP-ICDH, except that 15 m*M* potassium citrate substitutes for isocitrate and 0.5 U of NADP-ICDH is added (Patel, 1974a).

Procedure. After the addition of citrate to initiate the reaction, the rate of $NADP^+$ reduction is monitored by determining the increase in absorbance at 340 nm (Patel, 1974a).

2.2.6. Fumarase (Fum; Fumarate Hydratase, EC 4.2.1.2)

Principle. The assay relies on measuring the enzyme-catalyzed formation of fumarate from malate by determining the increase in absorbance at 240 nm (Racker, 1950; Lai and Clark, 1976). Compounds such as fumarate have an unsaturated C=C linkage and show a marked absorption in the ultraviolet (Racker, 1950).

Reagents. The reaction mixture consists of 0.1*M* potassium phosphate buffer, pH 7.4, 0.1% (v/v) Triton X-100, 50 m*M* L-malate (K or Na salt), and a small volume of tissue extract (Lai and Clark, 1976).

Procedure. The reaction is started with the addition of malate, and the increase in absorbance is measured at 240 nm (Racker, 1950; Lai and Clark, 1976).

2.2.7. NAD-Linked Malate Dehydrogenase (NAD-MDH; EC 1.1.1.37)

Principle. The rate of oxidation of NADH concomitant with the conversion of oxaloacetate to malate is determined by measuring the decrease in absorbance at 340 nm (Lai and Clark, 1976, 1978).

Reagents. The solution of oxaloacetate is freshly prepared just prior to the enzymatic assays and its pH adjusted to approximately 6.7 with sodium bicarbonate. The assay mixture contains: 0.1*M* potassium phosphate buffer, pH 7.4, 0.16 m*M* NADH, 0.16% (v/v) Triton X-100, 0.133 m*M* oxaloacetate, and a small volume of tissue extract (Lai and Clark, 1976, 1978).

Procedure. The reaction is started with the addition of oxaloacetate, and the initial velocity of NADH oxidation is measured by monitoring the decrease in absorbance at 340 nm (Lai and Clark, 1976, 1978).

2.2.8. NADP-Linked Malate Dehydrogenase [NADP-MDH; Malate Dehydrogenase (Oxaloacetate Decarboxylating) (NADP), EC 1.1.1.40]

Principle. The rate of oxidation of NADPH concomitant with the conversion of oxaloacetate to malate is measured (Lai and Clark, 1976).

Reagents. The assay mixture is the same as that for the assay of NAD-MDH, except that NADPH substitutes for NADH (Lai and Clark, 1976).

Procedure. The reaction is initiated with the addition of oxaloacetate, and the initial velocity of NADPH oxidation is assayed by determining the rate of decrease in absorbance at 340 nm (Lai and Clark, 1976).

2.2.9. Other Citric Acid Cycle Enzymes

Methods for the assay of TCA cycle enzymes not dicussed in the present review can be found in Lowenstein (1969).

2.3. Enzymes Related to the Citric Acid Cycle

2.3.1. Enzymes Associated with Glutamate Metabolism

2.3.1.1. NAD- AND NADP-LINKED GLUTAMATE DEHYDROGENASES (NAD- AND NADP-GDH; EC 1.4.1.3)

Principle. The rate of NADH (for NAD-GDH) or NADPH (for NADP-GDH) oxidation concomitant with the formation of glutamate from 2-oxoglutarate in the presence of ammonium ions is assayed (Lai and Clark, 1976). In the reverse direction, the rate of NAD^+ (for NAD-GDH) or $NADP^+$ (for NADP-GDH) reduction concomitant with the conversion of glutamate to 2-oxoglutarate is determined (Dennis et al., 1977). ADP is used as an allosteric activator since, in the absence of such a kind of activator, the detectable GDH activities in brain tissue extracts are very low (Dennis et al., 1977).

Reagents. The assay mixture for assaying enzymatic activity in the direction of glutamate formation consists of 0.1*M* phosphate-Tris, pH 7.7, 162 m*M* ammonium acetate, 1 m*M* EDTA (Na or K salt), 0.17 m*M* NADH or 0.33 m*M* NADPH, 1.7 m*M* ADP (usually Tris salt), 0.16% (v/v) Triton X-100, 10 m*M* 2-oxoglutarate (Tris or Na salt), and a small volume of tissue extract (Lai and Clark, 1976; Dennis et al., 1977).

For assaying enzymatic activities in the reverse direction (i.e., in the direction of 2-oxoglutarate formation), the assay mixture contains: 0.1*M* phosphate-Tris, pH 7.7, 1.7 m*M* ADP (usually Tris salt), 0.16% (v/v) Triton X-100, 4 m*M* NAD^+ (for NAD-GDH) or $NADP^+$ (for NADP-GDH), 8 m*M* glutamate, and a small volume of tissue extract (Dennis et al., 1977).

Procedure. For determining the enzymatic activities in the direction of glutamate formation, the reaction is initiated with the addition of 2-oxoglutarate, and the initial rate of NADH (or NADPH) oxidation is monitored by measuring the rate of decrease in absorbance at 340 nm (Lai and Clark, 1976). On the other hand, for assaying the enzymatic activities in the direction of 2-oxoglutarate formation, the initial rate of reduction of NAD^+ (or $NADP^+$) is measured by determining the rate of increase in absorbance at 340 nm after the addition of glutamate (Dennis et al., 1977).

2.3.1.2. Aspartate Aminotransferase (AAT: EC 2.6.1.1)

Principle. This enzyme catalyzes a fully reversible reaction in which the reactants are 2-oxoglutarate and aspartate, and the products are glutamate and oxaloacetate [*see* Gabay and Clarke (1983) and Cooper (1987) for discussions]. The rate of oxaloacetate formation is coupled to the reaction catalyzed by malate dehydrogenase (MDH). The concomitant rate of NADH oxidation is measured in the presence of excess exogenously added MDH (Dennis et al., 1976 and 1977). The AAT-catalyzed reaction can also be assayed in the reverse direction (i.e., in the direction of aspartate production) [*see* Dennis et al. (1976)].

Reagents. The reaction mixture contains: 88 m*M* HEPES-Tris pH 7.4, 8.3 m*M* 2-oxoglutarate, 16.7 m*M* L-aspartate, 0.23 m*M* NADH, 0.17% (v/v) Triton X-100, 10 U MDH, and a small volume of tissue extract (Dennis et al., 1977).

Procedure. The reaction is started with the addition of L-aspartate, and the rate of NADH oxidation is measured by monitoring the rate of decrease in absorbance at 340 nm (Dennis et al., 1976, 1977).

2.3.2. Enzymes of GABA Metabolism

2.3.2.1. GABA TRANSAMINASE (GABA-T; γ-AMINOBUTYRATE AMINOTRANSFERASE, EC 2.6.1.19)

Principle. GABA-T catalyzes the transamination of GABA with 2-oxoglutarate forming glutamate and succinic semialdehyde [*see* Baxter (1970) and Tapia (1983) for discussions]. In the assay procedure described below, the rate of glutamate formation is coupled to the reaction catalyzed by GDH. The concomitant rate of NAD^+ reduction is measured (Walsh and Clark, 1976).

Reagents. The 2-oxoglutarate solution is freshly made up just prior to the enzymatic assays, and its pH adjusted to 8.6 with triethanolamine. The reaction mixture consists of 0.1*M* triethanolamine hydrochloride pH 8.6, 1 m*M* NAD^+, 0.1% (v/v) Triton X-100, 2 U of GDH (in glycerol), 2.5 m*M* 2-oxoglutarate, 6 m*M* GABA, and a small volume of tissue extract.

Procedure. The reaction is initiated with the addition of GABA, and the linear rate of NADH formation is measured by monitoring the increase in absorbance at 340 nm after a brief lag (~1 min) (Walsh and Clark, 1976).

2.3.2.2. SUCCINIC SEMIALDEHYDE DEHYDROGENASE (SSADH; SUCCINATE-SEMIALDEHYDE DEHYDROGENASE, EC 1.2.1.16)

Principle. The assay involves the determination of the rate of NAD^+ reduction concomitant with the conversion of succinic semialdehyde to succinate (Walsh and Clark, 1976).

Reagents. The reaction mixture contains: 0.1*M* Tris-HCl, pH 8.8, 15 m*M* mercaptoethanol, 2.5 m*M* NAD^+, 0.1% (v/v) Triton X-100, 1 mM succinic semialdehyde, and a small volume of tissue extract (Walsh and Clark, 1976).

Procedure. The reaction is started with the addition of succinic semialdehyde, and the rate of NADH formation is monitored by measuring the rate of increase in absorbance at 340 nm (Walsh and Clark, 1976).

2.3.3. *Enzymes of Ketone Body Metabolism*

2.3.3.1. D-3-HYDROXYBUTYRATE DEHYDROGENASE (BOBDH; EC 1.1.1.30)

Principle. The assay measures the reduction of NAD^+ concomitant with the conversion of 3-hydroxybutyrate to acetoacetate (Leong et al., 1981).

Reagents. The reaction mixture consists of 100 m*M* Tris-HCl, pH 8.5, 400 m*M* hydrazine hydrate, 10 m*M* magnesium sulfate, 5 m*M* EDTA (K salt), 10 m*M* NAD^+, 0.001 m*M* rotenone, 20 m*M* DL-3-hydroxybutyrate, and a small volume of tissue extract (Leong et al., 1981).

Procedure. Just prior to the enzymatic assay, the tissue extract is frozen and thawed three times. The frozen and thawed tissue extract is then added to the assay mixture. The reaction is initiated with the addition of 3-hydroxybutyrate, and the rate of NAD reduction is monitored by determining the rate of increase in absorbance at 340 nm.

2.3.3.2. 3-OXO-ACID: COA TRANSFERASE (OCOA TR; 3-KETOACID COA-TRANSFERASE, EC 2.8.3.5)

Principle. The reaction is measured in the direction of succinyl-CoA formation from succinate concomitant with acetoacetyl-CoA disappearance (Booth et al., 1980). However, the reaction in the opposite direction can also be readily measured [*see* Williamson et al., (1971) for details].

Reagents. The assay mixture contains: 50 m*M* Tris-HCl, pH 8.5, 10 m*M* $MgCl_2$, 5 m*M* iodoacetamide, 0.1 m*M* acetoacetyl-CoA, 50 m*M* succinate (Na salt), 0.2% (v/v) Triton X-100, and a small volume of tissue extract.

Procedure. The assay mixture is prepared without succinate, tissue extract, and acetoacetyl-CoA. Upon the addition of acetoacetyl-CoA (molar extinction coefficient = 20.5×10^3/mol/cm), the decrease in absorbance at 303 nm (resulting from spontaneous hydrolysis of acetoacetyl-CoA) is measured for 2 min. Then, the tissue extract is added, and the decrease in absorbance at 303 nm (resulting from spontaneous acetoacetyl-CoA hydrolysis and acetoacetyl-CoA deacylase activity) is determined for 3 min. Finally, succinate is added to the reaction mixture, and the decrease in absorbance at 303 nm is recorded for a further 3 min [resulting from spontaneous acetoacetyl-CoA hydrolysis plus deacylase activity plus 3-oxo-acid:CoA transferase (OCoA Tr) activity]. From these changes in absorbance at 303 nm, the activity resulting from the 3-oxo-acid:CoA transferase (OCoA Tr) can be calculated using the molar extinction coefficient of acetoacetyl-CoA. [An alternative (but slightly more accurate) method for determining this enzymatic activity requires a dual wavelength recording spectrophotometer (e.g., an Aminco-Chance DW2 spectrophotometer). [*See* Booth et al. (1980) for details.] The rate of absorbance change resulting from succinate-dependent acetoacetyl-CoA hydrolysis is monitored at 303 nm, with the reference wavelength set at 373 nm.]

2.3.3.3. Acetoacetyl-CoA Thiolase (A-CoA Th; Acetyl-CoA Acetyl-Transferase, EC 2.3.1.9)

Principle. This enzymatic reaction catalyzes the formation of acetyl-CoA from acetoacetyl-CoA and CoA. The rate of hydrolysis of acetoacetyl-CoA is measured by the method of Middleton (1973a) as modifid by Booth et al. (1980). However, it is important that any thiol esterase activity present in the tissue extracts has to be substracted from the total apparent thiolase activity in order to accurately determine the activity resulting from acetoacetyl-CoA thiolase (A-CoA Th) [*see* Middleton (1973a) and Booth et al. (1980) for discussions]. Activities of other oxoacyl-CoA thiolases can also be determined using assay methods very similar to that for the acetoacetyl-CoA thiolase (A-CoA Th) [*see* Middleton (1973a,b) for a full discussion and experimental details].

Reagents. The assay mixture contains: 100 m*M* Tris-HCl, pH 8.1, 25 m*M* $MgCl_2$, 50 m*M* KCl, 0.01 m*M* acetoacetyl-CoA, 0.1% (v/v) Triton X-100, 0.05 m*M* CoA, and a small volume of tissue extract (Booth et al., 1980).

Procedure. The reaction is started by adding CoA. The CoA-dependent rate of decrease in absorbance at 303 nm (using 373 nm as the reference wavelength) resulting from acetoacetyl-CoA is monitored using a dual-wavelength recording spectrophotometer (e.g., an Aminco-Chance DW2). The rate of decrease in absorbance prior to adding the CoA, resulting from the hydrolysis (by thiol esterase) of the acetoacetyl-CoA, is substracted from the rate of decrease in absorbance after the addition of the CoA (Booth et al., 1980). Under the assay conditions employed herein, the apparent molar extinction coefficient of acetoacetyl-CoA is 16.9×10^3/mol/cm (Booth et al., 1980).

3. Brain Regional Variations of Glycolytic, Citric Acid Cycle and Related Enzymes

3.1. Phylogenetic Aspects and Sex Differences

3.1.1. Phylogenetic Aspects

Many (i.e., the rate-limiting and regulated as well as the nonrate-limiting) glycolytic, citric acid cycle and related enzymes in the brain show variable regional distribution (Tables 1 and 2). In general, the activities of these enzymes in phylogenetically younger regions (e.g., cerebral cortex and cerebellum) are consistently higher than corresponding enzymatic activities in phylogenetically older regions (e.g., pons and medulla) (Leong et al., 1981, 1984; Lai et al., 1982, 1984b, 1985b; Leong and Clark, 1984a, b, c; Sheu et al., 1984; Butterworth and Giguère, 1984; Butterworth et al., 1985, 1986; Lai and Cooper, 1986).

More recent studies indicate that the generalization regarding the phylogenetic differences in the regional distribution of the activities of enzymes of intermediary metabolism may not be universally applicable. For example, the olfactory bulb is exceptional in this respect (Sheu et al., 1984, 1985; Lai et al., 1985b; Lai and Cooper, 1986). The activities of hexokinase (HK) (Sheu et al, 1984; Lai et al., 1985b), lactate dehydrogenase (LDH) (Lai et al., 1985b), pyruvate dehydrogenase complex (PDHC) (Sheu et al., 1984, 1985; Lai et al., 1985b), pyruvate dehydrogenase kinase (Sheu et al., 1984; Lai et al., 1985b), pyruvate dehydrogenase phosphate phosphatase (Sheu et al., 1984), and NAD-isocitrate dehydrogenase (NAD-ICDH) (Lai et al., 1985b) are significantly lower than the corresponding enzymatic activities in the cerebral cortex. Consequently, in respect to the maximum potential activities of these enzymes of intermediary metabolism, the olfactory bulb resembles a phylogenetically older region such as the pons and medulla.

On the other hand, the activities of NADP-isocitrate dehydrogenase (NADP-ICDH), and NAD- and NADP-glutamate dehydrogenases (NAD-GDH and NADP-GDH) in the olfactory bulb are significantly higher than corresponding enzymatic activities in the cerebral cortex, suggesting that, as far as the regional distribution of these enzymes is concerned, the olfactory bulb resembles a phylogenetically younger (e.g., cerebral cortex) rather than a phylogenetically older (e.g., pons and medulla) region. The ontogenetic development of the olfactory bulb may influence the expression

Table 1
Brain Regional Distribution of Glycolytic Enzymes in Young Male (yM), Young Female (yF), and Old Female (oF) Rats*

Enz	Rats	Brain regions					
		Cb	PM	Hp	St	MB	CC
HK	yM	14.9 ± 1.4	7.9 ± 0.9	15.0 ± 1.2	13.9 ± 1.1	13.2 ± 1.1	17.1 ± 1.6
	yF	16.6 ± 2.3	9.0 ± 1.2	16.4 ± 2.0	14.5 ± 2.0	14.6 ± 1.8	18.6 ± 2.2
	oF	15.9 ± 0.7	7.9 ± 0.8	14.8 ± 1.6	13.7 ± 2.2	13.3 ± 1.3	19.1 ± 1.4
PFK	yM	16.3 ± 3.4[a]	10.9 ± 2.1	11.8 ± 5.1	15.6 ± 3.8	12.6 ± 3.7	20.1 ± 1.7[a]
	yF	11.2 ± 1.7	7.8 ± 2.0	7.8 ± 1.6	10.6 ± 1.8	10.7 ± 1.4	16.6 ± 2.2
	oF	7.7 ± 0.3[a]	6.3 ± 1.0	7.1 ± 1.3	11.2 ± 1.8	10.2 ± 0.9	11.8 ± 0.5[a]
AL	yM	8.2 ± 0.2	5.9 ± 0.2	7.2 ± 0.4	8.2 ± 0.4	7.3 ± 0.3	8.3 ± 0.5
	yF	8.3 ± 0.2	6.1 ± 0.2	7.1 ± 0.5	7.5 ± 0.3	7.6 ± 0.2	8.2 ± 0.1
	oF	7.0 ± 0.2[c]	4.9 ± 0.1[c]	6.1 ± 0.3[a]	6.9 ± 0.3[a]	6.3 ± 0.3[b]	7.5 ± 0.4[a]
PK	yM	123 ± 8	97 ± 7[a]	119 ± 17	142 ± 13	125 ± 7	126 ± 6
	yF	133 ± 1	107 ± 3	137 ± 10	144 ± 1	135 ± 5	133 ± 1
	oF	146 ± 14	106 ± 7	128 ± 12	161 ± 5[b]	141 ± 16	158 ± 17[a]
LDH	yM	97 ± 8	75 ± 9	101 ± 10	118 ± 9	101 ± 5	112 ± 6
	yF	106 ± 10	81 ± 3	105 ± 8	119 ± 9	109 ± 9	119 ± 15
	oF	92 ± 15	69 ± 9[a]	90 ± 9[a]	108 ± 20	94 ± 11[a]	111 ± 12

*Activities are in U/g wet weight of tissue (means ± SD of three or more experiments) and are taken from Leong et al. (1981); significant differences between yM and yF (superscript on yM values), and between y and o (superscript on oF): [a]$p < 0.05$; [b]$p < 0.005$; [c]$p < 0.0005$. The abbreviations are: Cb, cerebellum; PM, pons and medulla; Hp, hypothalamus; St, striatum; MB, midbrain; CC, cerebral cortex (+ hippocampus); Enz, enzymes; HK, hexokinase; PFK, phosphofructokinase; AL, aldolase; PK, pyruvate kinase; LDH, lactate dehydrogenase; yM, young male (90-d-old); yF, young female (90-d-old); oF, old female (723–768-d-old).

Table 2
Brain Regional Distribution of Citric Acid Cycle and Related Enzymes in Young Male, Young Female, and Old Female Rats*

Enz	Rats	Brain regions					
		Cb	PM	Hp	St	MB	CC
PDHC	yM	1.69 ± 0.18[a]	1.04 ± 0.17	1.33 ± 0.14	1.54 ± 0.32	1.31 ± 0.08	1.74 ± 0.30
	yF	1.40 ± 0.03	0.92 ± 0.07	1.19 ± 0.11	1.35 ± 0.13	1.35 ± 0.05	1.50 ± 0.30
	oF	1.35 ± 0.10	0.78 ± 0.18	1.08 ± 0.08	1.11 ± 0.20	1.19 ± 0.27[a]	1.44 ± 0.16
Cit Syn	yM	25.5 ± 0.9	19.0 ± 0.4[b]	24.3 ± 1.3[a]	26.6 ± 0.6[c]	24.9 ± 0.4[b]	28.2 ± 0.9[b]
	yF	24.2 ± 1.1	17.7 ± 0.1	21.0 ± 0.4	22.0 ± 0.6	22.7 ± 0.4	25.2 ± 0.7
	oF	20.7 ± 1.6[a]	13.6 ± 0.9[c]	18.3 ± 0.9[b]	19.8 ± 2.1	19.3 ± 1.8[a]	24.0 ± 2.9
NAD-ICDH	yM	4.9 ± 1.3	4.0 ± 0.6	5.3 ± 1.5	5.2 ± 0.8	5.4 ± 0.5	5.8 ± 0.2
	yF	5.6 ± 0.6	3.8 ± 0.4	4.8 ± 0.2	4.2 ± 0.5	5.0 ± 0.6	5.3 ± 0.7
	oF	4.3 ± 0.3[a]	2.3 ± 0.3[b]	3.2 ± 0.8[a]	2.9 ± 0.4[a]	3.5 ± 0.4[b]	4.8 ± 0.2
Fum	yM	27.7 ± 0.6[b]	17.5 ± 0.6[a]	16.1 ± 1.2	17.6 ± 0.4[b]	20.2 ± 0.4[a]	19.1 ± 0.3[b]
	yF	23.9 ± 1.2	15.8 ± 0.8	14.6 ± 0.6	15.0 ± 0.6	17.9 ± 0.8	16.7 ± 0.6
	oF	23.6 ± 1.2	12.9 ± 1.2[a]	13.5 ± 1.0	14.0 ± 1.3	16.1 ± 1.1	17.1 ± 1.0
NAD-MDH	yM	668 ± 31	512 ± 5[b]	559 ± 30	624 ± 8	607 ± 17[a]	651 ± 24
	yF	721 ± 41	561 ± 9	596 ± 11	648 ± 72	664 ± 15	695 ± 54
	oF	665 ± 35	464 ± 39[b]	540 ± 20[b]	603 ± 55	614 ± 47	732 ± 87
NAD-GDH	yM	20.6 ± 1.4[a]	31.3 ± 1.1[b]	30.6 ± 2.4	22.5 ± 1.7	30.6 ± 1.5	22.6 ± 0.9
	yF	23.8 ± 0.8	37.1 ± 1.6	33.7 ± 2.9	23.3 ± 0.9	33.7 ± 2.4	24.3 ± 1.8
	oF	22.2 ± 2.4	28.3 ± 4.0[a]	30.2 ± 2.2	22.5 ± 3.5	29.7 ± 2.9	24.2 ± 1.3

*Activities are in U/g wet weight of tissue (means ± SD of three or more experiments). The NAD-GDH data are from Leong and Clark (1984c), and the rest are from Leong et al. (1981). The abbreviations are: Enz, enzymes; PDHC, pyruvate dehydrogenase complex; Cit Syn, citrate synthase; NAD-ICDH, NAD-isocitrate dehydrogenase; Fum, fumarase: NAD-MDH, of the activities of these enzymes of intermediary metabolism in this brain region in the adult animal (Lai et al., 1985b).

As indicated by more recent studies, even within one brain region, the various subregions and/or nuclei may show differences in the enzyme distribution patterns. For example, Martin et al. (1987) have demonstrated in the rat that the activities of hexokinase (HK) are highest in the ventral medial hypothalamus (VMH), intermediate in ventral lateral hypothalamus (VLH), and lowest in the area postrema-nucleus of the tractus solitarius (AP-NTS). On the other hand, the activities of glucose-6-phosphate dehydrogenase (G6PDH), 3-hydroxybutyrate dehydrogenase (BOBDH), glutamate dehydrogenase (GDH), carnitine palmitoyl transferase [total as well as types I and II: *see* Bird et al. (1985) for a discussion of the subdivision of activities] are higher in AP-NTS than in the other two subregions (i.e., VMH and VLH) (Martin et al., 1987). The activities of pyruvate dehydrogenase complex (PDHC) are lower in VMH than in the other two subregions (i.e., VLH and AP-NTS) (Martin et al., 1987). Citrate synthase (Cit Syn) activities do not, however, significantly differ among the three subregions of the hypothalamus (Martin et al., 1987).

3.1.2. Sex Differences

Sex differences in the activities of numerous enzymes in various organs have been reported [*see* Pitot and Yatvin (1973) for a discussion], although the documentation of these differences in the brain is far from complete (Leong et al., 1981). In an extensive study covering 14 enzymes of intermediary metabolism and six brain regions, Leong et al. (1981) reported small (10–20%), region-specific differences in the activities of several enzymes.

For example, the activities of phosphofructokinase (PFK) (significant in cerebellum and cerebral cortex), pyruvate dehydrogenase complex (PDHC) (significant in cerebellum), citrate synthase (Cit Syn) (significant in pons and medulla, hypothalamus, striatum, midbrain, and cerebral cortex), and fumarase (Fum) (significant in cerebellum, pons and medulla, striatum, midbrain, and cerebral cortex) are generally lower in female than in male adult rat brain (Tables 1 and 2). By contrast, the activities of pyruvate kinase (PK) (significant in pons and medulla), NAD-malate dehydrogenase (NAD-MDH) (significant in pons and medulla and midbrain), glutamate dehydrogenase (GDH) (significant in cerebellum and pons and medulla), 3-hydroxybutyrate dehydrogenase (BOBDH) (significant in cerebellum, pons and medulla, hypothalamus, and striatum), and NADP-malate dehydrogenase (NADP-MDH) (significant in cerebellum and pons and medulla) are, in general, higher in the female than in the male adult rat brain (Tables 1–3). However, sex differences in the activities of hexokinase (HK), aldolase, lactate dehydrogenase (LDH), NAD-isocitrate dehydrogenase (NAD-ICDH), glucose-6-phosphate dehydrogenase (G6PDH), and NADP-isocitrate dehydrogenase (NADP-ICDH) are not apparent (Tables 1–3).

These observations of Leong and coworkers (Leong et al., 1981; Leong and Clark, 1984c) clearly indicate that there is a sexual dimorphism in the brain regional expression of the activities of enzymes of intermediary metabolism. Thus, the results of Leong and coworkers (Leong et al., 1981; Leong and Clark, 1984c) are consistent with the notion that the sexual dimorphism in enzymatic expressions may be the consequences of the influences of sex steroids on brain development. However, the underlying cellular and molecular mechanisms are unknown. This is certainly a worthwhile topic for further investigation.

3.2. Changes during Development

Until the early 1980s, most of the developmental studies (e.g., Cremer and Teal, 1974; Patel, 1974b; MacDonnell and Greengard, 1974; Land et al., 1977; Patel, 1979; Booth et al., 1980) on the enzymes associated with glucose utilization and tricarboxylic acid cycle were on age-related changes in whole brain or forebrain [*see* Clark and Nicklas (1984), Thurston and Hauhart (1985), and Miller (1985) for discussions]. The following discussion focuses, therefore, on the regional enzyme development in the rat brain.

3.2.1. Enzymes Associated with Glucose Utilization

Leong and Clark (1984a) studied the developmental patterns of the activities of hexokinase (HK), aldolase, lactate dehydrogenase (LDH), and glucose-6-phosphate dehydrogenase (G6PDH) in six brain regions (including cerebellum, pons and medulla, hypothalamus, striatum, midbrain, and cerebral cortex). In all the regions examined, the activities of the glycolytic enzymes develop as a single cluster, although the timing of this development varies from region to region (Leong and Clark, 1984a).

When the various regions are compared, it is evident that the glycolytic enzymatic potentials develop earliest in pons and medulla, somewhat later in hypothalamus, striatum, and midbrain, but much later in cerebellum and cerebral cortex (Leong and

Table 3
Brain Regional Distribution of 3-Hydroxybutyrate Dehydrogenase, Glucose-6-Phosphate Dehydrogenase, NADP-Isocitrate Dehydrogenase, and NADP-Malate Dehydrogenase in Young Male (yM), Young Female (yF), and Old Female (oF) Rats*

Enz	Rats	Brain regions					
		Cb	PM	Hp	St	MB	CC
BOBDH	yM	0.23 ± 0.02^a	0.14 ± 0.01^a	0.14 ± 0.02^a	0.18 ± 0.01^a	0.17 ± 0.00	0.22 ± 0.03
	yF	0.25 ± 0.01	0.16 ± 0.00	0.18 ± 0.03	0.23 ± 0.02	0.18 ± 0.01	0.24 ± 0.01
	oF	0.08 ± 0.02^c	0.06 ± 0.03^b	0.08 ± 0.03^b	0.11 ± 0.04^b	0.13 ± 0.04^a	0.13 ± 0.02^c
G6PDH	yM	2.81 ± 0.09	2.11 ± 0.19	2.09 ± 0.18	2.01 ± 0.09	2.32 ± 0.02	1.88 ± 0.02
	yF	2.84 ± 0.04	2.35 ± 0.06	2.13 ± 0.02	1.94 ± 0.06	2.44 ± 0.13	1.88 ± 0.06
	oF	2.73 ± 0.17	2.28 ± 0.20	2.30 ± 0.13^a	2.19 ± 0.23	2.48 ± 0.18	2.15 ± 0.16^a
NADP-ICDH	yM	2.31 ± 0.08	2.11 ± 0.15	2.34 ± 0.22	1.79 ± 0.15	2.24 ± 0.19	2.10 ± 0.15
	yF	2.48 ± 0.11	2.30 ± 0.08	2.44 ± 0.05	1.98 ± 0.10	2.40 ± 0.13	2.25 ± 0.09
	oF	2.39 ± 0.12	2.07 ± 0.21	2.65 ± 0.31	1.99 ± 0.40	2.38 ± 0.25	2.47 ± 0.18
NADP-MDH	yM	0.67 ± 0.08^a	0.47 ± 0.04^a	0.57 ± 0.05	0.67 ± 0.07	0.62 ± 0.06	0.68 ± 0.04
	yF	0.94 ± 0.19	0.69 ± 0.10	0.74 ± 0.16	0.79 ± 0.23	0.86 ± 0.20	0.86 ± 0.18
	oF	0.88 ± 0.12	0.50 ± 0.11^a	0.66 ± 0.14	0.71 ± 0.13	0.72 ± 0.11	0.90 ± 0.11

*Values are in U/g wet weight of tissue (means ± SD of three or more experiments) and are from Leong et al. (1981). The abbreviations are: Enz, enzymes; BOBDH, 3-hydroxybutyrate dehydrogenase; G6PDH, glucose-6-phosphate dehydrogenase; NADP-ICDH, NADP-isocitrate dehydrogenase; NADP-MDH, NADP-malate dehydrogenase. For other details, *see* the footnote to Table 1.

Clark, 1984a). On the other hand, during brain development, the activities of glucose-6-phosphate dehydrogenase (G6PDH), a key enzyme of the hexose monophosphate shunt pathway, decline relative to the glycolytic enzymatic activities (Leong and Clark, 1984a; Lai et al., 1984b).

3.2.2. Enzymes of Energy Metabolism

Leong and Clark (1984b) also studied the developmental patterns of the activities of several energy-metabolizing enzymes [including pyruvate dehydrogenase complex (PDHC), citrate synthase (Cit Syn), NAD-isocitrate dehydrogenase (NAD-ICDH), fumarase (Fum), and 3-hydroxybutyrate dehydrogenase (BOBDH)] in six rat brain regions (pons and medulla, hypothalamus, striatum, midbrain, cerebellum, and cerebral cortex).

Citrate synthase (Cit Syn), NAD-isocitrate dehydrogenase (NAD-ICDH), and pyruvate dehydrogenase complex (PDHC) apparently develop as a cluster in each brain region (Leong and Clark, 1984b). However, during the perinatal period, the development of the pyruvate dehydrogenase complex (PDHC) lags behind the other two enzymes (Leong and Clark, 1984b), a finding consistent with that of Land et al. (1977), who made a similar observation using isolated nonsynaptic mitochondria derived from the developing postnatal rat forebrain.

The timing of the development of the citric acid cycle and related enzymes varies from region to region (Leong and Clark, 1984b; Lai et al., 1984b; Butterworth and Giguère, 1984; Malloch et al., 1986). However, similar to the glycolytic enzyme cluster during development (Leong and Clark, 1984a), 50% of the adult activities of the TCA cycle enzymes are attained first in pons and medulla, then in hypothalamus, striatum, and midbrain, but last in cerebellum and cerebral cortex (Leong and Clark, 1984b).

The regional development of glutamate dehydrogenase (GDH) differs somewhat from the TCA cycle enzymes. In all six regions of the rat brain studied by Leong and Clark (1984c), glutamate dehydrogenase (GDH) activities develop rapidly from birth and reach a maximum around 30 d postpartum; thereafter, enzymatic activities either significantly decrease (as in pons and medulla) or level off to adult values. This enzymatic activity develops much earlier in the pons and medulla than in the other regions investigated (Leong and Clark, 1984c).

The 3-hydroxybutyrate dehydrogenase (BOBDH) also develops earlier in the pons and medulla than in the other regions

(Leong and Clark, 1984b). However, in all the regions examined, the activities of this enzyme attain a maximum at weaning, but proceed to decline from postnatal d 30 (Leong and Clark, 1984b). In this regard, the results of Leong and Clark (1984b) are compatible with those of Booth et al. (1980), who studied the age-related changes in 3-hydroxybutyrate dehydrogenase (BOBDH) activity in nonsynaptic mitochondria isolated from the postnatal rat forebrain. Moreover, it is interesting and pertinent to note that an earlier study of Land and Clark (1975) demonstrated that the 3-hydroxybutyrate-supported state-3 oxygen uptake rates by nonsynaptic mitochondria isolated from the postnatal rat forebrain show the same age-related changes, i.e., uptake rates increase till weaning and decline thereafter.

Land and Clark (1975) also noted that the age-related changes in the activities of pyruvate carboxylase in rat brain during postnatal development show the same developmental pattern as the 3-hydroxybutyrate-supported oxygen uptake rates by nonsynaptic mitochondria derived from the postnatal rat brain.

More recently, Bird et al. (1985) observed that the activities of both type I (the overt form) and type II (the latent form) carnitine palmitoyltransferases in nonsynaptic mitochondria isolated from the postnatal rat brain show maxima at d 20, but thereafter decline to lower values at d 50. The age-related increases in the type II enzymatic activities are particularly pronounced during the period of active myelination. These results led Bird et al. (1985) to propose the possibility that, in the rat brain, carnitine palmitoyltransferases may be more concerned with mitochondrial fatty acid elongation (which may make a contribution towards myelination) than with a role in energy homeostasis. Obviously, this is an interesting area for further investigation.

The metabolic significance of the development of the enzymes of intermediary metabolism in the brain in the light of the brain's unique dependence on glucose oxidation for energetic and synthetic purposes has been amply discussed elsewhere (Land et al., 1977; Leong and Clark, 1984a,b,c; Miller, 1985; Thurston and Hauhart, 1985; Lai and Clark, 1989) and will not be further elaborated upon.

3.2.3. Enzyme Development in Precocial vs Nonprecocial Species

Many histological and other morphological studies have provided a basis for the concept that certain animals are in a relatively "mature state" of neurological competence at birth [for a full discussion, *see*: Booth et al. (1980) and references cited therein]. These animals are called the precocial species (e.g., horse, sheep, and guinea pig). On the other hand, there are species that are born in a rather "immature state" of neurological competence, and these are the nonprecocial species (e.g., mouse, rat, rabbit, dog, and cat). The human neonate lies somewhere between these two extremes.

Booth et al. (1980) studied the development of ketone-body-metabolizing enzymes [3-hydroxybutyrate dehydrogenase (BOBDH), 3-oxo-acid: CoA transferase (OCoA Tr), and acetoacetyl-CoA thiolase (A-CoA-Th)] and several enzymes associated with aerobic glucose metabolism [hexokinase (HK), lactate dehydrogenase (LDH), pyruvate dehydrogenase complex (PDHC), and citrate synthase (Cit Syn)] in the cerebral cortex of the rat (a nonprecocial species) and the guinea pig (a precocial species). In the guinea pig brain, the activities of the ketone-body-metabolizing enzymes are low during fetal and neonatal development, with only small increases in enzymatic activities at birth (Booth et al., 1980). By contrast, in the rat, the activities of these enzymes increase three- to fourfold during the suckling period (1–22 d postpartum), but decrease in the adult (Booth et al., 1980).

In the rat brain, activities of hexokinase (HK), lactate dehydrogenase (LDH), pyruvate dehydrogenase complex (PDHC), and citrate synthase (Cit Syn) increase several-fold during the late suckling period (Booth et al., 1980). However, the activities of these enzymes in the guinea pig brain show similar increases only in the last 10–15 d before birth. In addition, at birth, these enzymatic activities in the brain of the guinea pig, a precocious species, are almost at adult levels (Booth et al., 1980). Thus the brain development of the enzymes of aerobic glycolysis in a precocial (i.e., the guinea pig) and a nonprecocial (i.e., the rat) species correlates well with the onset of neurological competence (Booth et al., 1980).

3.3. Changes during Aging

3.3.1. Enzymes Associated with Glucose Utilization

Glycolytic Enzymes. In aging, activities of aldolase are decreased in all rat brain regions, whereas the activities of lactate dehydrogenase (LDH) (decreases in pons and medulla, hypothalamus, and midbrain) and phosphofructokinase (PFK) (decreases in cerebellum and cerebral cortex) are decreased only in selected regions of the rat brain (Leong et al., 1981; *also see* Table 1). By contrast, the activities of pyruvate kinase (PK) are slightly increased in two regions (striatum and cerebral cortex) of the rat brain in aging (Leong et al., 1981; *also see* Table 1). However, the activities of hexokinase (HK) remained unaltered in the aged rat brain (Leong et al., 1981; *also see* Table 1).

Citric Acid Cycle and Related Enzymes. Activities of NAD-isocitrate dehydrogenase (NAD-ICDH) decrease in every region of the rat brain in aging (Leong et al., 1981; Lai et al., 1982; *also see* Table 2). Activities of citrate synthase (Cit Syn) (decrease in cerebellum, pons and medulla, hypothalamus, and midbrain), fumarase (Fum) (decrease in pons and medulla, and midbrain), and NAD-malate dehydrogenase (NAD-MDH) (decrease in pons and medulla, and hypothalamus) decrease in several regions of the rat brain in aging (Leong et al., 1981; *also see* Table 2). NADP-malate dehydrogenase (NADP-MDH) activities show a trend toward decreases in the rat brain in aging although the age-related decreases in enzymatic activity are statistically significant only in the pons and medulla (Leong et al., 1981; *also see* Table 3).

By contrast, the activities of pyruvate dehydrogenase complex (PDHC) (Table 2) and NADP-isocitrate dehydrogenase (NADP-ICDH) (Table 3) are not significantly decreased in the rat brain in aging (Leong et al., 1981). Similarly, Patel and coworkers (Deshmukh et al., 1980; Deshmukh and Patel, 1982) also did not find significant decreases in pyruvate dehydrogenase complex (PDHC) activity in mitochondria isolated from the aged rat brain. However, these workers (Deshmukh and Patel, 1982) did provide some evidence that the age-dependent decreases in pyruvate-supported oxygen uptake by synaptic and nonsynaptic mitochondria can be attributed to the decreases in pyruvate transport by the mitochondria isolated from the aged rat brain. More recently, working with 3-, 10- and 30-mo-old mice, Ksiezak-Reding et al. (1984) found that the activation state of pyruvate dehydrogenase complex (PDHC) in the mouse brain is not affected in aging, although there is a slight decrease (~17%) in the total activity of the complex in the aged mouse brain. Thus, the results of Ksiezak-Reding et al. (1984) obtained in the mouse confirm the earlier observations of Deshmukh et al. (1980) and Leong et al. (1981) in the rat.

Among all the energy-metabolizing enzymes so far studied, 3-hydroxybutyrate dehydrogenase (BOBDH) (Table 3) shows the most dramatic age-related changes: the activities of this enzyme significantly decrease (by 40–65%) in all regions of the rat brain in aging (Leong et al., 1981).

The activities of glucose-6-phosphate dehydrogenase (G6PDH) (Table 3) in the hypothalamus and cerebral cortex of the rat are slightly, but significantly, increased in aging (Leong et al., 1981).

4. Cellular and Subcellular Compartmentation of Glycolytic, Citric Acid Cycle and Related Enzymes

4.1. Neuronal and Glial Localization

Systematic studies on the neuronal and glial localization of glycolytic, citric acid cycle, and related enzymes are scarce. Most of the available studies are focused on the cellular localization of enzymes associated with the metabolism of putative amino acid neurotransmitters such as glutamate, aspartate, and GABA [*see* Hertz (1979 and 1982), Drejer et al. (1985), Tholey et al. (1985), Hertz et al. (1987), and references cited therein for discussions].

Despite the paucity of systematic studies on the cellular localization of enzymes of intermediary metabolism in different brain regions, some recent and current studies using immunocytochemical and histochemical techniques are quite revealing. Using a polyclonal antiserum to purified glutamate dehydrogenase (GDH), Aoki and coworkers (1987) studied the cellular and regional localization of GDH in the adult rat brain. These workers found that the localization of peroxidase immuno-reactivity for GDH in neurons vs glia is highly dependent on the methods of tissue preparation and upon whether or not the detergent Triton X-100 is employed. This caveat notwithstanding, Aoki et al. (1987) detected low to moderate levels of immunoreactivity for GDH in most neuronal perikarya without using Triton X-100. However, in the presence of 0.25% Triton X-100, intense labeling was selectively detected in glial cells and processes having an unequal distribution throughout the brain. GDH immunoreactivity was detected only in a subset of glial cells containing glial fibrillary acidic protein (GFAP). Most GDH-labeled glial processes were localized to regions (e.g., superficial layers of neocortex, dorsal neostriatum, nucleus accumbens, septohippocampal nucleus, intralaminar thalamic nuclei) that contained moderate to high densities of receptors for L-glutamate or L-aspartate (Aoki et al., 1987). By contrast, intense labeling for GDH was also detected in the central gray of the midbrain, the nuclei of the reticular formation, brainstem regions projecting to the cerebellum, and in the cranial nuclei of the trigeminal and vagal nerves, despite the fact that only some of these regions are known to receive but minor glutamatergic projections (Aoki et al., 1987). Aoki et al. (1987), therefore, concluded that their "findings are consistent with the existence of at least two forms of GDH differing in susceptibilities to solubilization by detergents." Additionally, the observations of Aoki et al. (1987) raise the possibility that GDH may be preferentially localized in certain populations of neurons and astrocytes. This interesting possibility requires further testing.

Recent studies on the light microscopic, peroxidase-antiperoxidase immunocytochemical localization of pyruvate dehydrogenase complex (PDHC) in the adult rat brain revealed that the levels of immunoreactivity in neuronal perikarya throughout the brain are higher than those in the surrounding neuropil (Milner et al., 1988). In addition, only selective populations of neurons have moderate to high accumulation of the reaction product: the intensely labeled perikarya are found in septal complex, dorsal and ventral striatum, entorhinal cortex, supraoptic hypothalamic nuclei, reticular nuclei of thalamus, lateral substantia nigra, tegmental nuclei, lateral nuclei of trapezoid body, raphé pontis and obscuris, and caudo-lateral reticular nuclei (Milner et al., 1988). The distribution of the heavily labeled cells appeared to be prominent in nuclei known to contain cholinergic neurons (as indicated by the distributions of the immunocytochemical labeling for the cholinergic markers, choline acetyltransferase and acetylcholinesterase) (Milner et al., 1988). (By contrast, catecholaminergic nuclei, identified by immunocytochemical labeling for tyrosine hydroxylase, contained few cell bodies that exhibited moderate degrees of labeling for PDHC.) Nonetheless, the perikaryal labeling for the cholinergic markers and for PDHC did not show any overlap in several brain regions (e.g., primary olfactory cortex, pyramidal layer of regio inferior of hippocampus, Purkinje layer of cerebellum) (Milner et al., 1988). Thus, the findings of Milner et al. (1988) are compatible with the notion that brain pyruvate metabolism may play a key role in the regulation of acetylcholine synthesis [*see* Clark et al. (1982) and Gibson and Blass (1983) for a discussion].

4.1.1. Cellular Localization of the Enzymes from the Perspective of the Metabolic Compartmentation Concept

As discussed earlier [*see* Section 1.5 above and Lai and Clark (1989)], the metabolic compartmentation concept stipulates that brain glucose metabolism occurs in at least two biochemically distinct compartments—the "large" and the "small" compartments [*see* Balázs and Cremer (1973) and Berl et al. (1975), and articles therein]. There is a "concensus that the 'large' and the 'small' compartments may be localized, respectively, in neurons and glia (especially astrocytes) [*see* Balázs and Cremer (1973) and Berl et al. (1975), and articles therein]. Consequently, one would anticipate that the glycolytic and citric acid cycle enzymes are more enriched in neurons than in glia. However, the proposal that the "large" and the "small" compartments are localized in neurons and glia, respectively, may be too *simplistic* and cannot fully account for *all* the enzymatic and metabolic data obtained in recent and current studies (*see* Section 4.1.2. below for examples of these studies) using primary cultures of neurons and glial cells (both astrocytes and oligodendrocytes) derived from the developing brain. It is, therefore, reasonable to conclude that additional studies are needed to clearly delineate the morphological correlates of the two—or more—metabolic compartments [see Lai and Clark (1989) for an additional discussion].

4.1.2. Cellular Localization of the Enzymes Based on Studies Using Primary Cultures of Neurons and Glia Derived from Perinatal Brain Tissue

Glycolytic Enzymes. Lai and coworkers (Table 4) studied the activities of several enzymes of intermediary metabolism in primary cultures of neurons and astrocytes, and in mixed cultures (of neurons on astrocytes) derived from fetal rat hypothalamus, and compared them with the corresponding enzymatic activities in the homogenates of the 6-d-old hypothalamus. The hypothalamic cells were cultured in vitro for 10–12 d prior to enzymatic assays (*see* footnotes to Table 4 for details).

Similar to those in the 6-d-old hypothalamic homogenates, the activities of hexokinase (HK) in astrocytes are slightly higher than in neurons or in the mixed cultures (Table 4). Thus, the cellular distribution of hexokinase (HK) activity suggests that hexokinase (HK) is likely to be found in most—if not all—cell types. Similarly, using immunohistochemical techniques Wilkin and Wilson (1977) and Kao-Jen and Wilson (1980) also localized hexokinase (HK) in different kinds of neurons and in astrocytes in the adult rat brain. However, it is interesting to note that Kao-Jen and Wilson (1980) did not find immunoreactive staining in oligodendrocytes in the adult rat cerebellar cortex.

Table 4
Activities of Enzymes of Intermediary Metabolism in Cultured Rat Hypothalamic Neurons (N), Astrocytes (G), and Neurons on Astrocytes (M), and in 6-Day-Old Rat Hypothalamic Homogenates (H)*

Enzymes	Activity, mU/mg protein			
	N	G	M	H
HK	45 ± 2	56 ± 4	47 ± 4	68 ± 0
LDH	599 ± 24	1343 ± 23	721 ± 52	630 ± 1
G6PDH	33 ± 2	56 ± 0	43 ± 5	18 ± 1
NAD-GDH	75 ± 4	93 ± 7	74 ± 7	82 ± 3

*Values (means ± SD of 3–6 determinations) are unpublished data of J.C.K. Lai, T. K. C. Leung, S. Whatley, and L. Lim. Abbreviations are: N, cultured hypothalamic neurons; G, cultured hypothalamic astrocytes; M, mixed cultures of hypothalamic neurons on astrocyte monolayer; H, homogenates of 6-d-old (postnatal) hypothalamus; HK, hexokinase; LDH, lactate dehydrogenase; G6PDH, glucose-6-phosphate dehydrogenase; NAD-GDH, NAD-glutamate dehydrogenase. The primary cultures of rat hypothalamic N, G, and M (i.e., N on G) were prepared by the procedures of Whatley et al. (1981) and were maintained in vitro for 10–12 d prior to the enzymatic assays.

The activities of lactate dehydrogenase (LDH) in cultured hypothalamic astrocytes are double those in cultured hypothalamic neurons, in mixed hypothalamic cells, or in homogenates of the 6-d-old postnatal hypothalamus (Table 4). [In this regard, this enzyme is interesting in that its activity in primary cultures of astrocytes and oligodendrocytes can be induced by treatment of the cultured cells with dibutyryl cyclic AMP (McCarthy and de Vellis, 1980).]

Taken together, the results of Lai and coworkers (*see* Table 4) suggest that–at least in the rat hypothalamus–none of the neural cell types is likely to be totally lacking in glycolytic enzymes. This conclusion is in accord with the notion that even the developing brain is dependent, to some extent, on the aerobic oxidation of glucose for energy [*see* Thurston and Hauhart (1985) and Miller (1985) for discussions].

Hexose Monophosphate Shunt. Activities of glucose-6-phosphate dehydrogenase (G6PDH), a key enzyme of the hexose monophosphate shunt, in cultured hypothalamic astrocytes are almost double the activities in cultured hypothalamic neurons (Table 4). The enzymatic activities in the mixed hypothalamic cell cultures are midway between the corresponding activities in cultured hypothalamic neurons and astrocytes (Table 4). However, the activities of this enzyme in the homogenates of the 6-d-old hypothalamus are considerably lower than the activities of this enzyme in all three kinds of hypothalamic cells in culture (Table 4).

The results of Lai and coworkers (Table 4) suggest that the enzymes of the hexose monophosphate shunt are likely to occur in all cell types, although the overall catalytic capacities of these enzymes in one cell type (e.g., astroglia) may be higher than the capacities in another cell type (e.g., neurons). Indeed, it is evident from some ongoing studies by Edmond et al. (1985, 1986) that this kind of metabolic compartmentation *does* occur. For example, primary cultures of rat neurons, astrocytes, and oligodendrocytes can utilize glucose via the hexose monophosphate shunt (Edmond et al., 1985, 1986). Moreover, based on studies using [1-^{14}C]- and [6-^{14}C]glucose, Edmond and coworkers (Edmond et al., 1985, 1986) concluded that both oligodendrocytes and astrocytes in culture utilize glucose preferentially via the hexose monophosphate shunt pathway, whereas cultured neurons are more dependent on pathway(s) other than the hexose monophosphate shunt for glucose utilization.

Citric Acid Cycle and Related Enzymes. The activities of total pyruvate dehydrogenase complex (PDHC) and 3-oxoacid:CoA transferase (OCoA Tr) are highest in cultured rat neurons, intermediate in cultured rat astrocytes, but lowest in cultured oligodendrocytes (Edmond et al., 1986). However, flux studies employing [3-^{14}C]-labeled acetoacetate and 3-hydroxybutyrate indicate that, although all three cell types can utilize these two ketones, the rate of $^{14}CO_2$ production from the labeled ketones by cultured neurons and oligodendrocytes are approximately three times the corresponding rates by cultured astrocytes (Edmond et al., 1986). These data led Edmond et al. (1986) to conclude that ketone bodies are preferred substrates for cultured neurons and oligodendroglia [*see* Edmond et al. (1985 and 1986) for further discussions].

In fractions of cell bodies isolated from 8-d-old rat cerebella, the activities of glutamate dehydrogenase (GDH) are two- and fivefold higher in astrocytes than in Purkinje and granule cells, respectively (Patel et al., 1982). Similarly, glutamate dehydrogenase (GDH) activities are higher in cultured rat cerebellar astrocytes than in cultured rat cerebellar interneurons (Patel et al., 1982). However, in contrast to the data obtained with rat cerebella

(Patel et al., 1982), the activities of this enzyme in cultured mouse cerebro-cortical interneurons are threefold higher than activities in cultured mouse cerebro-cortal astrocytes (Larsson et al., 1985; Drejer et al., 1985). More recent studies using primary cultures of rat hypothalamic neurons, astrocytes, and mixed cultures of neurons on astrocytes (*see* Table 4) indicate that glutamate dehydrogenase (GDH) activities are higher in cultured astrocytes than in cultured neurons, cultured neurons on astrocytes, or in hypothalamic homogenates. Clearly, further studies are necessary to resolve these contrasting data. However, it is possible that species differences may partially account for such dissimilar results. On the other hand, it is also possible that the neural cell populations studied in various laboratories may significantly differ (and thus lead to apparently contrasting results) because of the dissimilar procedures used for preparing the primary cell cultures.

4.2. Subcellular Localization

4.2.1. Glycolytic Enzymes

It is generally acknowledged that, in mammalian tissues, glycolytic enzymes are primarily—if not exclusively—localized in the cytosolic compartment. However, in the 1960s, it was reported that there are substantial amounts of glycolytic activity in crude brain mitochondrial preparations: those early observations led to the conclusion that brain mitochondria contain glycolytic enzymes [for discussions, *see* Beattie et al. (1964), Clark and Nicklas (1970), and references cited therein]. This erroneous conclusion was later proven to be the result of inadequate mitochondrial isolation methodology [*see* Lai and Clark (1989) for a critical evaluation of this issue].

On the other hand, evidence is accumulating that several kinases (including creatine kinase [Booth and Clark, 1978b] and at least two [i.e., hexokinase (HK) and phosphofructokinase (PFK)] from the glycolytic pathway) are bound to mitochondria [*see* Beattie et al. (1964), Craven and Basford (1974), Land et al. (1977), Wilson (1980, 1983), Lai and Barrow (1984), Lai and Blass (1985), and Lai and Clark (1989) for discussions]. Moreover, there are some indications that the activities of these kinases do differ depending on whether the kinases are localized in synaptic or nonsynaptic mitochondria [*see* Clark and Nicklas (1984), Lai and Clark (1989), and references cited therein for discussions].

Recently, Lim et al. (1983) observed that some of the neuron-specific enolase and pyruvate kinase (PK) activities are associated with the synaptic plasma membranes. These and other results led Lim et al. (1983) to propose that the close association of these glycolytic enzymes with the synaptic plasma membranes constitutes an enzyme assembly capable of generating ATP *in situ*.

4.2.2. Citric Acid Cycle and Related Enzymes

Citric Acid Cycle Enzymes. Extensive studies using the structure-and-function approach have established the mitochondrial and submitochondrial localization of citric acid cycle enzymes in several mammalian tissues, including brain, although most of these studies are primarily focused on liver and/or heart [*see* Srere (1985) and references cited therein]. However, although several studies [*see* Clark and Nicklas (1984), Lai and Clark (1989) and references cited therein] using subcellular fractions derived from the homogenates of whole brain or a brain region (including purified preparations of synaptic and nonsynaptic mitochondria) have yielded results that are consistent with a mitochondrial localization of these enzymes in the brain, the precise submitochondrial localizations of these enzymes in the brain remain to be fully elucidated [*see* Lai and Cooper (1986) for a discussion]. Nonetheless, there are some indications that citric acid cycle enzymes may show submitochondrial compartmentation (Lai and Cooper, 1986; J. C. K. Lai, unpublished observations).

Enzymes Related to the Citric Acid Cycle. Again, based on studies using primarily the liver and heart, it is generally accepted that many of these enzymes are either exclusively or preferentially localized in mitochondria [*see* Tzagoloff (1982) for a discussion]. For example, based on subcellular studies of liver enzymes, it has been accepted for some time that glutamate dehydrogenase (GDH) is exclusively localized in mitochondria; consequently, glutamate dehydrogenase (GDH) has been extensively employed as a mitochondrial marker enzyme [*see* Lai et al. (1986) and references cited therein for discussions]. However, there is recent evidence that this enzyme is also found in purified nuclear fractions derived from several tissues, including brain [*see* Lai et al. (1986) for a discussion and references].

Aspartate aminotransferase (AAT), another enzyme important in glutamate metabolism (Dennis et al., 1977), shows a bimodal distribution. In fact, there is evidence that there are cytosolic and mitochondrial isozymes [*see* Cooper (1988) for a discussion].

GABA transaminase (GABA-T) and succinic semialdehyde dehydrogenase (SSADH), the enzymes of GABA metabolism, are primarily mitochondrial enzymes (Walsh and Clark, 1976; Lai et al., 1977).

Most of the key enzymes of ketone body metabolism are bimodally distributed, being localized in cytosol and mitochondria (Williamson et al., 1971; Middleton, 1973a,b). However, 3-hydroxybutyrate dehydrogenase (BOBDH) is primarily a mitochondrial enzyme [*see* Tzagoloff (1982) for a discussion].

Distributions of Enzymes in Heterogeneous Populations of Synaptic and Nonsynaptic Brain Mitochondria. There is a substantial body of evidence indicating that brain mitochondria are very heterogeneous, and the distributions of citric acid cycle and related enzymes in the various populations of synaptic and nonsynaptic mitochondria are not uniform. Furthermore, the heterogeneity of mitochondria is found not only in the whole brain, but also in brain regions. For a detailed discussion of this topic, the chapter by Lai and Clark in this volume should be consulted.

5. Pathophysiological States and Changes in Glycolytic, Citric Acid Cycle and Related Enzymes

It is impossible to comprehensively discuss the substantial literature [*see*, for example, Benzi (1983)] on this topic in the present review. Nonetheless, a few examples will be discussed to illustrate how the enzyme assays described in the present review can be usefully exploited in pathophysiological studies.

5.1. *Deficiency States*

5.1.1. *Nutritional Deficiencies*

Nutritional deficiencies can give rise to disease states and altered metabolism. One can assess the extent of altered metabolism and use that as a metabolic indicator of a nutritional deficiency by determining the activities of enzymes that are predicted to be affected by such a deficiency. Studies on thiamine deficiency serve to illustrate this approach very well.

Thiamine deficiency in humans and animals leads to the development of neurological signs and symptoms that either partially or totally disappear upon treatment with thiamine [*see* Gibson et al.

(1984), McCandless (1985), and Butterworth et al. (1985, 1986) for discussions].

Thiamine-dependent enzymes are good candidates to be employed as such biochemical indices [*see* Gibson et al. (1984), Butterworth et al. (1985, 1986), and references cited therein]. For example, in symptomatic pyrithiamin-induced thiamine-deficient rats, the brain activities of two TPP-dependent enzymes, 2-oxoglutarate dehydrogenase complex (KGDHC) and transketolase (EC 2.2.1.1), are decreased by 36 and 60%, respectively (Gibson et al., 1984). Upon treatment of the thiamine-deficient animals with thiamine, the neurological symptoms in these animals disappear concomitant with a full restoration of the brain 2-oxoglutarate dehydrogenase complex (KGDHC) activity, although the brain transketolase activity is only partially restored (Gibson et al., 1984). However, the brain activities of pyruvate dehydrogenase complex (PDHC), another TPP-dependent enzyme complex, and pyruvate dehydrogenase phosphate phosphatase are unaltered in the pyrithiamin-induced thiamine-deficient rats (Gibson et al., 1984).

In a recent study on brain regional activities of pyruvate dehydrogenase complex (PDHC) (Butterworth et al., 1985), the observations of Gibson et al. (1984) are confirmed. Nonetheless, in rats fed chronically with a thiamine-deficient diet, the activities of pyruvate dehydrogenase complex (PDHC) are selectively decreased by 15–30% in the midbrain and pons (lateral vestibular nucleus) in parallel with the appearance of ataxia and the loss of the righting reflex (Butterworth et al., 1985). When the chronically thiamine-deficient animals are treated with thiamine, the enzymatic and neurological abnormalities are no longer detected (Butterworth et al., 1985).

5.1.2. Enzyme Deficiencies and Inborn Errors of Metabolism

Enzyme deficiencies can arise either directly or indirectly from inborn error(s) of metabolism. Enzymatic activity measurement provides a convenient, accessible, and sensitive approach for unraveling the molecular mechanism(s) underlying an enzyme deficiency and/or inborn error(s) of metabolism.

Since the normal adult brain is strongly dependent upon glucose as a metabolic fuel, any enzyme deficiencies and/or inborn errors of metabolism occurring in the pathways concerned with the oxidative metabolism of glucose will have devastating effects on brain function. Indeed, not only will oxidative metabolism be affected, but the syntheses of putative neurotransmitters (e.g., acetylocholine, glutamate, aspartate, GABA), whose metabolism is closely associated with the oxidative metabolism of glucose, will also be affected [*see* Clark et al. (1982) and Gibson and Blass (1983) for a discussion]. Thus, the approach of enzymatic activity measurement will help identify the biochemical lesions in the enzyme deficiencies [*see* Burman et al. (1980) and articles therein].

Although detailed discussions on the variety of neurological syndromes in which enzyme deficiencies have been identified using this enzymological approach are beyond the scope of the present review, it is worthwhile to mention one example in which extensive investigations in several laboratories have identified the enzyme deficiencies that may be the direct or indirect consequences of inborn error(s) of metabolism. The case in point is the so-called "disorders of pyruvate metabolism" [*see* Blass (1980) and Butterworth (1985), and references cited therein].

5.2. Toxicological States

Measurements of the activities of glycolytic, citric acid cycle, and related enzymes can be—and have been—used as indices of neurotoxic effects of several different kinds of toxins. Since the normal adult brain is dependent on aerobic glucose oxidation for energy, any detectable effects of the toxins on these enzymatic activities imply that alteration of energy metabolism may be one of the biochemical mechanisms that underlie the neurotoxic effects of the toxins.

Sabri and coworkers [*see* Sabri et al. (1979), Sabri and Spencer (1980), Sabri (1983), and references cited therein] found that several neurotoxins (including 2,5-hexanedione, methyl *n*-butyl ketone, acrylamide, carbon disulfide) inhibit glyceraldehyde-3-phosphate dehydrogenase and phosphofructokinase (PFK) activities in brain homogenates. These and other observations led these workers to hypothesize that "certain substances producing distal axonopathies may act at nearby or related sites in metabolic pathways, and specifically, by impairing energy transformation" (Sabri and Spencer, 1980).

Derangement of oxidative metabolism of glucose may also be an important pathophysiological and/or pathogenetic mechanism underlying the neurotoxicity of metals. For example, Lai and coworkers observed that the inorganic salts of several neurotoxic metals (including Cr, Al, Hg, Cd, Cu, Pb, Zn, Fe) are potent inhibitors of brain hexokinase (HK), the rate-limiting enzyme that regulates the entry of glucose-C into glycolysis (Lai and Blass, 1984a,b; Lai and Barrow, 1984; Lai et al., 1984a, 1985a). Using lactate production from glucose as an indicator of glycolytic flux, Lai and his associates found a good correlation between the inhibition of brain hexokinase (HK) by toxic metal ions and the inhibition of glycolytic flux in brain postnuclear supernatants by the same metal ions (Lai and Blass, 1984a,b; J. C. K. Lai, unpublished data). These observations led Lai and coworkers to hypothesize that one of the major neurotoxic mechanisms of metals may be the inhibition of brain glycolysis through the selective inhibition of HK (Lai and Blass, 1984a,b; Lai et al., 1985a).

In diseases associated with various kinds of inborn errors of metabolism, "toxic metabolites" may accumulate in various tissues and alter intermediary metabolism. Assay of enzymatic activities may help identify key step(s) in metabolic pathway(s) that could be influenced by the accumulated toxic metabolites [*see* Lai and Clark (1989) for an additional discussion]. For example, in several diseases (including hepatic encephalopathy, Reye's disease, and inborn errors of urea metabolism) in which hyperammonemia is a persistent feature, ammonia is the "toxic metabolite" that accumulates in tissues. Lai and Cooper (1986) recently observed that pathophysiological levels of ammonia can inhibit brain 2-oxoglutarate dehydrogenase complex (KGDHC) activity. Since 2-oxoglutarate dehydrogenase complex (KGDHC) is a rate-limiting and regulated enzyme complex, these observations led Lai and Cooper (1986) to postulate that one of the biochemical mechanisms underlying the neurotoxic effects of ammonia may be an inhibition of brain citric acid cycle flux through the inhibition of KGDHC.

6. Conclusions and Prospects for Future Studies

6.1. Concluding Remarks

Although measurements of V_{max}-like enzymatic activities have several recognized limitations [*see* Lai and Clark (1989) for a discussion], such measurements—particularly of rate-limiting and/or regulated enzymes—may provide some useful clues as to the metabolic and pathophysiological roles of enzymes in different pathways. This approach can also provide insights and information for formulating the right premise for the subsequent elucidation of the underlying molecular mechanisms. Furthermore, this enzymological approach may help delineate cases and situations that may or may not be amenable to investigations employing molecular biological approaches.

6.2. Prospects for Future Studies

Several areas that merit further studies have already been alluded to in the previous sections. Nonetheless, there are two areas that deserve priority considerations.

1. Although data are accumulating regarding the region-specific changes in the activities of the enzymes of intermediary metabolism during brain development and aging (Leong et al., 1981, 1984; Lai et al., 1982 and 1984a; Leong and Clark, 1984a,b,c), additional studies will illuminate how far the age-related, region-specific changes conform to the "enzyme cluster" concept [*see* Leong and Clark (1984a,b,c) for a discussion].
2. With the availability of primary cultures of different neural cell types, it may be enlightening to systematically address metabolic compartmentation from the *cellular* perspective. In particular, it will be revealing to determine the activities of enzymes of intermediary metabolism in various neural cell types in primary culture and ascertain to what extent these activity measurements may reflect the actual metabolic fluxes through the different pathways in the cultured cells.

7. Summary

The importance of brain glucose metabolism and the enzymological aspects of the regulation of glycolysis and citric acid cycle have been briefly reviewed. The assay procedures for the determination of the activities of glycolytic, citric acid cycle, and related enzymes have been detailed. The brain regional variations of these enzymes have been discussed in terms of sex-related differences and changes during development and aging. The cellular and subcellular compartmentation of these enzymes was evaluated from the structural and functional viewpoints despite the fact that the available data are far from complete. Pertinent examples were discussed in order to illustrate how the enzymological methodologies described in the present review can be usefully exploited to address mechanistic questions in pathophysiological studies. The need for future studies to address (i) the change in the enzymes of intermediary metabolism in brain regions during development and aging, and (ii) the localization of these enzymes in neurons and glia was emphasized.

Acknowledgments

The authors wish to thank Dr. Arthur J. L. Cooper for a critical reading of the present review and helpful comments, and to the authors' past and present collaborators for their contributions over the years. J.B. Clark's research was supported, in part, by the Medical and Science and Engineering Research Councils (UK), British Diabetic Association and NATO. J. C. K. Lai's research was supported, in part, by the Medical Research Council (UK), the Worshipful Company of Pewterers (UK), and the National Institutes of Health (USA).

References

Aoki C., Milner T. A., Sheu K.-F. R., Blass J. P., and Pickel V. M. (1987) Regional distribution of astrocytes with intense immunoreactivity for glutamate dehydrogenase in rat brain: implication for neuronal-glial interactions in glutamate transmission. *J. Neurosci.* **7**, 2214–2231.

Bachelard H. S. (1970) Control of carbohydrate metabolism, in *Handbook of Neurochemistry*, Vol. 4 (1st Ed.) (Lajtha A., ed.), pp. 1–12, Plenum, New York.

Bachelard H. S. (1983) Transport of hexoses and monocarboxylic acids, in *Handbook of Neurochemistry*, Vol. 5 (2nd Ed.) (Lajtha A., ed.), pp. 339–354, Plenum, New York.

Balázs R. (1970) Carbohydrate metabolism, in *Handbook of Neurochemistry*, Vol. 3 (1st Ed.) (Lajtha A., ed.), pp. 1–36, Plenum, New York.

Balázs R. and Cremer J. E. (1973) *Metabolic Compartmentation in the Brain.* MacMillan, London.

Baxter C. F. (1970) The nature of gamma-aminobutyric acid, in *Handbook of Neurochemistry*, Vol. 3 (1st Ed.) (Lajtha A., ed.), pp. 289–353, Plenum, New York.

Beattie D. S., Sloan H. R., and Basford R. E. (1964) Brain mitochondria II. The relationship of brain mitochondria to glycolysis. *J. Cell Biol.* **19,** 309–316.

Benzi G. (1983) Drug-induced changes in some cerebral enzymatic activities related to energy transduction, in *Handbook of Neurochemistry,* Vol. 4 (2nd Ed.) (Lajtha A., ed.), pp. 531–542, Plenum, New York.

Berl S., Clarke D. D., and Schneider D. (1975) *Metabolic Compartmentation and Neurotransmission.* Plenum, New York.

Bird M. I., Munday L. A., Saggerson E. D., and Clark J. B. (1985) Carnitine acetyltransferase activities in rat brain mitochondria. Bimodal distribution, kinetic constants, regulation by malonyl-CoA and development. *Biochem. J.* **226,** 323–330.

Blass J. P. (1980) Pyruvate dehydrogenase deficiencies, in *Inherited Disorders of Carbohydrate Metabolism* (Burman D., Holton J. B., and Pennock C. A., eds.). pp. 239–268, MTP, Lancaster.

Booth R. F. G. and Clark J. B. (1978a) The control of pyruvate dehydrogenase in isolated brain mitochondria. *J. Neurochem.* **30,** 1003–1008.

Booth R. F. G. and Clark J. B. (1978b) Studies on the mitochondrially bound form of rat brain creatine kinase. *Biochem. J.* **170,** 145–151.

Booth R. F. G., Patel T. B., and Clark J. B. (1980) The development of enzymes of energy metabolism in the brain of a precocial (Guinea pig) and non-precocial (Rat) species. *J. Neurochem.* **34,** 17–25.

Burman D., Holton J. B., and Pennock C. A. (eds.) (1980) *Inherited Disorders of Carbohydrate Metabolism.* MTP, Lancaster.

Butterworth R. F. (1985) Pyruvate dehydrogenase deficiency disorders, in *Cerebral Energy Metabolism and Metabolic Encephalopathy* (McCandless D. W., ed.), pp. 121–141. Plenum, New York.

Butterworth R. F. (1989) Enzymes of the pyruvate dehydrogenase complex of mammalian brain, in *Neuromethods* Vol. 11 (Boulton, A. A., Baker, G. B., and Butterworth R. F., eds.), pp. 283–307, Humana, Clifton, N.J.

Butterworth R. F. and Giguère J.-F. (1984) Pyruvate dehydrogenase activity in regions of the rat brain during postnatal development. *J. Neurochem.* **43,** 280–282.

Butterworth R. F., Giguère J.-F., and Besnard A.-M. (1985) Activities of thiamine-dependent enzymes in two experimental models of thiamine-deficiency encephalopathy: 1. The pyruvate dehydrogenase complex. *Neurochem. Res.* **10,** 1417–1428.

Butterworth R. F., Giguère J.-F., and Besnard A.-M. (1986) Activities of thiamine-dependent enzymes in two experimental models of thiamine-deficiency encephalopathy. 2. α-Ketoglutarate dehydrogenase. *Neurochem. Res.* **11,** 567–577.

Cheng S.-C. (1971) The tricarboxylic acid cycle, in *Handbook of Neurochemistry,* Vol. 5, Pt. A (1st Ed.) (Lajtha A., ed.), pp. 283–315, Plenum, New York.

Clark J. B. and Land J. M. (1974) Differential effects of 2-oxo acids on pyruvate utilization and fatty acid synthesis in rat brain. *Biochem. J.* **140,** 25–29.

Clark J. B. and Nicklas W. J. (1970) The metabolism of rat brain mitochondria. *J. Biol. Chem.* **245,** 4724–4731.

Clark J. B. and Nicklas W. J. (1984) Brain mitochondria, in *Handbook of Neurochemistry* Vol. 7 (2nd Ed.) (Lajtha A., ed.), pp. 135–159, Plenum, New York.

Clark J. B., Booth R. F. G., Harvey S. A. K., Leong S. F., and Patel T. B. (1982) Compartmentation of the supply of the acetyl moiety for acetylcholine synthesis, in *Neurotransmitter Interaction and Compartmentation* (Bradford H. F., ed.), pp. 431–460, Plenum, New York.

Cooper A. J. L. (1988) *L*-glutamate (2-oxoglutarate) aminotransferases, in *Glutamine and Glutamate in Mammals* **Vol. I** (Kvamme E., ed.), CRC, Boca Raton, pp. 123–152.

Craven P. A. and Basford R. E. (1974) ADP-induced binding of phosphofructokinase to the brain mitochondrial membrane. *Biochim. Biophys. Acta* **354,** 49–56.

Cremer J. E. and Teal H. M. (1974) The activity of pyruvate dehydrogenase in rat brain during postnatal development. *FEBS Lett.* **39,** 17–20.

Dennis S. C., Lai J. C. K., and Clark J. B. (1977) Comparative studies on glutamate metabolism in synaptic and non-synaptic rat brain mitochondria. *Biochem. J.* **164,** 727–736.

Dennis S. C., Land J. M., and Clark J. B. (1976) Glutamate metabolism and transport in rat brain mitochondria. *Biochem. J.* **156,** 323–331.

Deshmukh D. R. and Patel M. S. (1982) Age-dependent changes in pyruvate uptake by nonsynaptic and synaptic mitochondria from rat brain. *Mech. Ageing Dev.* **20,** 343–351.

Deshmukh D. R., Qwen O. E., and Patel M. S. (1980) Effect of aging on the metabolism of pyruvate and 3-hydroxybutyrate in nonsynaptic and synaptic mitochondria from rat brain. *J. Neurochem.* **34,** 1219–1224.

Drejer J., Larsson O. M., Kvamme E., Svenneby G., Hertz L., and Schousboe A. (1985) Ontogenetic development of glutamate metabolizing enzymes in cultured cerebellar granule cells and in cerebellum *in vivo*. *Neurochem. Res.* **10,** 49–62.

Edmond J., Auestad N., Robbins R. A., and Bergstrom J. D. (1985) Ketone body metabolism in the neonate: development and the effect of diet. *Fed. Proc.* **44,** 2359–2364.

Edmond J., Robbins R. A., Bergstrom J. D., Cole R. A., and de Vellis J. (1986) Capacity for substrate utilization in oxidative metabolism by neurons, astrocytes, and oligodendrocytes from developing brain in primary culture, in *Proceedings of the American Society for Neurochemistry Special Workshop on Cell Cultures and Cell Markers in Neurochemistry.*

Gabay S. and Clarke C. C. (1983) Aminotransferases, in *Handbook of Neurochemistry* Vol. 4 (2nd Ed.) (Lajtha A., ed.), pp. 67–83, Plenum, New York.

Gibson G. E. and Blass J. P. (1983) Metabolism and neurotransmission, in *Handbook of Neurochemistry* Vol. 3 (2nd Ed.) (Lajtha A.), pp. 633–651, Plenum, New York.

Gibson G. E., Ksiezak-Reding H., Sheu K.-F. R., Mykytyn V., and Blass J. P. (1984) Correlation of enzymatic, metabolic and behavioral defects in thiamin deficiency and its reversal. *Neurochem. Res.* **9,** 803–814.

Hansford R. G. (1980) Control of mitochondrial substrate oxidation, in *Current Topics in Bioenergetics* Vol. 10 (Sanadi D. R., ed.), pp. 217–278, Academic, New York.

Hawkins R. A. (1985) Cerebral energy metabolism, in *Cerebral Energy Metabolism and Metabolic Encephalopathy* (McCandless D. W., ed.), pp. 3–23, Plenum, New York.

Hawkins R. A. and Mans A. M. (1983) Intermediary metabolism of carbohydrates and other fuels, in *Handbook of Neurochemistry* Vol. 3 (2nd Ed.) (Lajtha A., ed.), pp. 259–294, Plenum, New York.

Hers H. G. and Hue L. (1983) Gluconeogenesis and related aspects of glycolysis. *Ann. Rev. Biochem.* **52,** 617–653.

Hertz L. (1979) Functional interactions between neurons and astrocytes I. Turnover and metabolism of putative amino acid transmitters, *Prog. Neurobiol.* **13,** 277–323.

Hertz L. (1982) Astrocytes, in *Handbook of Neurochemistry* Vol. 1 (2nd Ed.) (Lajtha A., ed.), pp. 319–355, Plenum, New York.

Hertz L., Murthy Ch. R. K., Lai J. C. K., Fitzpatrick S. M., and Cooper A. J. L. (1987) Some metabolic effects of ammonia on astrocytes and neurons in primary cultures. *Neurochem. Pathol.*, **6,** 97–129.

Hinman L. M., Ksiezak-Reding H., Baker A. C., and Blass J. P. (1986) Pigeon liver phosphoprotein phosphatase: an effective activator of pyruvate dehydrogenase in tissue homogenates. *Arch. Biochem. Biophys.* **246,** 381–390.

Kao-Jen J. and Wilson J. E. (1980) Localization of hexokinase in neural tissue: electron microscopic studies of rat cerebellar cortex. *J. Neurochem.* **35,** 667–678.

Ksiezak-Reding H., Blass J. P., and Gibson G. E. (1982) Studies on the pyruvate dehydrogenase complex in brain with the arylamine acetyltransferase-coupled assay. *J. Neurochem.* **38,** 1627–1636.

Ksiezak-Reding H., Peterson C., and Gibson G. E. (1984) The pyruvate dehydrogenase complex during aging. *Mech. Ageing Dev.* **26,** 67–73.

Kvamme E. (1983a) Glutamine, in *Handbook of Neurochemistry* Vol. 3 (2nd Ed.) (Lajtha A., ed.), pp. 405–422, Plenum, New York.

Kvamme E. (1983b) Deaminases and amidases, in *Handbook of Neurochemistry* Vol. 4 (2nd Ed.) (Lajtha A., ed.), pp. 85–110, Plenum, New York.

Lai J. C. K. (1975) Ph.D. Thesis, University of London, England.

Lai J. C. K. and Barrow H. N. (1984) Comparison of the inhibitory effects of mercuric chloride on cytosolic and mitochondrial hexokinase activities in rat brain, kidney and spleen. *Comp. Biochem. Physiol.* **78C,** 81–87.

Lai J. C. K. and Blass J. P. (1984a) Inhibition of brain glycolysis by aluminum. *J. Neurochem.* **42,** 438–446.

Lai J. C. K. and Blass J. P. (1984b) Neurotoxic effects of copper: inhibition of glycolysis and glycolytic enzymes. *Neurochem. Res.* **9,** 1713–1724.

Lai J. C. K. and Blass J. P. (1985) Differences in the responses of brain cytosolic and mitochondrial hexokinases to three essential divalent metal ions. *Comp. Biochem. Physiol.* **80C,** 285–290.

Lai J. C. K. and Clark J. B. (1976) Preparation and properties of mitochondria derived from synaptosomes. *Biochem. J.* **154,** 423–432.

Lai J. C. K. and Clark J. B. (1978) Isocitric dehydrogenase and malate dehydrogenase in synaptic and non-synaptic rat brain mitochondria: a comparison of their kinetic constants. *Biochem. Soc. Trans.* **6,** 993–995.

Lai J. C. K. and Clark J. B. (1979) Preparation of synaptic and nonsynaptic mitochondria from mammalian brain, in *Methods in Enzymology* 55, Pt. F (Fleischer S. and Packer L., eds.), pp. 51–60, Academic, New York.

Lai J. C. K. and Clark J. B. (1989) Isolation and characterization of synaptic and nonsynaptic mitochondria from mammalian brain, in *Neuromethods* Vol. 11 (Boulton, A. A., Baker, G. B. and Butterworth R. F., eds). pp. pp. 43–98, Humana, Clifton, N.J.

Lai J. C. K. and Cooper A. J. L. (1986) Brain α-ketoglutarate dehydrogenase complex: kinetic properties, regional distribution, and effects of inhibitors. *J. Neurochem.* **47,** 1376–1386.

Lai J. C. K. and Sheu K.-F. R. (1985) Relationship between activation state of pyruvate dehydrogenase complex and rate of pyruvate oxidation in isolated cerebro-cortical mitochondria: effects of potassium ions and adenine nucleotides. *J. Neurochem.* **45,** 1861–1868.

Lai J. C. K. and Sheu K.-F. R. (1987) The effects of 2-oxoglutarate or 3-hydroxybutyrate on pyruvate dehydrogenase complex in isolated cerebro-cortical mitochondria. *Neurochem. Res.* **12,** 715–722.

Lai J. C. K., Baker A., Carlson Jr. K. C., and Blass J. P. (1985a) Differential effects of monovalent, divalent and trivalent metal ions on rat brain hexokinase. *Comp. Biochem. Physiol.* **80C,** 291–294.

Lai J. C. K., Sheu K.-F. R., and Carlson Jr. K. C. (1985b) Differences in some of the metabolic properties of mitochondria isolated from cerebral cortex and olfactory bulb in the rat. *Brain Res.* **343,** 52–59.

Lai J. C. K., Barrow H. N., Carlson Jr. K. C., and Blass J. P. (1984a) Effects of mercury, lead and tin on brain hexokinase. *Fed. Proc.* **43,** 581.

Lai J. C. K., Leung T. K. C., and Lim L. (1984b) Differences in the neurotoxic effects of manganese during development and aging: some observations on brain regional neurotransmitter and non-neurotransmitter metabolism in a developmental rat model of chronic manganese encephalopathy. *Neurotoxicology* **5,** 37–48.

Lai J. C. K., Leung T. K. C., and Lim L. (1982) Activities of the mitochondrial NAD-linked isocitric dehydrogenase in different regions of the rat brain. Changes in ageing and the effect of chronic manganese chloride administration. *Gerontology* **28,** 81–85.

Lai J. C. K., Leung T. K. C., Marr W., and Lim L. (1980) The activity of glucose-6-phosphate dehydrogenase in liver and hypothalamus of the female rat: effects of administration of ethinyl oestradiol and the progestogens norethisterone acetate and *d*-norgestrel. *Biochem. Soc. Trans.* **8,** 64.

Lai J. C. K., Sheu K.-F. R., Kim Y. T., Clarke D. D., and Blass J. P. (1986) The subcellular localization of glutamate dehydrogenase (GDH): is GDH a marker for mitochondria in brain? *Neurochem. Res.* **11,** 733–744.

Lai J. C. K., Walsh J. M., Dennis S. C., and Clark J. B. (1975) Compartmentation of citric acid cycle and related enzymes in distinct populations of rat brain mitochondria, in *Metabolic Compartmentation and Neurotransmission* (Berl S., Clarke D. D. and Schneider D., eds.), pp. 487–496, Plenum, New York.

Lai J. C. K., Walsh J. M., Dennis S. C., and Clark J. B. (1977) Synaptic and non-synaptic mitochondria from rat brain: isolation and characterization. *J. Neurochem.* **28,** 625–631.

Land J. M. and Clark J. B. (1975) The changing pattern of brain mitochondrial substrate utilization during development, in *Normal and Pathological Development of Energy Metabolism* (Hommes F. A. and Van den Berg C. J., eds.), Academic, New York-London, pp. 155–167.

Land J. M., Booth R. F. G., Berger R., and Clark J. B. (1977) Development of mitochondrial energy metabolism in rat brain. *Biochem. J.* **164,** 339–348.

Larsson O. M., Drejer J., Kvamme E., Svenneby G., Hertz L., and Schousboe A. (1985) Ontogenetic development of glutamine and GABA metabolizing enzymes in cultured cerebral cortex interneurons and in cerebral cortex in vivo. *Int. J. Develop. Neurosci.* **3,** 177–185.

Leong S. F. and Clark J. B. (1984a) Regional enzyme development in rat brain. Enzymes associates with glucose utilization. *Biochem. J.* **218,** 131–138.

Leong S. F. and Clark J. B. (1984b) Regional enzyme development in rat brain. Enzymes of energy metabolism. *Biochem. J.* **218,** 139–145.

Leong S. F. and Clark J. B. (1984c) Regional development of glutamate dehydrogenase in the rat brain. *J. Neurochem.* **43,** 106–111.

Leong S. F., Lai J. C. K., Lim L., and Clark J. B. (1981) Energy-metabolizing enzymes in brain regions of adult and aging rats. *J. Neurochem.* **37,** 1548–1556.

Leong S. F., Lai J. C. K., Lim L., and Clark J. B. (1984) The activities of some energy-metabolizing enzymes in nonsynaptic (free) and synaptic mitochondria derived from selected brain regions. *J. Neurochem.* **42,** 1306–1312.

Lim L., Hall C., Leung T., Mahadevan L., and Whatley S. (1983) Neurone-specific enolase and creatine phosphokinase are protein components of rat brain synaptic plasma membranes. *J. Neurochem.* **41,** 1177–1182.

Lowenstein J. M. (ed.) (1969) *Methods in Enzymology* Vol. 13. Academic, New York–London.

MacDonnell P. C. and Greengard O. (1974) Enzymes in intracellular organelles of adult and developing rat brain. *Arch. Biochem. Biophys.* **163,** 644–655.

Malloch G. D. A., Munday L. A., Olson M. S., and Clark J. B. (1986) Comparative development of the pyruvate dehydrogenase complex and citrate synthase in rat brain mitochondria. *Biochem. J.* **238,** 729–736.

Martin R. J., Bird M. I., Saggerson E. D., Munday L. A., and Clark J. B. (1987) Enzyme activities in regions of the hypothalamus. *J. Neurochem.* **48,** 738–740.

McCandless D. W. (1985) Thiamine deficiency and cerebral energy metabolism, in *Cerebral Energy Metabolism and Metabolic Encephalopathy* (McCandless D. W., ed.), pp. 335–351, Plenum, New York.

McCarthy K. D. and de Vellis J. (1980) Preparation of separate astroglial and oligodendroglial cell cultures from rat cerebral tissue. *J. Cell Biol.* **85,** 890–902.

Middleton B. (1973a) The oxoacyl-coenzyme A thiolase of animal tissues. *Biochem. J.* **132,** 717–730.

Middleton B. (1973b) The acetoacetyl-coenzyme A thiolases of rat brain and their relative activities during development. *Biochem. J.* **132,** 731–737.

Miller A. L. (1985) The tricarboxylic acid cycle, in *Developmental Neurochemistry* (Wiggins R. C., McCandless D. W., and Enna S. J., eds.), pp. 127–159, University of Texas, Austin.

Milner T. A., Aoki C., Sheu K.-F. R., Blass J. P., and Pickel V. M. (1988) Light microscopic immunocytochemical localization of pyruvate dehydrogenase complex in rat brain: topographical distribution and relation to cholinergic and catecholaminergic nuclei, in *Neural Systems* (Spear P. D. and Humphrey D. R., eds.), in press.

Newsholme E. A. and Leech A. R. (1983) *Biochemistry for the Medical Sciences,* John Wiley & Sons, Chichester–New York.

Newsholme E. A. and Start C. (1973) *Regulation in Metabolism.* John Wiley & Sons, London.

Ogasawara N., Watanabe T. and Goto H. (1973) Bilirubin: a potent inhibitor of NAD^+–linked isocitrate dehydrogenase. *Biochim. Biophys. Acta* **327,** 233–237.

Patel A. J., Hunt A., Gordon R. D., and Balázs R. (1982) The activities in different neural cell types of certain enzymes associated with the metabolic compartmentation of glutamate. *Develop. Brain Res.* **4,** 3–11.

Patel M. S. (1974a) The relative significance of CO_2-fixing enzymes in the metabolism of rat brain. *J. Neurochem.* **22,** 717–724.

Patel M. S. (1974b) Inhibition by the branched-chain 2-oxoacids of the 2-oxoglutarate dehydrogenase complex in developing rat and human brain. *Biochem. J.* **144,** 91–97.

Patel M. S. (1979) Influence of neonatal hypothyroidism on the development of ketone-body-metabolizing enzymes in rat brain. *Biochem. J.* **184,** 169–172.

Pevzner L. (1983) Multiple forms of enzymes, in *Handbook of Neurochemistry* Vol. 4 (2nd Ed.) (Lajtha A., ed.), pp. 461–491, Plenum, New York.

Pitot H. C. and Yatvin M. B. (1973) Interrelationships of mammalian hormones and enzyme levels *in vivo. Physiol. Rev.* **53,** 228–325.

Racker E. (1950) Spectrophotometric measurements of the enzymatic formation of fumaric and cis-aconitic acids. *Biochim. Biophys. Acta* **4,** 211–214.

Sabri M. I. (1983) In vitro and in vivo inhibition of glycolytic enzymes by acrylamide. *Neurochem. Pathol.* **1,** 179–191.

Sabri M. I. and Spencer P. S. (1980) Toxic distal axonopathy: biochemical studies and hypothetical mechanisms, in *Experimental and Clinical Neurotoxicology* (Spencer P. S. and Schaumburg H. H., eds.), pp. 206–219, Williams and Wilkins, Baltimore.

Sabri M. I., Moore C. L., and Spencer P. S. (1979) Studies on the biochemical basis of distal axonopathies—I. Inhibition of glycolysis by neurotoxic hexacarbon compounds. *J. Neurochem.* **32,** 683–689.

Sacks W. (1983) Cerebral metabolism in vivo, in *Handbook of Neurochemistry* Vol. 3 (2nd Ed.) (Lajtha A., ed.), pp. 321–351, Plenum, New York.

Sheu K.-F. R., Lai J. C. K., and Blass J. P. (1983) Pyruvate dehydrogenase phosphate (PDH_b) phosphatase in brain: activity, properties and subcellular localization. *J. Neurochem.* **40,** 1366–1372.

Sheu K.-F. R., Lai J. C. K., and Blass J. P. (1984) Properties and regional distribution of pyruvate dehydrogenase (PDH_a) kinase in rat brain. *J. Neurochem.* **42,** 230–236.

Sheu K.-F. R., Lai J. C. K., Kim Y. T., Bagg J., and Dorant G. (1985) Immunochemical characterization of pyruvate dehydrogenase complex in rat brain. *J. Neurochem.* **44,** 593–599.

Sokoloff L. (1983) Measurement of local glucose utilization in the central nervous system and its relationship to local functional activity, in *Handbook of Neurochemistry* Vol. 3 (2nd Ed.) (Lajtha A., ed.), pp. 225–257, Plenum, New York.

Srere P. A. (1985) Organization of proteins within the mitochondrion, in *Organized Multienzyme Systems: Catalytic Properties* (Welch G. R., ed.). pp. 1–61, Academic, Orlando.

Tapia R. (1983) gamma-Aminobutyric acid: metabolism and biochemistry of synaptic transmission, in *Handbook of Neurochemistry* Vol. 3 (2nd Ed.) (Lajtha A., ed.), pp. 423–466, Plenum, New York.

Tholey G., Ledig M., Bloch S., and Mandel P. (1985) Glutamine synthetase and energy metabolism enzymes in cultured chick glial cells: modulation by dibutyryl cyclic AMP, hydrocortisone, and trypsinization. *Neurochem. Res.* **10,** 191–200.

Thurston J. H. and Hauhart R. E. (1985) Ketone bodies, lactate, glucose: metabolic fuels for developing brain, in *Developmental Neurochemistry* (Wiggins R. C., McCandless D. W., and Enna S. J., eds.), pp. 100–126, University of Texas, Austin.

Tzagoloff A. (1982) *Mitochondria,* Plenum, New York—London.

Van den Berg C. J. (1970) Glutamate and glutamine, in *Handbook of Neurochemistry* Vol. 3 (1st Ed.) (Lajtha A., ed.), pp. 355–379, Plenum, New York.

Walsh J. M. and Clark J. B. (1976) Studies on the control or 4-aminobutyrate metabolism in 'synaptosomal' and free rat brain mitochondria. *Biochem. J.* **160,** 147–157.

Whatley S. A., Hall C., and Lim L. (1981) Hypothalamic neurons in dissociated cell culture: the mechanism of increased survival times in the presence of non-neuronal cells. *J. Neurochem.* **36,** 2052–2056.

Wilkin G. P. and Wilson J. E. (1977) Localization of hexokinase in neural tissue: light microscopic studies with immunofluorescence and histochemical procedures. *J. Neurochem.* **29,** 1039–1051.

Williamson D. H., Bates M. W., Page M. A., and Krebs H. A. (1971) Activities of enzymes involved in acetoacetate utilization in adult mammalian tissues. *Biochem. J.* **121** 41–47.

Willson V. J. C. and Tipton K. F. (1979) Purification and characterization of ox brain NAD^+-dependent isocitrate dehydrogenase. *J. Neurochem.* **33,** 1239–1247.

Willson V. J. C. and Tipton K. F. (1980) Allosteric properties of ox brain nicotinamide-adenine dinucleotide dependent isocitrate dehydrogenase. *J. Neurochem.* **34,** 793–799.

Wilson J. E. (1980) Brain hexokinase, the prototype ambiquitous enzyme, in *Current Topics in Cellular Regulation* Vol. 16 (Horecker B. L. and Stadtman E. R., eds.), pp. 1–44, Academic, New York.

Wilson J. E. (1983) Hexokinase, in *Handbook of Neurochemistry* Vol. 4 (2nd Ed.) (Lajtha A., ed.), pp. 151–172, Plenum, New York.

Wood W. A. (ed.) (1982a) *Methods in Enzymology* Vol. 89. Academic, New York.

Wood W. A. (ed.) (1982b) *Methods in Enzymology* Vol. 90. Academic, New York.

Zomzely-Neurath C. E. (1983) Enolase, in *Handbook of Neurochemistry* Vol. 4 (2nd Ed.) (Lajtha A., ed.), pp. 403–433, Plenum, New York.

Enzymes of the Pyruvate Dehydrogenase Complex of Mammalian Brain

Roger F. Butterworth

1. Introduction

Under normal physiological conditions, cerebral function requires a continuous supply of glucose for its energetic and biosynthetic needs. Pyruvate oxidation via the pyruvate dehydrogenase enzyme complex (PDHC) represents a step of key importance in cerebral glucose utilization.

The pyruvate dehydrogenase complex (EC 1.2.4.1) is a five-component enzyme system responsible for the irreversible transformation of pyruvate into acetyl CoA according to the reaction:

$$\text{pyruvate} + NAD^{+} + \text{CoASH} \rightarrow \text{acetyl CoA} + \text{NADH} + CO_2$$

The acetyl CoA formed is then either further oxidized in the tricarboxylic acid cycle or is used for biosynthetic purposes. PDHC exists in all mammalian tissues, including brain, as a complex of three catalytic enzymes and two regulatory enzymes. The catalytic enzymes are pyruvate decarboxylase (E_1, EC 4.1.1.1), dihydrolipoyl transacetylase (E_2, EC 2.3.1.12), and dihydrolipoyl dehydrogenase (E_3, EC 1.6.4.3). E_1, E_2, and E_3 separately catalyze the decarboxylation of pyruvate, the formation of acetyl, CoA, and the production of NADH, respectively), as shown in Fig. 1. E_1 functions irreversibly, whereas E_2 and E_3 are reversible. Consequently, lipoate may be reduced or acetylated (by NADH or acetyl CoA, respectively).

The two regulatory enzymes are PDHC kinase (EC 2.7.1.99), which catalyzes the phosphorylation of E_1 with subsequent inactivation of the complex, and PDHC phosphate phosphatase (EC 3.1.3.43), which is responsible for the dephosphorylation of the phosphorylated enzyme leading to activation of PDHC (Fig. 2). The kinase is tightly bound to E_2, whereas the phosphatase is more weakly bound to the complex (Barrera et al., 1972). In mammals,

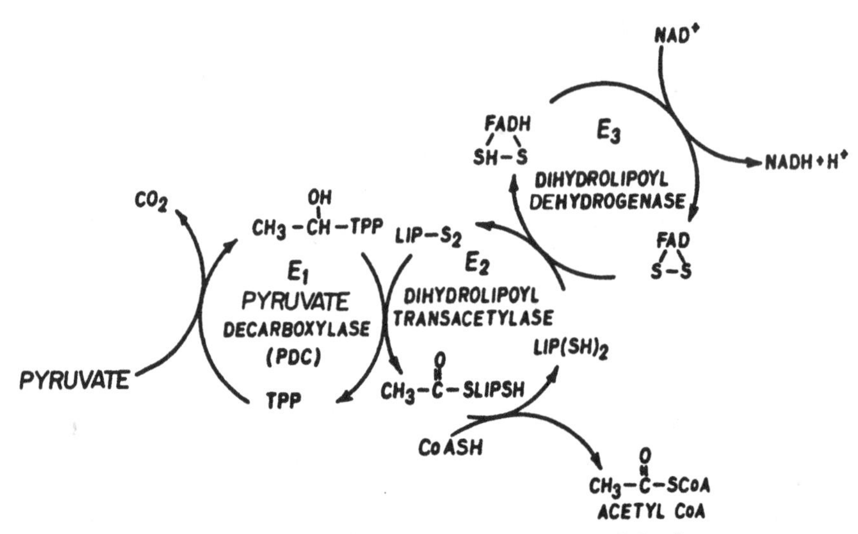

Fig. 1. Constituent enzymes of the pyruvate dehydrogenase complex.

the entire PDH complex including regulatory enzymes is localized within the inner mitochondrial membrane in all tissues studied to date, including brain (Lai and Clarke 1979; Szutowicz et al., 1981; Ksiezak-Reding et al., 1982). PDHC of mammalian brain has a pH optimum around 7.8 (Blass and Lewis, 1973).

2. Regulation of PDHC in Mammalian Brain

As is the case in other tissues, cerebral PDHC is subject to inhibition by the products of the enzymic reaction, namely acetyl CoA and NADH. The sites of end-product inhibition by acetyl CoA and NADH are E_2 and E_3, respectively. Regulation of PDHC in vivo depends on acetyl CoA/CoASH and NADH/NAD^+ ratios. Other regulatory mechanisms involve the kinase and phosphatase enzymes. It has been estimated that the "active form" of PDHC in brain may represent 60–70% of total under normal physiological conditions (Cremer and Teal, 1974; Jope and Blass, 1976; Booth and Clarke, 1978). Starvation results in inactivation of PDHC in most tissues, but not in brain. One suggested mechanism involved in this latter phenomenon lies in the ability of the substrate (pyruvate) to protect cerebral PDHC against inactivation. The observation that brain mitochondria are able to accumulate pyruvate to high con-

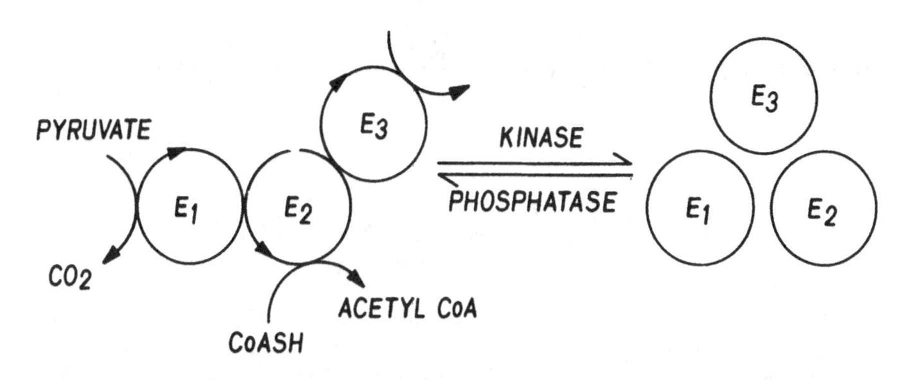

Fig. 2. Interconversion of active and inactive forms of PDHC.

centrations may thus facilitate the maintenance of PDHC in its active form, and in so doing, help to ensure a continuous supply of acetyl CoA to the tricarboxylic acid cycle. It has been suggested that such a mechanism may form the basis of a metabolic adaptation to ensure adequate glucose supply to brain (Booth and Clarke, 1978).

The proportion of cerebral PDHC in the active form changes inversely with the adenylate energy charge or ATP/ADP ratio (Jope and Blass, 1976). The kinase is an ATP-dependent enzyme, and increased ATP production results in increased phosphorylation of PDHC. Ischemia, on the other hand, leads to a rapid decline in the energy charge, subsequent inhibition of the kinase, and increase in the activation state of PDHC. The divalent cations Mg^{2+} and Ca^{2+} are required for PDHC activation. Fluoride ion stimulates the kinase and leads to inactivation of PDHC. Recent studies have shown that the hypoglycemic agent dichloroacetate leads to activation of PDHC by inhibition of the kinase (Abemayor et al., 1984).

3. Methods of PDHC Assay in Brain

3.1. Measurement of PDHC (Total Complex)

3.1.1. The Arylamine Acetyltransferase (ArAT) Coupled Assay

The ArAT coupled assay consists of the transfer of the acetyl moiety of acetyl CoA formed by PDHC to *p*-(*p*-aminophenyl-azo)benzene sulfonic acid (AABS) catalyzed by arylamine acetyltransferase (ArAT) according to the coupled reactions:

$$\text{pyruvate} + \text{NAD}^+ + \text{CoASH} \xrightarrow{\text{PDHC}} \text{acetyl CoA} + \text{NADH} + \text{CO}_2 \quad (1)$$

$$\text{acetyl CoA} + \text{AABS} \xrightarrow{\text{ArAT}} \underset{460\ \text{nm}}{\text{acetyl AABS}} + \text{CoASH} \quad (2)$$

Reaction (2) is monitored as a decrease in absorbance at 460 nm.

3.1.1.1. PREPARATION OF ArAT FROM PIGEON LIVER The following steps are carried out at 0–4°C (Ksiezak-Reding et al., 1982): 10 g of commercial pigeon liver acetone powder (Sigma) is suspended in 90 mL of 10% saturated ammonium sulfate and mixed (vortex) for 5 min. After allowing to settle (15 min), the suspension is centrifuged (40,000*g*, 3 min). The supernatant is then further fractionated with ammonium sulfate (20, 40, 60 and 92% saturation) by serial addition of appropriate quantities of saturated ammonium sulfate (saturated at room temperature). Following a 15-min precipitation period, the mixture is centrifuged (40,000*g*, 30 min), and the protein precipitate is resuspended in the volume indicated in Table 1 of a buffered solution containing 50 m*M* potassium phosphate buffer (pH 7.8), 1 m*M* potassium EDTA, 1 m*M* 2-mercaptoethanol, and Triton X-100 (0.1% w/v). The yield and specific activity of ArAT is found to be highest in the fraction precipitated between 40 and 60% saturation (Table 1). The 40–60% fraction is than dialyzed against 5 m*M* phosphate buffer, pH 7.8 (Stumpf and Kraus, 1979), and stored in 1-mL aliquots at –20°C. Under these conditions, the preparation is stable for at least 4 mo (Ksiezak-Reding et al., 1982). ArAT activity is assayed by acetylation of *p*-nitroaniline as described by Tabor et al. (1953) and Ksiezak-Reding et al. (1982). The decrease in absorbance of *p*-nitroaniline is monitored (molar extinction coefficient $6 \times 10^3/M/\text{cm}$). The reaction mixture contains 50 m*M* potassium phosphate buffer, pH 7.8, 5 m*M* 2-mercaptoethanol, 5 m*M* potassium EDTA, 12 m*M* lithium acetyl phosphate, phosphotransacetylase (4 IU/mL), 0.12 m*M* CoA, and 0.1 m*M* *p*-nitroaniline. ArAT units are expressed as nmol of *p*-nitroaniline acetylated per min at 30°C.

3.1.1.2. PDHC ASSAY. Brain tissue is homogenized at 4°C in 10 vol of a buffered medium containing 50 m*M* potassium phosphate buffer (pH 7.8), 1 m*M* potassium EDTA, 1 m*M* mercaptoethanol, and 0.1% (w/v) Triton X-100. PDHC activity is measured by the coupling of acetyl CoA production to the acetylation of *p*-(*p*-

Table 1
Purification of ArAT

Ammonium sulfate saturation %	Volume, mL	Total protein yield, g	Specific activity, U/mg protein
10	60	0.942 ± 0.108	1.09 ± 0.16
10–20	4	0.022 ± 0.003	1.22 ± 0.33
20–40	20	0.377 ± 0.012	0.78 ± 0.19
40–60	20	0.698 ± 0.028	2.40 ± 0.27
60–92	20	0.224	0.36
Total		2.263	

aminophenylazo) benzene sulfonic acid (AABS, Pfalz and Bauer, USA), as previously described (Ksiezak-Reding et al., 1982). The reaction mixture contains, in a final volume of 1 mL, the following constituents:

50 m*M* Tris-HCl buffer, pH 8.0
0.5 m*M* potassium EDTA
1 m*M* $MgCl_2$
5 m*M* mercaptoethanol
0.3 m*M* dithiothreitol
Triton X-100, 0.2% (w/v)
0.3 m*M* thiamine pyrophosphate (TPP)*
2 m*M* NAD*
0.1 m*M* CoASH*
0.05 mg of AABS
5 m*M* sodium pyruvate
10 μg of crystalline lactate dehydrogenase (LDH) (approximately 3IU)
3 U of purified, dialyzed (40–60% ammonium sulfate fraction) ArAT brain homogenate (up to 0.5 mg of protein equivalent)

*The enzyme cofactors (TPP, NAD, CoASH) are best added separately to the cuvet just prior to measurement since premixing of these substances reportedly leads to lower activities of PDHC (Ksiezak-Reding et al., 1982).

The decrease in absorbance at 460 nm is monitored using a temperature-controlled double-beam recording spectrophotometer. The reference sample is identical to that of the test sample with the exclusion of the substrate (pyruvate). Following baseline stabilization, pyruvate (test) or water (reference) are added.

Following a short lag period of 3–4 min (Leong and Clark, 1984), the assay is allowed to run for approximately 10–15 min, during which time rates are found to be linear with respect to time. The difference in molar extinction coefficient of AABS ($8.75 \times 10^3/M/cm$) and *N*-acetyl AABS ($1.64 \times 10^3/M/cm$) is used to compute the enzyme activity that is expressed as nmol AABS formed per min. Thus, for a light path of 1 cm and final volume of 1 mL

$$\text{nmol/min/mg protein} = \frac{\Delta \text{ optical density/min} \times 1000}{7.11 \times \text{mg protein in cuvet}}$$

The assay described has an absolute requirement for NAD, CoASH, $MgCl_2$, and pyruvate and a 95% requirement for TPP. Optimal substrate and cofactor concentrations are shown in Fig. 3. Omission of Triton X-100 yields enzyme activities about 50% less than optimal. Thus, the arylamine acetyltransferase coupled assay requires several time-consuming steps involving purification and assay of the coupling enzyme. However, once this step is completed, aliquots of ArAT are stable, and the enzyme assay is relatively robust compared to methods involving ^{14}C-pyruvate. It should be noted that the arylamine acetyltransferase coupled assay as described above cannot be used for measurement of the "active form" of PDHC in brain since the ArAT preparations contain substantial phosphatase activity. In order to measure PDHC (active), a lengthier ArAT purification procedure is required (Ksiezak-Reding et al., 1982).

Using the coupled assay, PDHC activities in rat brain homogenates in which Triton X-100 is included in the assay mixture are found to be in the range of 40–50 nmol/min/mg protein (Ksiezak-Reding et al., 1982; Sheu et al., 1983; Butterworth and Giguere, 1984).

3.1.2. The Citrate Synthetase Coupled Assay

Citrate synthetase (CS; EC 4.1.3.7) catalyzes the condensation reaction of acetyl CoA with oxaloacetate, the equilibrium being markedly in the direction of citrate formation. This reaction has therefore found application in the assays of acetyl CoA-producing enzymes such as the pyruvate dehydrogenase complex. The assay is based on the coupled reactions:

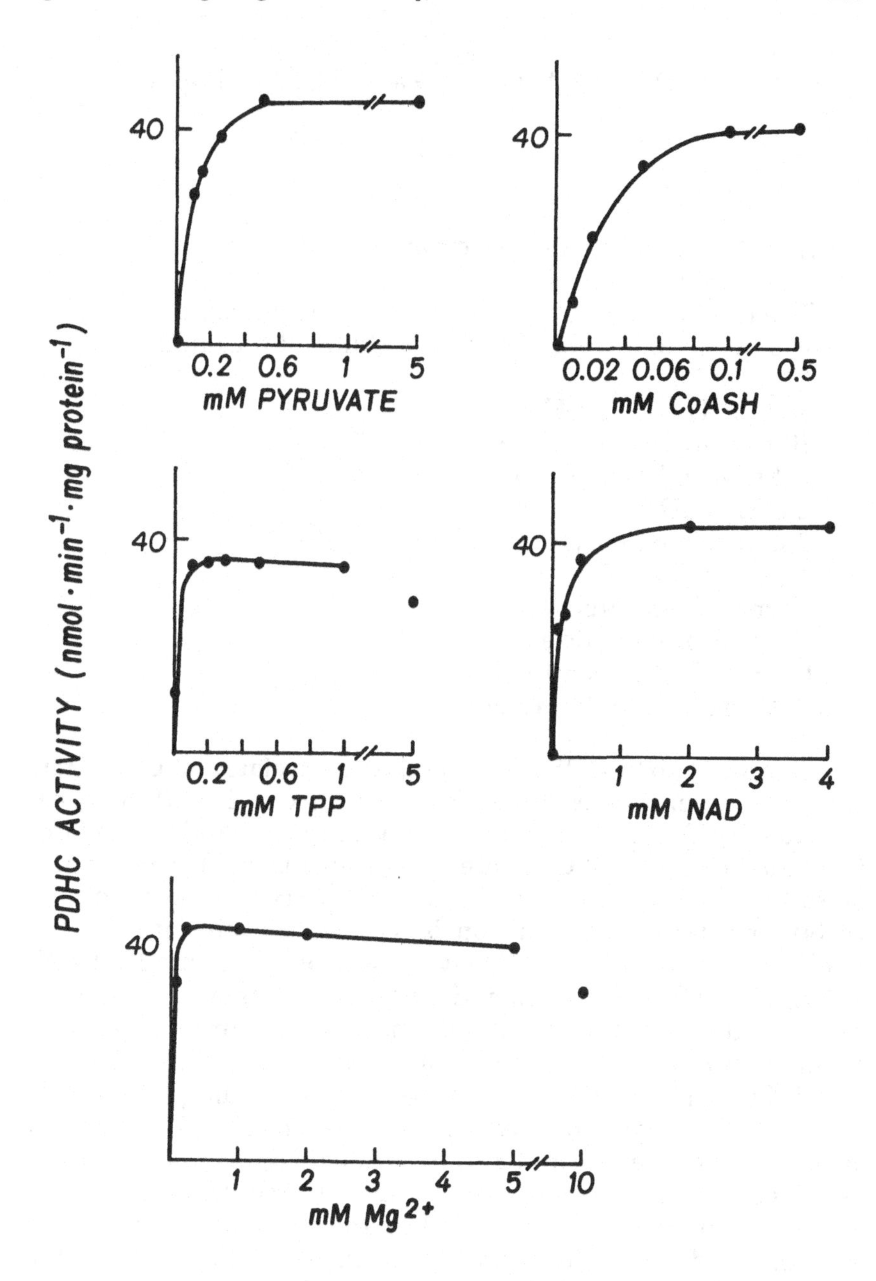

Fig. 3. Optimal substrate and cofactor concentrations in arylamine acetyltransferase coupled assay.

$$\text{pyruvate} + NAD^+ + \text{CoASH} \xrightarrow{\text{PDHC}} \text{acetyl CoA} + \text{NADH} + CO_2 \quad (1)$$

$$\text{acetyl CoA} + \text{oxaloacetate} \xrightarrow{\text{CS}} \text{citrate} + \text{CoASH} \quad (3)$$

The reaction mixture, as described by Szutowicz et al. (1981), consists of the following, in a final volume of 0.25 mL:

50 m*M* Tris-HCl, pH 8.0
10 m*M* sodium pyruvate
0.2 m*M* sodium CoASH
2 m*M* NAD^+
2 m*M* thiamine pyrophosphate
1 m*M* $MgCl_2$
10 m*M* dithiothreitol
2.5 m*M* oxaloacetic acid
0.5 IU of citrate synthetase
0.05 mg of sample protein

Reaction mixture (150 μL) (minus oxaloacetate and CoASH) is placed in Eppendorf test tubes. Freshly prepared 25 m*M* oxaloacetic acid (25 μL) and 50 μL of test sample are added and the mixture incubated 10 min at 37°C in order to activate PDHC. The reaction is started by the addition of 25 μL of 2m*M* CoASH or an equal volume of buffer (blank). The reaction is stopped 30 min later by the addition of 50 μL of $HC10_4$ (20% w/v); samples are neutralized with 2*M* K_2CO_3 (0.02–0.025 mL), and portions of clear supernatant are taken for determination of citrate. The assay mixture for measurement of citrate contains, in a final volume of 0.5 mL: 50 m*M* Tris-HCl, pH 7.6, 0.5–0.8 m*M* NADH (depending on sample volume). Unreacted oxaloacetate is removed by addition of 2 IU of malate dehydrogenase and 2 IU of lactate dehydrogenase. To this is added 0.1 IU of citrate lyase, and the absorbance change is monitored at 334 nm using a recording spectrophotometer. Results are corrected for citrate formation in blank samples. Using the citrate synthetase coupled assay, PDHC activities in homogenates of whole rat brain are reportedly of the order of 23 nmol/min/mg protein (Szutowicz et al., 1981).

3.1.3. Measurement of PDHC by $^{14}CO_2$ Formation from ^{14}C-Pyruvate

The assay is carried out with constant shaking at 37°C in sealed 25-mL Erlenmeyer flasks or test tubes equipped with serum caps and plastic center wells containing filter paper wicks soaked in hyamine hydroxide (0.2 mL). The reaction mixture contains, in a final volume of 1 mL, 48 m*M* phosphate buffer, pH 7.8, 6 m*M* NAD^+, 2 m*M* thiamine pyrophosphate, 0.1 m*M* CoASH, 2 m*M* $CaCl_2$, 9 m*M* dithiothreitol (alternatively, mercaptoethanol has been used) (Jope and Blass, 1976), 125 µg of lactate dehydrogenase, and brain homogenate. The reaction is started by the addition of 10 m*M* (0.2 µCi) 1-^{14}C-pyruvate. *Note:* Aqueous solutions of ^{14}C-pyruvate are unstable, so solutions should be freshly prepared and kept at room temperature; otherwise, treatment with HCl or EDTA is required. It is recommended that radiopurity of ^{14}C-pyruvate be summarily checked prior to use in this assay (Silverstein and Boyer, 1964).

The reaction is stopped after 15 min by injection of 0.2 mL of 6*N* H_2SO_4 into the reaction mixture. Following a further incubation for 30 min, center wells are cut into liquid scintillation vials containing 10 mL of Instagel (Packard) or other appropriate scintillation liquid.

It has been suggested that preincubation with 10 m*M* $MgCl_2$ and 1 m*M* $CaCl_2$ leads to activation of PDHC (Cremer and Teal, 1974; Jope and Blass, 1976). However, others report that such preincubations may result in proteolysis and reduced enzyme activity (Ksiezak-Reding et al., 1982). Preincubation with 50 m*M* dichloroacetate for 30 min reportedly leads to increased PDHC activities in brain slices (Abemayor et al., 1984).

Measurement of PDHC activities by this assay yields, in the absence of added Triton X-100, values for rat brain in the region of 18–24 nmol ^{14}C-pyruvate decarboxylated per min per mg protein (Cremer and Teal, 1974; Jope and Blass, 1976).

3.1.4. PDHC Assays Based on NADH Production

Assay systems based on production of CO_2 from labeled pyruvate or coupled assays in which acetyl CoA production is monitored do not, in the strict sense, measure activities of the total PDHC complex since CO_2 is produced by E_1, and acetyl CoA formation is catalyzed by E_2 (*see* Fig. 1). Measurement of NADH production, on the other hand, does reflect activity of the three-enzyme complex, and monitoring of NADH has been used as an assay for PDHC in brain preparations. In order to measure activity of PDHC in brain homogenates, NADH formed [by reaction (1)] is generally coupled to reduction of a tetrazolium dye with the use of intermediate electron carriers (Hochela and Weinhouse, 1965; Hinman and Blass, 1981). Such a coupling procedure is necessary for NADH measurement in crude homogenates since the latter contain substantital quantities of lactate dehydrogenase, which readily utilize any NADH formed in order to reduce pyruvate to lactate. It has been reported that transfer of reducing equivalents from NADH to the dye in the presence of electron carriers such as phenazine methosulfate or lipoamide dehydrogenase (diaphorase) is both rapid and irreversible even in the presence of pyruvate and lactate dehydrogenase (Hinman and Blass, 1981).

The reaction mixture for assay of PDHC contains:

0.05*M* potassium phosphate buffer, pH 7.8
2.5 m*M* NAD
0.2 m*M* thiamine pyrophosphate
0.1 m*M* CoASH
0.3 m*M* dithiothreitol
1 m*M* $MgCl_2$
0.6 m*M* 2(*p*-iodophenyl)-3-*p*-nitrophenyl-5 phenyl-tetrazolium chloride (INT)
6.5 μ*M* phenazine methosulfate (or lipoamide dehydrogenase 0.1 mg/mL)

The enzymic reaction is initiated by addition of pyruvate (5 m*M*). Absorption at 500 nm is measured using a double-beam recording spectrophotometer at 25°C. Activity is calculated using an extinction coefficient for reduced INT of 12.4/n*M*/cm (Hinman and Blass, 1981). Enzyme activity is expressed as nmol of NADH produced per min per mg protein.

3.2. Measurement of PDHC (Constituent Enzymes)

3.2.1. Pyruvate Dehydrogenase (E_1, EC 1.2.4.1)

Pyruvate dehydrogenase, sometimes also referred to as pyruvate decarboxylase, the first catalytic enzyme of the PDH complex, may be assayed either by measurement of $^{14}CO_2$ production from ^{14}C-pyruvate (*see* section 3.1.3), but with omission of NAD and coenzyme A, or by the colorimetric determination of ferrocyanide produced by oxidative decarboxylation of pyruvate (Reed and Willms, 1966), as represented by the reaction:

$$\text{pyruvate} + 2Fe(CN)_6^{-3} + H_2O \rightarrow \text{acetate} + CO_2 + 2Fe(CN)_6^{-4} + 2H^+ \quad (4)$$

Necessary reagents include the following:

potassium phosphate buffer 1.0*M*, pH. 6.0
thiamine pyrophosphate (TPP) 2 m*M*
$MgSO_4$ 3 m*M*
potassium pyruvate 0.5*M*
potassium ferricyanide 0.25*M*
trichloroacetic acid 10%
sodium lauryl sulfate 4%
ferric ammonium sulfate (Prussian Blue) reagent, prepared by dissolving 1.7 g of ferric ammonium sulfate in 10 mL of distilled water, filtering, and adding the filtrate to 20 mL of a solution containing 1.5 g of sodium lauryl sulfate in 20 mL of water. To this solution is added 27 mL of 85% H_3PO_4, and the mixture is diluted to 140 mL with water.

A mixture containing 0.15 mL of phosphate buffer, 0.1 mL of TPP, 0.1 mL of $MgCl_2$, 0.1 mL of potassium pyruvate, 0.1 mL of potassium ferricyanide, and appropriate quantities of homogenate is diluted to 1.4 mL with water. Control mixtures contain all ingredients except homogenate. The mixture is incubated at 30°C for 30 min and the reaction is terminated by addition of 1 mL of 10% trichloroacetic acid. The mixture is centrifuged, and 0.2 mL of supernatant is added to a mixture of 1 mL of trichloroacetic acid and 0.1 mL of potassium ferricyanide followed by water to a final volume of 2.4 mL. To this solution is added 1 mL of 4% sodium lauryl sulfate and 0.5 mL of ferric ammonium sulfate reagent. The mixture is allowed to stand 30 min and optical density then determined at 540 nm. Standard curves for potassium ferrocyanide are prepared. Enzyme activity is expressed as the amount of enzyme required to produce 2 μmol of ferrocyanide per min per mg protein.

The ferrocyanide method has also been used for measurement of the total PDH complex (with inclusion of NAD and CoASH). However, the assay cannot be used in the presence of reducing agents such as dithiothreitol because of nonenzymatic reduction of ferricyanide (Kresze, 1979).

3.2.2. Dihydrolipoyl Transacetylase (E_2, EC 2.3.1.12)

The assay for E_2 is generally based on determination of acetyl dihydrolipoamide generated by the enzyme according to the reaction:

$$\text{Lip (SH)}_2 + \text{acetyl CoA} \xrightarrow{E_2} \text{acetyl-S-Lip SH} + \text{CoASH} \qquad (5)$$

Acetyl CoA is generated *in situ* from acetyl phosphate, CoASH, and phosphotransacetylase (Reed and Willms, 1966), and the acetyl group is transferred to dihydrolipoamide by E_2. The S-acetyldihydrolipoamide product is then determined as the ferric acethydroxamate complex.

The reaction mixture contains 0.1 mL of Tris buffer (1*M*), pH 7.0, 0.1 mL of acetyl phosphate (0.1*M*), 5 IU of phosphotransacetylase, 0.1 mL of coenzyme A, the enzyme source or homogenate (omit for blanks), and distilled water to a final volume of 0.95 mL. A 0.05-mL aliquot of substrate (dihydrolipoamide 0.2*M*) is added, and mixtures are incubated for 30 min at 30°C. The reaction is stopped by the addition of 0.1 mL of HCl (1N), and the mixture heated to boiling for 5 min to destroy unreacted acetyl phosphate. After cooling, 1 mL of hydroxylamine solution (2*M*), pH 6.4, is added, and the mixture stored at room temperature for 10 min. A 3mL $FeCl_3$ solution containing equal volumes of 3N HCl, trichloroacetic acid (12% v/v), and 5% $FeCl_3$ in 0.1N HCl are added, and the mixture is centrifuged. Optical density of the supernatant is read at 540 nm. Enzyme activity is expressed as the amount required to produce 1 μmol of acetyl-S-LipSH per min per mg protein.

3.2.3. Dihydrolipoyl Dehydrogenase (E_3, EC 1.6.4.3)

This third catalytic enzyme of the PDH complex, often referred to as lipoamide dehydrogenase, is generally assayed spectrophotometrically by monitoring loss of NADH at 340 nm produced in the reaction:

$$\text{lipoamide} + \text{NADH} + \text{H}^+ \rightarrow \text{dihydrolipoamide} + \text{NAD}^+ \qquad (6)$$

Reactions are typically performed in phosphate buffer 0.3*M*, pH 8.1, containing NADH (0.2 m*M*) at 37°C. The decrease in absorbance at 340 nm is measured in a double-beam spectrophotometer following addition of substrate (lipoamide, 2.5 m*M*) to the test cuvet (Reed and Willms, 1966; Robinson et al., 1980). Enzyme activity is expressed as the amount of enzyme required to catalyze oxidation of 1 μmol NADH per min per mg protein.

3.2.4. PDHC Phosphate Phosphatase (EC 3.1.3.43)

Few attempts have been made to directly measure the activity of the regulatory PDHC phosphate phosphatase in mammalian brain. One such assay depends on measurement of the rate of activation of exogenous beef kidney phospho-PDHC (Sheu et al., 1983). In order to do this, PDHC is first purified to near homogeneity from bovine kidney as described by Machicao and Wieland (1980). Phospho-PDHC is then prepared by phosphorylation with MgATP as described by Sheu et al. (1983). The incubation mixture contains in 1 mL, 50 m*M* Tris HCl (pH 7.8), 1 m*M* dithiothreitol, 0.5 m*M* $MgCl_2$, 0.5 m*M* ATP, and 30–50 U of PDHC. Inactivation is allowed to take place at room temperature and is monitored by withdrawal of aliquots of the mixture at various times for measurement of residual PDHC activity. After 30 min, when >98% of PDHC is inactivated, the reaction is stopped by adding 1 m*M* EDTA followed by dialysis in 20 m*M* 3-(*N*-morphilino)propanesulfonic acid (MOPS), pH 7.2, 0.5 m*M* 2-mercaptoethanol, and 0.5 m*M* EDTA for 4 h. Aliquots of phospho-PDHC thus prepared are stable at –80°C for 2 wk or more (Sheu et al., 1983).

PDHC phosphate phosphatase is then assayed in brain homogenates or mitochondrial preparations by measurement of the rate of activation of added phospho-PDHC. The reaction mixture at 30°C contains, in a final volume of 30 μL, 75 m*M* MOPS (Tris), pH 7.2, 10 m*M* $MgCl_2$, 0.67 m*M* $CaCl_2$, 0.33 m*M* 2-mercaptoethanol, 0.12% (w/v) Triton X-100, and 0.1–0.5 mg of homogenate protein/mL. To this mixture, following preincubation for 2.5 min, is added 6–9 U/mL of phospho-PDHC to initiate the reaction. Aliquots (2–5 μL) are withdrawn at various times and assayed for PDHC activity. The rate of phospho-PDHC activation is then determined from the slope of the linear regression of PDHC activity versus incubation time and is expressed as mU PDHC activated per min per mg protein.

3.2.5. PDHC Kinase (EC 2.7.2.99)

PDHC kinase may be assayed in brain homogenates by monitoring the time course of the ATP-dependent inactivation of

PDHC after the latter has been fully activated by treatment with 5 m*M* $MgCl_2$ and 0.1 m*M* $CaCl_2$ as described by Sheu et al. (1984). Brain samples are homogenized in 9 vol of 20 m*M* ice-cold Tris buffer, pH 7.2, containing 50 μM leupeptin and 0.2 m*M* mercaptoethanol. To this is added (final conc.) 5 m*M* $MgCl_2$, 0.1 m*M* $CaCl_2$, and 0.1% Triton X-100. Homogenates are incubated 10–15 min at room temperature to activate PDHC and then stored in aliquots at –80°C. Quantities (20 μL) of homogenate are then pipeted into spectrophotometer cuvets previously equilibrated at 30°C and the PDHC kinase reaction initiated by addition of 50 μL of the following mixture: 40 m*M* Tris containing MOPS (pH 7.2, 2 m*M* ATP, 0.5 m*M* EGTA, 40 m*M* KCl, 2 m*M* NaF, 1.4 m*M* $MgCl_2$, 0.03 m*M* $CaCl_2$, 0.1% bovine serum albumin, and 0.1% Triton X-100. After a measured time (usually of the order of a few minutes or less), the reaction is quenched by addition of 0.8 mL of PDHC assay mixture as described in section 3.1.1.2. (or as described by Sheu et al., 1984).

The residual PDHC activity is then monitored at 460 nm to follow acetylation of AABS as described in section 3.1.1. The activity of PDHC kinase is defined as the first-order rate constant (K_{ATP})dependent inactivation of PDHC thus:

$$K_{ATP} = \frac{\ln (Ao/At)}{t}$$

where A_o and A_t are the initial and final PDHC activities at time = 0 and t seconds, respectively (Sheu et al., 1984).

4. Regional and Subcellular Localization of PDHC in Brain

4.1. Regional Distribution of PDHC Enzymes

PDHC activities in hippocampus and cerebral cortex have consistently been found to be higher than in phylogenetically older brain structures such as medulla oblongata and pons (Szutowicz et al., 1981; Butterworth and Giguere, 1984; Leong and Clarke, 1984; Sheu et al., 1984; *see* Table 2). PDHC activities in white matter are substantially less than those observed in gray matter structures, reflecting the relatively greater cerebral glucose utilization of gray matter structures. The regulatory kinase and phosphatase enzymes show similar distribution patterns in the CNS with cerebral

Table 2
Distribution of PDHC Activities in Regions of Rat CNS

CNS region	PDHC activity, (nmol/min/mg protein)
Olfactory bulb	26.93 ± 2.38
Cerebral cortex	46.73 ± 1.65
Hippocampus	52.24 ± 1.54
Striatum	46.13 ± 1.60
Midbrain	40.16 ± 1.03
Hypothalamus	39.70 ± 1.39
Cerebellum	37.81 ± 1.66
Medulla oblongata	26.91 ± 1.33
Pons	32.01 ± 1.99
Spinal cord (gray)	27.02 ± 2.48
Spinal cord (white)	7.50 ± 1.12
(Whole brain)	(45.00 ± 0.91)

cortical and hippocampal values being in excess of enzyme activities in other brain regions (Sheu et al., 1984).

Maximal values for PDHC activity in whole rat brain obtained in vitro with the use of Triton X-100 have generally been found to be in the range of 45 nmol/min/mg protein (Ksiezak-Reding et al., 1982; Butterworth and Giguere, 1984; Table 2). Such values for PDHC activity are approximately equal to (or slightly superior to) maximal rates of pyruvate flux in whole brain observed during seizures (i.e., 41 nmol pyruvate/min/mg protein; Siesjo, 1978). It has been suggested that the value of PDHC activity of 45 nmol/min/mg protein may therefore represent the enzyme's true maximum activity in rat brain (Ksiezak-Reding et al., 1982).

Results of a previous study suggested that the percent PDHC in its active form under normal physiological conditions was of the order of 14 nmol/min/mg protein, i.e., just in excess of normal pyruvate flux through PDHC, suggesting that PDHC may be rate-limiting (Cremer and Teal, 1974; Jope and Blass, 1976; Stumpf and Kraus, 1979; Butterworth, 1985). Furthermore, there is evidence to suggest regional variations in the ratio of PDHC to pyruvate flux in the CNS (Reynolds and Blass, 1976), thus rendering certain brain structures particularly vulnerable to inherited and acquired disorders of PDHC (*see* section 6).

4.2. Subcellular Localization of PDHC in Brain

Highest PDHC activities have consistently been found in mitochondrial fractions from mammalian brain. Cytosolic or synaptoplasmic fractions show very low or undetectable activities of PDHC (Szutowicz et al., 1981; Ksiezak-Reding et al., 1982; Sheu et al., 1983). Enzyme activities expressed per milligram of mitochondrial protein are higher in nonsynaptic mitochondria than in mitochondria from synaptosomes, and the significance of these findings has been discussed in terms of the different metabolic activities of the two mitochondrial populations and compartmentation of cerebral metabolism (Leong et al., 1984).

5. PDHC and Brain Development

Developmental changes in PDHC activity in regions of the rat brain are shown in Fig. 4. Thus, 3 d after birth, PDHC activities are of the order of 10% adult values in cerebral cortex, hippocampus, midbrain, straitum, and cerebellum, but somewhat higher, of the order of 25%, in hypothalamus and medulla-pons. Enzyme activity in medulla-pons continues to be more highly developed at all times up to weaning, being some 5–15 d ahead of other brain structures in this regard. The order of development of PDHC parallels the neurophylogenetic development of brain structures, which proceeds from the medulla rostrally, and it has been suggested that the development of morphological and neurological maturity in the different brain regions may relate directly to the development of the potential for aerobic glycolysis and, particularly, to development of PDHC (Leong and Clark, 1984). Thus, cerebral PDHC activities in species born neurologically mature (such as guinea pig) are fully developed at birth, whereas PDHC activities in species born neurologically immature (such as rat) are poorly developed (Booth et al., 1980). The delayed developmental profile for PDHC may also relate to the changeover of energy substrate for brain during the developmental period from one of glucose and ketone bodies to one of glucose alone that occurs in rat and human brain (Leong and Clark, 1984). This change of energy substrate and the role of PDHC is depicted schematically in Fig. 5.

Of particular interest is the confirmed observation of delayed maturation of PDHC in rat cerebellum (Butterworth and Giguere, 1984; *see* Fig. 4; Leong and Clark, 1984), which by the time of

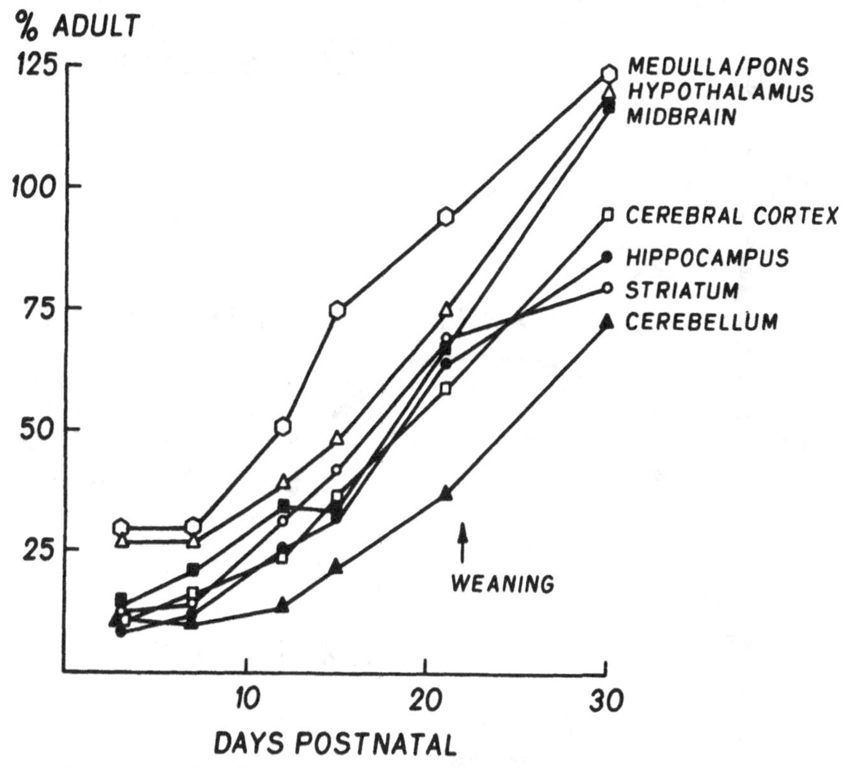

Fig. 4. Regional development of PDHC in rat brain.

weaning has attained only some 40% of adult levels of activity. It has been suggested that this developmental lag in cerebellar PDHC may relate to delayed maturation of cerebellar function in the rat (Butterworth and Giguere, 1984). In support of such a contention, studies have shown that maturation of cerebellar GABA neurons is not complete until 60 d of age in the rat (Gilad and Kopin, 1979), and establishment of parallel fiber synapses and associated high-affinity glutamate binding sites in rat cerebellum is not complete until the fourth postnatal week (deBarry et al., 1980).

6. PDHC Deficiency Disorders

6.1. Inherited Disorders of PDHC

In the last two decades, inherited neurological disorders have been identified in which there is convincing evidence for genetic

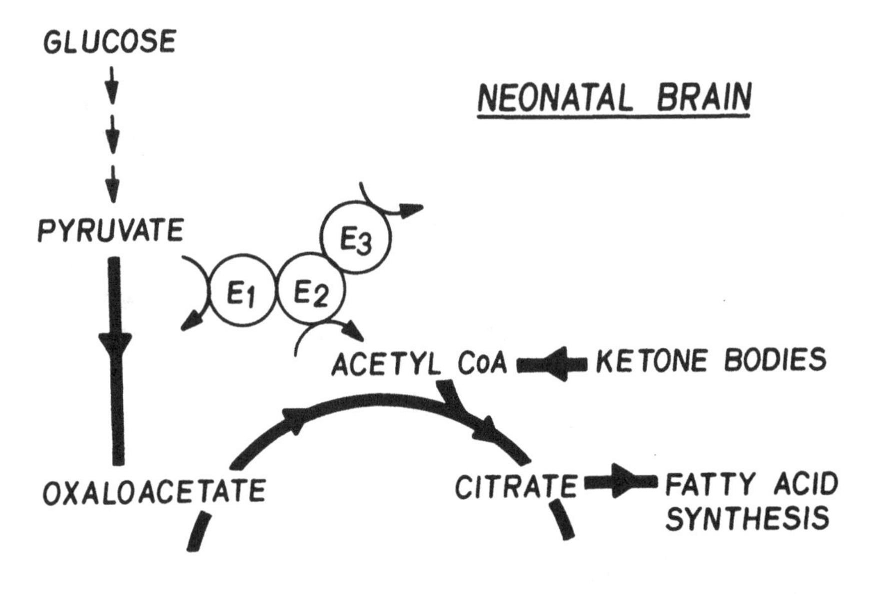

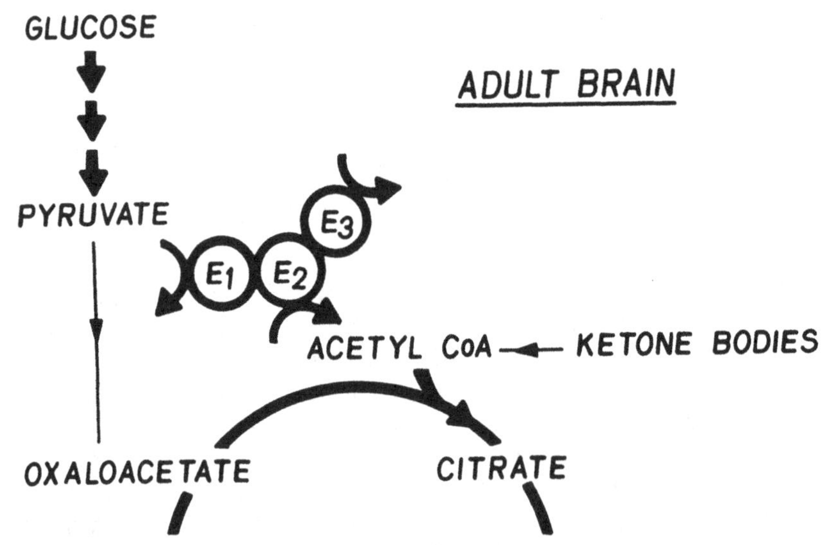

Fig. 5. Role of PDHC in the changeover of cerebral energy substrate during development.

abnormalities of PDHC or one of its constituent enzymes (*see* reviews by Blass, 1981; Butterworth, 1985). Generally, the more severe the PDHC defect is, the more severe is the accompanying clinical illness, the earlier its onset, the more rapid its progression, and the more widespread the CNS damage. Thus, patients with less than 15–20% of normal PDHC activity generally present with "lactic acidosis of infancy" and severe neurological deficits, whereas patients with 40–50% residual PDHC activity have a milder illness in which intermittent ataxia is the most promonent neurologial symptom. Specific, biochemically proven abnormalities of the E_1, E_2, or E_3 components of PDHC have been identified (*see* review by Butterworth, 1985). In addition, inherited defects of PDHC *regulation* have been reported in association with Leigh disease (subacute necrotizing encephalomyelopathy) (Devivo et al., 1979; Butterworth, 1982; Sorbi and Blass, 1982; Toshima et al., 1982).

6.2. Acquired Disorders of PDHC

Measurement of PDHC in liver homogenates from arsenic-intoxicated rats leads to decreased activities of PDHC both before and after in vitro activation with Mg^{2+}. It was suggested that the toxic effects of arsenic may be mediated by the regulatory enzymes PDHC kinase or phosphatase (Schiller et al., 1977). There is also indirect evidence to suggest that impairment of cerebral PDHC may mediate the neurotoxic effects of methyl mercury and of certain trialkyltin compounds (Cremer, 1970; *see* review by Butterworth, 1985).

Until recently it was generally accepted that thiamine deficiency resulted in decreased activity of PDHC in brain and that this played an important pathogenic role in thiamine-deficiency encephalopathy. However, although blood pyruvate levels are frequently increased as a result of thiamine deficiency, altered cerebral PDHC activities have been more difficult to demonstrate. Alterations of PDHC in the brain of chronically thiamine-deprived rats are confined to brainstem and are relatively small (of the order of 25%). Furthermore such changes follow rather than precede the neurological signs and neuropathological damage associated with chronic thiamine deprivation (Butterworth et al., 1985; Butterworth, 1986). Treatment of rats with the central thiamine antagonist pyrithiamine does not lead to alterations of cerebral PDHC (Gibson et al., 1984; Butterworth et al., 1985).

6.3. Pathophysiology of PDHC Deficiency Disorders

Normal cerebral function in the adult requires, except during prolonged starvation or ketoacidosis, an adequate supply of glucose for ATP synthesis. Since currently available evidence suggests that PDHC may be rate-limiting for cerebral glucose utilization,

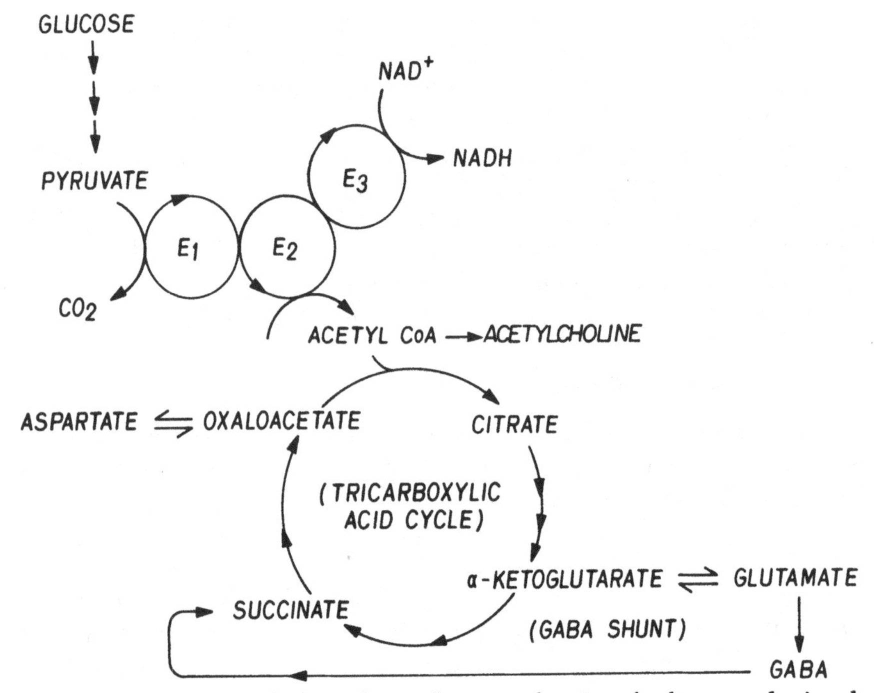

Fig. 6. Role of PDHC in the synthesis of glucose-derived neurotransmitters (acetylcholine, GABA, glutamate, aspartate).

inherited or acquired deficiencies of PDHC, if sufficiently severe, are likely to have an adverse effect on cerebral energy metabolism. Furthermore, synthetic pathways involving synthesis of glucose-derived neurotransmitters may be particularly susceptible to mild impairments of PDHC. The acetyl moiety of acetylcholine is normally derived from pyruvate (Fig. 6). Thus, impaired pyruvate oxidation might be expected to cause a reduction in acetylcholine synthesis. Indeed, studies have shown that a number of substances that inhibited conversion of ^{14}C-pyruvate to $^{14}CO_2$ in rat brain preparations also led to a corresponding decrease in conversion of ^{14}C-pyruvate to acetylcholine (Gibson et al., 1975). Included in the list of substances tested were the irreversible PDHC inhibitor 3-bromopyruvate and the competitive PDHC inhibitor 2-oxobutyrate.

There is evidence to suggest that the neurotransmitter pools of GABA as well as glutamate and aspartate are synthesized from glucose via pyruvate (Fig. 5). Incorporation of glucose into these amino acids is sensitive to changes in physiological function, whereas incorporation from other precursors is not. For example, anesthetic doses of sodium pentobarbitone lead to decreased incorporation of ^{14}C-glucose into amino acids, but have no effect on incorporation of ^{14}C-acetate or ^{14}C-butyrate (Cremer and Lucas, 1971). Electrical stimulation of guinea pig cortical slices evokes a selective tetrodotoxin-sensitive release of amino acids newly synthesized from ^{14}C-glucose (Potashner, 1978). Impairment of cerebral glucose metabolism results in reductions of GABA, glutamate, and aspartate in brain (*see*, for example, Butterworth, 1982).

Another mechanism by which PDHC deficiency may affect cerebral function is by alteration of calcium homeostasis. There is considerable evidence available to suggest that mitochondrial metabolism can influence neurotransmitter release by regulation of calcium levels in nerve terminals (Alnaes and Rahaminoff, 1975). Calcium accumulation by brain mitochondria appears to be tightly linked to PDHC activity, the process being mediated by the phosphorylation state of the enzyme complex. Dichloroacetate, the PDHC kinase inhibitor, stimulates pyruvate-supported calcium accumulation at concentrations at which it stimulates cerebral PDHC activity (Browning et al., 1981). Yet another recent report suggests that PDHC regulation may play an important role in the modulation of synaptic plasticity involved in learning and memory function (Morgan and Routtenberg, 1981).

7. Summary

PDHC is a multienzyme complex responsible for the irreversible transformation of pyruvate into acetyl CoA. The enzyme complex consists of three catalytic and two regulatory enzymes. The activation state of PDHC depends in the brain, as in other mammalian tissues, on the energy charge, but, in contrast with peripheral tissues, it is unaffected by starvation. Measurement of cerebral PDHC may be performed by:

1. coupling of acetyl CoA formed to either citrate synthesis or acetylation of AABS,
2. measurement of $^{14}CO_2$ formed from ^{14}C-pyruvate,
3. measurement of NADH formed.

Modifications to these techniques afford assay systems for the constituent enzymes of PDHC. With inclusion of detergents such as Triton X-100, maximal activities of PDHC in homogenates of adult whole rat brain are found to be of the order of 45 nmol/min/mg protein.

Maximal cerebral PDHC activities measured in vitro are of a similar magnitude to estimates of maximal pyruvate oxidation, suggesting that PDHC may be rate-limiting in mammalian brain.

Cerebral PDHC activities increase during the postnatal period, reflecting the changeover of cerebral energy substrate from ketone bodies to glucose. PDHC develops in a region-selective manner, in parallel with the establishment of adult patterns of brain metabolism and regional maturation of cerebral function.

Inherited neurological diseases involving genetic mutations of each of the constituent enzymes of PDHC have been described.

Decreases of cerebral PDHC may be associated not only with impairments of energy metabolism, but also with reduced synthesis of the glucose-derived neurotransmitters acetylcholine, GABA, glutamate, and aspartate. In addition, there is evidence to suggest that PDHC may play a role in neurotransmitter release mechanisms as well as in synaptic plasticity.

8. Acknowledgments

The author is grateful to Françoise Trotier and Diane Mallette for their assistance with the preparation of this manuscript. The author's research laboratory is funded by The Medical Research Council of Canada, Fonds de la Recherche en Santé du Québec, The Savoy Foundation, and La Fondation de l'Hôpital Saint-Luc.

References

Abemayor E., Kovachich G. B., and Haugaard N. (1984) Effects of dichloroacetate on brain pyruvate dehydrogenase. *J. Neurochem.* **42,** 38–42.

Alnaes E. and Rahaminoff R. (1975). On the role of mitochondria in trasmitter release from motor nerve terminals. *J. Physiol.* **248,** 285–306.

Barrera C. R., Namihira G., Hamilton L., Munk P., Eley M. H., Linn T. C., and Reed L. J. (1972) α-Ketoacid dehydrogenase complexes from bovine kidney and heart. *Arch. Biochem. Biophys.* **148,** 343–358.

Blass J. P. (1981) Hereditary Ataxias, in *Current Neurology* vol. III (Appel S. H., ed.) John Wiley, New York.

Blass J. P. and Lewis C. A. (1973) Kinetic properties of the partially purified pyruvate dehydrogenase complex from ox brain. *Biochem. J.* **131,** 31–37.

Booth R. F. G. and Clark J. B. (1978) The control of pyruvate dehydrogenase in isolated brain mitochondria. *J. Neurochem.* **30,** 1003–1008.

Booth R. F. G., Patel T. B., and Clark J. B. (1980) The development of enzymes of energy metabolism in the brain of a precocial (guinea pig) and non-precocial (rat) species. *J. Neurochem.* **34,** 17–25.

Browning M., Beaudry M., Bennett W. F., and Lynch G. (1981) Phosphorylation mediated changes in pyruvate dehydrogenase activity influence pyruvate-supported calcium accumulation by brain mitochondria. *J. Neurochem.* **36,** 1932–1940.

Butterworth R. F. (1982) Neurotransmitter function in thiamine-deficiency encepahlopathy. *Neurochem. Int.* **4,** 449–464.

Butterworth R. F. (1985) Pyruvate Dehydrogenase Deficiency Disorders, in *Cerebral Energy Metabolism and Metabolic Encephalopathy* (McCandless D. W., ed.) Plenum, New York.

Butterworth R. F. (1986) Cerebral thiamine-dependent enzyme changes in experimental Wernicke's encephalopathy. *Metab. Brain Dis.* **1,** 165–175.

Butterworth R. F. and Giguere J. F. (1984) Pyruvate dehydrogenase activity in regions of the rat brain during postnatal development. *J. Neurochem.* **43,** 280–282.

Butterworth R. F., Giguere J. F., and Besnard A. M. (1985) Activities of thiamine-dependent enzymes in two experimental models of thiamine-deficiency encephalopathy. 1. The pyruvate dehydrogenase complex. *Neurochem. Res.* **10,** 1417–1428.

Cremer J. E. (1970) Selective inhibition of glucose oxidation by triethyltin in rat brain in vivo. *Biochem. J.* **119,** 95–102.

Cremer J. E. and Lucas H. M. (1971) Sodium pentobarbitone and metabolic compartments in rat brain. *Brain Res.* **35,** 619–621.

Cremer J. E. and Teal H. M. (1974) The activity of pyruvate dehydrogenase in rat brain during postnatal development. *FEBS Lett.* **39,** 17–20.

deBarry J., Vincedon G., and Gombos G. (1980) High affinity glutamate binding during postnatal development of rat cerebellum. *FEBS Lett.* **109,** 175–179.

DeVivo D. C., Haymond M. W., Obert K. A., Nelson J. S., and Pagliara A. S. (1979) Defective activation of the pyruvate dehydrogenase complex in subacute necrotizing encephalomyelopathy (Leigh disease). *Ann. Neurol.* **6,** 483–494.

Gibson G. E., Jope R., and Blass J. P. (1975) Decreased synthesis of acetylcholine accompanying impaired oxidation of pyruvic acid in rat brain minces. *Biochem. J.* **148,** 17–23.

Gibson G. E., Ksiezak-Reding H., Sheu K.F.R., Mykytyn V., and Blass J. P. (1984) Correlation of enzymatic, metabolic and behavioural deficits in thiamine deficiency and its reversal. *Neurochem. Res.* **9,** 803–814.

Gilad G. M. and Kopin I. J. (1979) Neurochemical aspects of neuronal ontogenesis in the developing rat cerebellum: Changes in neurotransmitter and polyamine synthesizing enzymes. *J. Neurochem.* **33,** 1195–1204.

Hinman L. M. and Blass J.P. (1981) An NADH-linked spectrophotometric assay for pyruvate dehydrogenase complex in crude tissue homogenates. *J. Biol. Chem.* **256,** 6583–6586.

Hochella N J. and Weinhouse S. (1965) Automated assay of lactate dehydrogenase in urine. *Anal. Biochem.* **13,** 322–335.

Jope R. and Blass J. P. (1976) The regulation of pyruvate dehydrogenase in brain *in vivo*. *J. Neurochem.* **26,** 709–714.

Kresze G. B. (1979) An improved procedure for the assay of pyruvate dehydrogenase. *Anal. Biochem.* **98,** 85–88.

Ksiezak-Reding H., Blass J. P., and Gibson G. E. (1982) Studies on the pyruvate dehydrogenase complex in brain with the arylamine acetyltransferase coupled assay. *J. Neurochem.* **38,** 1627–1636.

Lai J.C.K. and Clark J. B. (1979) Preparation of synaptic and non-synaptic mitochondria from mammalian brain. *Meth. Enzymol.* **55,** 51–60.

Leong S. F. and Clark J. B. (1984) Regional enzyme development in rat brain. Enzymes of energy metabolism. *Biochem. J.* **218,** 139–145.

Leong S. F., Lai J.C.K., Lim L., and Clark J. B. (1984) The activities of some energy-metabolizing enzymes in non-synaptic (free) and synaptic mitochondria derived from selected brain regions. *J. Neurochem.* **42,** 1306–1312.

Machicao F. and Wieland O. H. (1980) Subunit structure of dihydrolipoamide acetyltransferase component of pyruvate dehydrogenase complex from bovine kidney. *Hoppe Seylers Z. Physiol. Chem.* **361,** 1093–1106.

Morgan D. G. and Routtenberg A. (1981) Brain pyruvate dehydrogenase: Phosphorylation and enzyme activity altered by a training exercise. *Science* **214,** 470–471.

Potashner S. J. (1978) Effects of tetrodotoxin, calcium and magnesium on the release of amino acids from slices of guinea pig cerebral cortex. *J. Neurochem.* **31,** 187–195.

Reed L. J. and Willms C. R. (1966) Purification and resolution of the pyruvate dehydrogenase complex *(Escherichia coli)*. *Meth. Enzymol.* **9,** 247–278.

Reynolds S. F. and Blass J. P. (1976) A possible mechanism for selective cerebellar damage in partial pyruvate dehydrogenase deficiency. *Neurology* **26,** 625–628.

Robinson B. H., Taylor J., and Sherwood W. G. (1980) The genetic heterogeneity of lactic acidosis: occurrence of recognizable inborn errors of metabolism in a pediatric population with lactic acidosis. *Pediatr. Res.* **14,** 956–962.

Schiller C. M., Fowler B. A., and Woods J. S. (1977) Effects of arsenic on pyruvate dehydrogenase activation. *Environ. Health Perspec.* **19,** 205–207.

Sheu K.F.R., Lai J.C.K., and Blass J. P. (1983) Pyruvate dehydrogenase phosphate (PDH_b) phosphatase in brain: Activity, properties and subcellular localization. *J. Neurochem.* **40,** 1366–1372.

Sheu K.F.R., Lai J.C.K., and Blass J.P. (1984) Properties and regional distribution of pyruvate dehydrogenase kinase in rat brain. *J. Neurochem.* **42,** 230–236.

Siesjo B. K. (1978) Epileptic Seizures, in *Brain Energy Metabolism* (Siesjo B. K., ed.) John Wiley, New York.

Silverstein E. and Boyer P. D. (1964) Instability of pyruvate-^{14}C in aqueous solution as detected by enzymic assay. *Anal. Biochem.* **8,** 470–476.

Sorbi S. and Blass J. P. (1982) Abnormal activation of pyruvate dehydrogenase in Leigh disease fibroblasts. *Neurology* **32,** 555–558.

Stumpf B. and Kraus H. (1979) Regulatory aspects of glucose and ketone body metabolism in infant rat brain. *Pediat. Res.* **13,** 585–590.

Szutowicz A., Stepien M., and Piec G. (1981) Determination of pyruvate dehydrogenase and acetyl CoA synthetase activities using citrate synthetase. *Anal. Biochem.* **115,** 81–87.

Tabor H., Mehler A. H., and Stadtman E. R. (1953) The enzymatic acetylation of amines. *J. Biol. Chem.* **204,** 127–138.

Toshima K., Kuroda Y., and Hashimoto T. (1982) Enzymologic studies and therapy of Leigh disease associated with pyruvate decarboxylase deficiency. *Pediatr. Res.* **16,** 430–435.

CO_2-Fixing Enzymes

Mulchand S. Patel

1. Introduction

CO_2 fixation by the nervous system has been demonstrated with the use of ^{14}C-labeled bicarbonate in the retina (Crane and Ball, 1951), in the brain of 1-d-old mice (Moldave et al., 1953), and in the adult cat brain (Berl et al., 1962a,b). When cat brain was perfused with labeled bicarbonate and the specific radioactivity of $^{14}CO_2$ was maintained at a constant level, most of the radioactivity fixed in the brain was found in free aspartate, glutamate, and glutamine, with a small amount in lactate (Berl et al., 1962a,b; Otsuki et al., 1963; Waelsch et al., 1964). Aspartate had the highest specific radioactivity, which approached about 10% of that of infused radioactive bicarbonate. When ammonia was administered together with $H^{14}CO_3^-$ to cats, the total amount of $^{14}CO_2$-fixation into the acid-soluble fraction of the brain was increased, and the specific radioactivity of glutamine increased without any change in the specific radioactivities of aspartate and glutamate (Waelsch et al., 1964). It was suggested that the fixation of $H^{14}CO_3^-$ in vivo occurred at the level of oxalacetate or malate (Otsuki et al., 1963), and that the $^{14}CO_2$-fixation in the brain played an important role in the replenishment of dicarboxylic acid needed for the formation of glutamine as a result of stress caused by high concentrations of ammonia (Waelsch et al., 1964). By comparing the relative amounts of radioactivity incorporated from [1-^{14}C]pyruvate and [2-^{14}C]pyruvate in aspartate and citrate in rat striatum slices, it was estimated that the relative proportion of pyruvate utilization via the CO_2-fixation pathway compared with that via the acetyl-CoA pathway (pyruvate decarboxylation pathway) was approximately 1 to 10 (Cheng et al., 1967). The fixation of CO_2 in peripheral nerves has also been demonstrated (Waelsch et al., 1965; Naruse, et al., 1966a,b; Cote et al., 1966; Cheng et al., 1967).

It has been demonstrated that CO_2-fixation in the mammalian brain can be mediated by both biotin-dependent and biotin-independent enzymatic reactions (Table 1; Fig. 1). The purpose of this chapter is to describe these "CO_2-fixing enzymes" in the brain in some detail and to discuss their possible significance in brain

Table 1
CO_2-Fixing Enzymes in Mammalian Brain

Enzyme	Reaction catalyzed	Cellular location	Metabolic process
Biotin-dependent CO_2 fixation			
Pyruvate carboxylase	Pyruvate → oxalacetate	Mitochondria	Anaplerotic reaction and lipogenesis[a]
Acetyl-CoA carboxylase	Acetyl-CoA → malonyl-CoA	Cytosol	Lipogenesis[a]
Propionyl-CoA carboxylase	Propionyl-CoA → methylmalonyl-CoA	Mitochondria	Propionate metabolism, catabolism of several amino acids
β-Methylcrotonyl-CoA carboxylase	β-Methylcrontonyl-CoA → β-methylglutaconyl-CoA	Mitochondria	Catabolism of leucine
Biotin-independent CO_2 fixation			
NADP-Malate dehydrogenase	Malate ⇌ pyruvate	Cytosol, mitochondria	Lipogenesis[a]
NADP-Isocitrate dehydrogenase	Isocitrate ⇌ α-ketoglutarate	Cytosol, mitochonrial	Unknown (possibly lipogenesis[a])
Phosphoenolpyruvate carboxykinase	Oxalacetate ⇌ phosphoenolpyruvate	Mitochondria	Unknown

[a]During the developmental period.

metabolism. Since these seven enzymes in the brain are not extensively investigated as compared with these enzymes from other tissues, relevant information on noncerebral enzymes is also presented when necessary.

2. Biotin-Dependent CO_2 Fixation

2.1. *Pyruvate Carboxylase*

Pyruvate carboxylase (pyruvate:CO_2 ligase[ADP], E.C. 6.4.1.1) catalyzes the formation of oxalacetate from pyruvate involving two partial reactions as shown below:

$$\text{(i) E-biotin} + \text{MgATP} + HCO_3^- \underset{\text{acetyl-CoA}}{\overset{Mg^{2+}}{\rightleftharpoons}} \text{E-biotin-}CO_2 + \text{MgADP} + \text{Pi}$$

$$\text{(ii) E-biotin-}CO_2 + \text{pyruvate} \rightleftharpoons \text{oxalacetate} + \text{E-biotin}$$

$$\text{Sum: Pyruvate} + \text{MgATP} + HCO_3^- \rightleftharpoons \text{oxalacetate} + \text{MgADP} + \text{Pi}$$

Pyruvate carboxylase is assayed by measuring the fixation of $H^{14}CO_3^-$ in the presence of pyruvate, ATP, $MgCl_2$, and acetyl-CoA (Utter and Keech, 1963; Ballard and Hanson, 1967b). Rat brain is homogenized (10 or 20% w/v) in ice-cold buffered sucrose medium (0.25*M* sucrose, 5 m*M* triethanolamine buffer, pH 7.4, or another suitable isotonic buffered solution) at 500 rpm for 1 min in a glass homogenizer fitted with a Teflon pestle. Aliquots of homogenates, or the crude mitochondrial fraction obtained by differential centrifugation, can be used to assay mitochondrial enzyme activities. Since pyruvate carboxylase is localized in the matrix of mitochondria, these organelles must be disrupted prior to the measurement of enzyme activity. This can be easily achieved by freeze-drying followed by resuspension in water, sonification, or treatment with Triton X-100 (final concentration up to 0.5%). The solubilized preparation may then be centrifuged at 100,000*g* for 30 min, and the resulting supernatant is assayed for enzyme activities. The reaction mixture for the measurement of pyruvate carboxylase activity contains: 25 m*M* of Tris, pH 7.4, 10 m*M* of sodium pyruvate, 2.5 m*M* of sodium ATP, 0.75 m*M* of acetyl-CoA, 50 m*M* of $NaH^{14}CO_3$ (2 μCi), 5 m*M* of $MgCl_2$, and suitable amounts of sample in a total volume of 1 mL. Omission of acetyl-CoA serves as blank. The entire procedure is then carried out in the hood. The

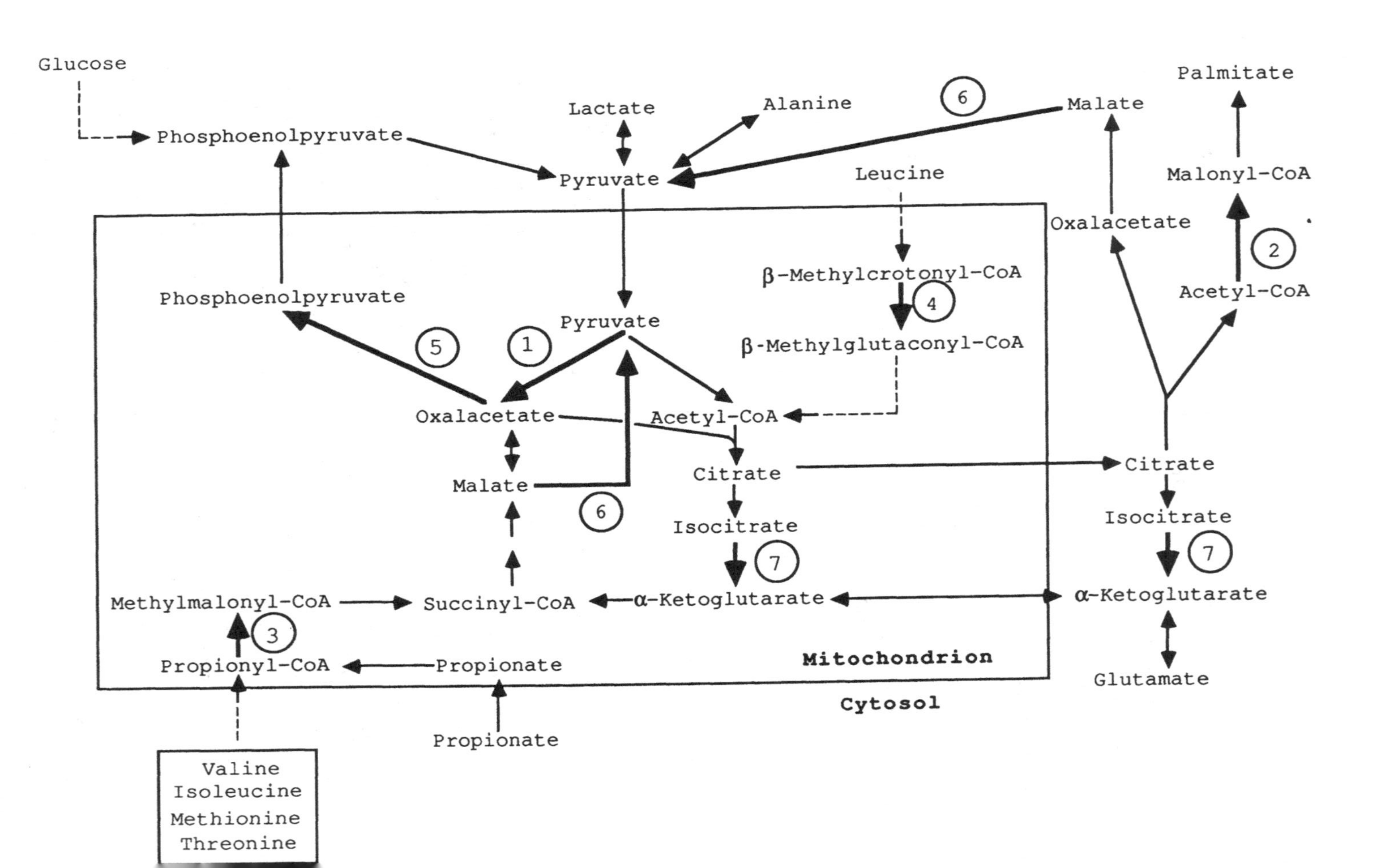

Glucose
Phosphoenolpyruvate
Lactate
Alanine
6
Malate
Palmitate
Pyruvate
Leucine
Malonyl-CoA
Oxalacetate
2
Acetyl-CoA
β-Methylcrotonyl-CoA
4
β-Methylglutaconyl-CoA
Phosphoenolpyruvate
Pyruvate
5
1
Oxalacetate
Acetyl-CoA
Malate
6
Citrate
Isocitrate
7
Citrate
Isocitrate
7
Methylmalonyl-CoA
3
Succinyl-CoA
α-Ketoglutarate
α-Ketoglutarate
Propionyl-CoA
Propionate
Mitochondrion
Glutamate
Cytosol
Propionate
Valine
Isoleucine
Methionine
Threonine

Fig. 1. A schematic representation of the involvement of the "CO_2-fixing enzymes" in intermediary metabolism in the brain cell. The enzymes are: (1) pyruvate carboxylase; (2) acetyl-CoA carboxylase; (3) propionyl-CoA carboxylase; (4) β-methylcrotonyl-CoA carboxylase; (5) phosphoenolpyruvate carboxykinase; (6) NADP-malate dehydrogenase; and (7) NADP-isocitrate dehydrogenase.

reaction is started by the addition of small volumes of samples (up to 0.1 mL) to the prewarmed reaction mixture at 37°C for 5–30 min. The reaction is stopped by adding 0.5 mL of 10% trichloroacetic acid in the hood to remove released $^{14}CO_2$ from the working area. To release any $^{14}CO_2$ radioactivity that is still present in the solution, CO_2 gas is bubbled in the tubes for at least 3 min. If CO_2 gas is not available, repeated addition of very small pieces of dry ice to the tubes placed in a 37°C waterbath to maintain bubbling (avoiding freezing) for 3 min provides an alternate approach for gassing with CO_2. Acid-stable ^{14}C-radioactivity in an aliquot of the solution is then determined using an appropriate scintillation fluid and a liquid scintillation spectrometer.

In this assay, ^{14}C-oxalacetate formed from pyruvate carboxylation is then utilized for the formation of ^{14}C-citrate by endogenous citrate synthase in the presence of added acetyl-CoA as shown below. Since acetyl-CoA is added in excess (0.75 m*M* acetyl-CoA in the reaction mixture compared with a K_a of 5–25 μ*M* for acetyl-CoA for pyruvate carboxylase) (*see* Table 2), the utilization of acetyl-CoA for the formation of citrate does not affect the activation of the enzyme.

(iii) pyruvate + MgATP + $H^{14}CO_3^-$ $\rightarrow$ ^{14}C-oxalacetate + MgADP + Pi
pyruvate carboxylase

(iv) ^{14}C-oxalacetate + acetyl-CoA $\rightarrow$ ^{14}C-citrate + CoA
citrate synthase

Alternatively, ^{14}C-oxalacetate formed from pyruvate carboxylation can be reduced to ^{14}C-malate by endogenous (or excess exogenous) NAD-malate dehydrogenase in the presence of added NADH (Scrutton et al., 1969), as shown in reaction (v).

(v) ^{14}C-Oxalacetate + NADH + H^+ $\rightarrow$ ^{14}C-Malate + NAD^+

This coupled system is also used to assay partially purified preparations of pyruvate carboxylase by measuring the oxidation of NADH spectrophotometrically and replacing nonradioactive HCO_3^- for radioactive $H^{14}CO_3$ in the assay mixture (Scrutton et al., 1969).

Utter and Keech (1963) first demonstrated the presence of pyruvate carboxylase in chicken liver mitochondria. In subsequent studies, pyruvate carboxylases from chicken liver (Utter et al., 1964; Utter and Scrutton, 1969; Scrutton et al., 1969) rat liver (McClure et al., 1971a,b; Seufert et al., 1971), bovine liver (Utter

Table 2
Kinetic Parameters for Pyruvate Carboxylase from Different Avian and Mammalian Tissues

Parameters	Chicken liver[a]	Rat liver[b]	Rat brain[c]
Mol wt (tetrameric form)	6.6×10^5	5.2×10^5	—
S^o_{20W}	14.7–17.0	15	14.8
MgATP (ATP)	(0.028 m*M*)	0.082 m*M*	[d]
HCO_3^-	0.98 m*M*	2.5 m*M*	[d]
Pyruvate (I)	—	0.049 m*M*	[d]
Pyruvate (II)	0.44 m*M*	0.53 m*M*	[d]
K_a for acetyl-CoA	0.002 m*M*	0.025 m*M*	0.017 m*M*
Hill coefficient (n)	2.85	2.0–2.1	2.0
Cold sensitivity	Labile	Nonlabile	Nonlabile
K_i			
Malonyl-CoA	0.073 m*M*		
Methylmalonyl-CoA	0.32 m*M*		
Fluoropyruvate	0.17 m*M*		
Phenylpyruvate	0.48 m*M*		

[a]Data taken from Utter et al. (1964); Utter and Scrutton (1969); Scrutton et al. (1969).
[b]Data taken from McClure et al. (1971a,b).
[c]Data taken from Mahan et al. (1975).
[d]Reportedly similar to that of rat liver pyruvate carboxylase (Mahan et al., 1975).

and Scrutton, 1969), and sheep kidney (Ling and Keech, 1966) have been purified, and their structural and kinetic properties have been extensively studied. The findings of these and many other studies pertaining to pyruvate carboxylase have been reviewed periodically (Utter et al., 1964; Utter and Scrutton, 1969; Scrutton and Young, 1972; Achuta Murthy and Mistry, 1977; Wood and Barden, 1977), and readers are referred to two recent reviews by Attwood and Keech (1984) and Keech and Wallace (1985). The tetrameric (active) form of pyruvate carboxylase from chicken liver has a molecular weight of about 660 kdalton, and contains 4 mol of biotin and 4 g atoms of bound manganese per mole of enzyme (Scrutton and Utter, 1965; Scrutton et al., 1966). Manganese is tightly bound to the enzyme and participates in the second partial reaction (*see* reaction ii) of pyruvate carboxylase, possibly through mediation of the enzyme–manganese–pyruvate bridge complex (Scrutton et al., 1966; Mildvan et al., 1966; Utter and Scrutton, 1969; Scrutton and Mildvan, 1970). Manganese is also present in liver pyruvate carboxylases isolated from turkey, calf, and rat. The K_m for pyruvate of chicken liver pyruvate carboxylase is 0.44 mM (Table 2) (Scrutton et al., 1968). Phenylpyruvate inhibits the enzyme from chicken liver (K_i of 0.48 mM) (Scrutton et al., 1968) and rat brain (Patel, 1972; Mahan et al., 1975), and the inhibition occurs at the transcarboxylation step of the pyruvate carboxylase reaction. Acetyl-CoA is an obligatory allosteric activator for all mammalian pyruvate carboxylases, and the K_a values for acetyl-CoA range from 2 to 25 μM for pyruvate carboxylase from different sources (*see* Table 2). Malonyl-CoA and methylmalonyl-CoA inhibit chicken liver pyruvate carboxylase with the K_i values of 73 and 320 μM, respectively (Table 2). A divalent cation (e.g., Mg^{2+}) is required in excess of that complexed with ATP. K^+ is most effective among other cations in stimulating pyruvate carboxylase. In contrast to pyruvate carboxylase from rat liver (McClure et al., 1971a) and brain (Mahan et al., 1975), the enzyme from chicken liver, turkey liver, and sheep kidney is inactivated at 2°C (Table 2). Using electron microscopic studies and sedimentation studies of pyruvate carboxylase, it was shown that inactivation of the enzyme at 2°C is brought about by a conversion of the active tetramer to inactive monomer subunits (Valentine et al., 1966). The loss of catalytic activity of chicken liver pyruvate carboxylase at low temperatures can be restored by rewarming the enzyme preparation to 23°C.

Pyruvate carboxylase is present in all rat tissues, with considerable activities in gluconeogenic tissues such as liver and kidney (103 and 90 mU/mg of cellular protein, respectively) and in tissues with high rates of lipogenesis such as white adipose tissue and lactating mammary gland (103 and 50 mU/mg of cellular protein, respectively) (Ballard et al., 1970). Significant activity (10 mU/mg cellular protein) is also present in rat brain (Table 3). Immunological studies established that pyruvate carboxylase is localized only in the mitochondria, and the antibody raised against rat liver pyruvate carboxylase immunoreacted with pyruvate carboxylase present in other tissues of the rat (Ballard et al., 1970).

Table 3
Pyruvate Carboxylase Activity in the Brain of Various Animal, in Brain Mitochondrial Populations, and in Neuronal Cells

Source	Pyruvate carboxylase		Reference
	U/g tissue	mU/mg protein	
Rabbit brain	—	10	Keech and Utter (1963)
Rabbit brain cortex mitochondria	—	54	Felicioli et al. (1966)
Guinea pig brain cortex mitochondria	—	119	Felicioli et al. (1966)
Cat brain cortex mitochondria	—	118	Felicioli et al. (1966)
Ox brain cortex mitochondria	—	362	Felicioli et al. (1966)
Rat brain cortex mitochondria	—	107	Felicioli et al. (1966)
Rat brain	0.88	9.7	Ballard et al. (1970)
Rat brain	0.80	—	Wilbur and Patel (1974)
Rat brain	0.2[a]	—	Salganicoff and Koeppe (1968)
Rat brain nonsynaptic mitochondria	—	66	Patel and Tilghman (1973)
Mouse brain	1.24	—	Yu et al. (1983)
Astrocytes (1 wk in culture)	—	0.14	Yu et al. (1983)
Astrocytes (3 wk in culture)	—	1.75	Yu et al. (1983)
Cerebral cortex neurons	—	<0.05	Yu et al. (1983)
Cerebellar granule cells	—	<0.05	Yu et al. (1983)
Astrocyte-enriched cerebellar cells	—	2.5	Shank et al. (1985)
Granule cell-enriched cells	—	0.8	Shank et al. (1985)

[a]At 25°C.

Pyruvate carboxylase activity in the brain of newborn rats is very low and increases by about 15-fold during the first month of postnatal life (Wilbur and Patel, 1974). Aging has no effect on pyruvate carboxylase activity in brains of 1- and 2-yr-old rats compared to 3-mo-old animals (Patel, 1977). The distribution of pyruvate carboxylase in rat brain nerve endings and glial plus neuronal bodies was 75 and 25% of the total rat brain activity, respectively (Salganicoff and Koeppe, 1968). Immunocytochemical studies and the measurement of pyruvate carboxylase activity in cultured astrocytes or astrocyte-enriched cell populations clearly show that pyruvate carboxylase is an astrocytic enzyme (Yu et al., 1983; Shank et al., 1985 *also see* Table 2).

Biotin deficiency induced in rats by feeding a diet containing 20% egg white for several weeks caused a marked reduction in pyruvate carboxylase activity in liver (95%) and kidney (75%), whereas the reduction in brain pyruvate carboxylase activity was only 35% after prolonged deficiency (40 d) (Chiang and Mistry, 1974). The differential changes in the activities of biotin-containing enzymes in the liver, kidney, heart, and brain of biotin-deficient rats (Arinze and Mistry, 1971; Chiang and Mistry, 1974) may reflect the differences in the turnover rates of these enzymes in different tissues or their relative metabolic importance within the tissue.

Deficiency of pyruvate carboxylase has been reported in several children with severe developmental neurological abnormalities (psychomotor retardation, poor head control, and hypotonia) and who at autopsy showed marked reduction in cerebral white matter, depletion of cerebral cortical neurons, numerous ectopic neurons, gliosis, and presence of other degenerative changes (Blass, 1983; Barritt, 1985). Pyruvate carboxylase deficiency has been claimed to be a cause of Leigh's necrotizing encephalomyelopathy (Hommes et al., 1968); however, careful analyses have failed to establish this relationship (Grover et al., 1972; Atkin et al., 1979; Murphy et al., 1981; Gilbert et al., 1983).

2.2. Acetyl-CoA Carboxylase

Acetyl-CoA carboxylase (acetyl-CoA:CO_2 ligase[ADP], E.C.6.4.1.2) catalyzes the carboxylation of acetyl-CoA to malonyl-CoA involving two partial reactions as shown below:

(vi) E-biotin + MgATP + $HCO_3^- \rightleftharpoons$ E-biotin-CO_2 + MgADP + Pi

(vii) E-biotin-CO_2 + acetyl-CoA $\rightleftharpoons$ E-biotin + malonyl-CoA

Sum: acetyl-CoA + MgATP + $HCO_3^- \rightleftharpoons$ malonyl-CoA + MgADP + Pi

Brain acetyl-CoA carboxylase activity is assayed by measuring ATP-dependent carboxylation of acetyl-CoA (Gross and Warshaw, 1974) by a modification of the method described by Greenspan and Lowenstein (1967). Since the enzyme undergoes polymerization and hence activation in the presence of citrate, as described below, an activation step is essential prior to assaying the enzyme activity. The enzyme is activated by incubating for 30 min at 37°C in 0.25 mL medium containing: 40 m*M* Tris-Cl, pH 7.4; 20 m*M* $MgCl_2$, 4 m*M* EDTA, 1 mg of fatty acid-free bovine serum albumin, 8 m*M* sodium citrate, and suitable amounts of enzyme preparation (Gross and Warshaw, 1974). After a 30-min activation of the enzyme in the presence of citrate, 0.25 mL of the assay mixture is added to the activation medium to initiate the enzyme assay and further incubated at 37°C for a desirable period (varying from 5 to 30 min). The reaction is terminated by the addition of 0.1 mL of 7*N* trichloroacetic acid in the hood. The assay mixture (0.25 mL) contains: 40 m*M* Tris-Cl, pH 7.4, 20 m*M* $MgCl_2$, 4 m*M* ATP, 0.8 m*M* acetyl-CoA, 160 m*M* $NaH^{14}CO_3$ (approximately 0.5 μCi/0.25 mL), and 20 m*M* sodium citrate (Gross and Warshaw, 1974). Omission of acetyl-CoA serves as blank. To measure acid-stable radioactivity in the solution, the samples are gassed with CO_2, and the radioactivity is determined as described above for the assay of pyruvate carboxylase.

Acetyl-CoA carboxylase is considered to be the rate-limiting step in the biosynthesis of the long-chain fatty acids. Since acetyl-CoA carboxylase is subjected to both short-term (allosteric and covalent modifications) and long-term (changes in enzyme synthesis) controls, this enzyme has been extensively studied and has been reviewed periodically (Alberts and Vagelos, 1972; Lane et al., 1974; Numa and Yamashita, 1974; Volpe and Vagelos, 1976; Kim, 1983). This enzyme is widely distributed in animals, plants, and microorganisms (Alberts and Vagelos, 1972). Acetyl-CoA carboxylase has been purified from chicken liver, rat liver, mammary gland, and adipose tissue, and bovine adipose tissue (Waite and Wakil, 1962; Matsuhashi et al., 1964; Moss et al., 1972; Vagelos et al., 1963; Miller and Levy, 1969). Citrate activation of acetyl-CoA carboxylase is accompanied by polymerization of the enzyme. The equilibrium between the active polymeric form and the inactive protomers of acetyl-CoA carboxylase is influenced by citrate, isocitrate, coenzyme A, covalent phosphorylation, fatty acyl-CoAs, and ATP-Mg^{2+} plus HCO_3^-. Both partial reactions of acetyl-CoA carboxylase (*see* above vi and vii) are stimulated by citrate, an allosteric positive effector. Citrate activation of the enzyme results in an increase in the V_{max} without affecting the K_m values for its substrates (Lane et al., 1974). Inactivation of acetyl-CoA carboxylase through phosphorylation results in depolymerization in the absence of HCO_3^-, and depolymerization occurs even in the presence of citrate (Lent et al., 1978). Covalent modification of acetyl-CoA carboxylase results in lowering of citrate concentration required for activation by about one-tenth of that needed for the phosphorylated enzyme. Similarly, CoA activation of acetyl-CoA carboxylase results in lowering the K_m for acetyl-CoA (Kim, 1983). CoA activation of acetyl-CoA carboxylase can occur with or without polymerization of the enzyme. CoA and citrate binding sites are different, and activation of acetyl-CoA carboxylase by these compounds is synergistic (Yeh and Kim, 1980). However, binding of CoA to acetyl-CoA carboxylase is inhibited by palmityl-CoA. Acetyl-CoA carboxylase is also inhibited by long-chain fatty acyl-CoAs that cause depolymerization of the enzyme.

Covalent modification of acetyl-CoA carboxylase by phosphorylation mediated by cAMP-dependent protein kinase, the Ca^{2+}- and phospholipid-dependent protein kinase, has been described (Kim, 1983; Brownsey and Denton, 1982; Witters et al., 1983; Holland and Hardie, 1985; Hardie et al., 1986). The cAMP-dependent, the calmodulin-dependent, and the phospholipid-dependent protein kinases phosphorylate distinct sites on rat mammary acetyl-CoA carboxylase. One of the three phosphorylated tryptic peptides derived from phosphorylated acetyl-CoA carboxylase in the presence of phospholipid-dependent kinase is identical with the major phosphopeptide generated from enzyme treated with cAMP-dependent protein kinase, and both protein kinases inactivate acetyl-CoA carboxylase in a very similar manner (Hardie et al., 1986). Although the calmodulin-dependent protein kinase phosphorylates acetyl-CoA carboxylase, no effect has been detected on the enzyme activity.

The biosynthesis of long-chain fatty acids via the *de novo* pathway in developing rat brain has been observed both in vivo (Dhopeshwarkar et al., 1969) and in vitro (Majno and Karnovsky, 1958; Brady, 1960; Aeberhard et al., 1969; Patel and Tonkonow, 1974). The activity of acetyl-CoA carboxylase was high in rat brain during late fetal life (Gross and Warshaw, 1974) and during the first 10–15 d of postnatal life (Kelley and Joel, 1973; Patel and Tonkonow, 1974; Gross and Warshaw, 1974), and declined slowly within the next 2 wk, reaching 30–50% of the value observed in the newborn rat brain (approximately 3 and 1.6 mU/mg of soluble protein in brains of 2-d-old and adult rats (Gross and Warshaw, 1974). The K_m values for acetyl-CoA of acetyl-CoA carboxylase from brains of newborn and adult rats were 59 and 240 μM, respectively (Gross and Warshaw, 1974), whereas a K_m value of only 9 ± 2 μM was observed by Kelley and Joel (1973) for the enzyme in adult rat brain. This difference may be attributed to the phosphorylation state of the enzyme preparation. There is a paucity of information of the structure, kinetic properties, and regulation of purified acetyl-CoA carboxylase from the brain.

2.3. Propionyl-CoA Carboxylase

The presence of propionyl-CoA carboxylase was first demonstrated in pig heart by Flavin and Ochoa (1957) and was obtained in crystalline form by Kaziro et al. (1961). Propionyl-CoA carboxylase (propionyl CoA:CO_2 ligase-[ADP], EC 6.4.1.3), a mitochondrial enzyme, catalyzes ATP-dependent carboxylation of propionyl-CoA to D-methylmalonyl-CoA involving two partial reactions as shown below:

(viii) E-biotin + MgATP + HCO_3^- $\rightleftharpoons$ E-biotin-CO_2 + MgADP + Pi

(ix) E-biotin-CO_2 + propionyl-CoA $\rightleftharpoons$ methylmalonyl-CoA + E-biotin

Sum: Propionyl-CoA + MgATP + HCO_3^- $\rightleftharpoons$ methylmalonyl-CoA + MgADP + Pi

The enzyme is assayed by measuring the fixation of ^{14}C-bicarbonate in the presence of propionyl-CoA and ATP. The reaction mixture (in a final volume of 0.75 mL) contains: 66.7 mM Tris-Cl, pH 8.5, 10 mM $KH^{14}CO_3$ (approximately 200,000 dpm/μmol), 2.67 mM ATP, 2.67 mM $MgCl_2$, 3.33 mM reduced glutathione, 0.67 mM propionyl-CoA, and enzyme sample (Halenz et al, 1962). Blank tubes contain no propionyl-CoA. The mixture is preincubated for 3–5 min at 37°C, and the reaction is started by the addition of an aliquot of enzyme preparation. After incubation for 10–30 min, the reaction is terminated by acidification in the hood, followed by gassing with CO_2 and the measurement of acid-stable radioactivity as described for pyruvate carboxylase.

In nonruminant mammals propionyl-CoA is mostly derived from the catabolism of isoleucine, valine, methionine, threonine, cholesterol (side chain only), and odd chain fatty acids (Rosenberg, 1983). Propionyl-CoA carboxylase has also been purified to homogeneity from bovine kidney (Lau et al., 1979) and human liver (Kalousek et al., 1980). The native enzyme has a molecular weight of 540 kdalton with an $(\alpha\beta)_4$ quaternary structure and contains 4 mol of biotin per mole of enzyme. The native enzyme is a tetramer $(\alpha\beta)_4$ of protomers, each protomer containing a single α (mol wt, 72,000) and a single β (mol wt, 56,000) subunit. Biotin is bound only to the α subunit (Lau et al., 1979; Kalousek et al., 1980). The presence of propionyl-CoA carboxylase in tissues of several species has been reviewed by Achuta Murthy and Mistry (1977). In rat brain, the activity of propionyl-CoA carboxylase is 0.27 U/g of tissue (Chiang and Mistry, 1974). Biotin deficiency caused a marked reduction in the activity of propionyl-CoA carboxylase in chicken liver and rat tissue (Arinze and Mistry, 1971; Chiang and Mistry, 1974). However, the reduction in the activity was less marked in the brain compared with other rat tissues. The administration of biotin in a single dose to biotin-deficient animals restored the activities of propionyl-CoA carboxylase and other biotin-containing carboxylases in different tissues within 2–12 h (Arinze and Mistry, 1971; Chiang and Mistry, 1974). Deficiency of propionyl-CoA carboxylase has been extensively studied in affected children in recent years, and these studies have been reviewed by Rosenberg (1983).

2.4. 3-Methylcrotonyl-CoA Carboxylase

Lynen and associates were first to identify and isolate 3-methyl-crotonyl-CoA carboxylase from bacterial sources (Knappe et al., 1961). Mammalian 3-methylcrotonyl-CoA carboxylase (3-methylcrotonyl-CoA:CO_2 ligase[ADP], EC 6.4.1.4) has been isolated from bovine kidney and extensively characterized (Lau et al., 1979, 1980). The native enzyme has a molecular weight of 835 kdalton, and its protomer is composed of a biotin-free subunit (A subunit; mol wt, 61,000) and a biotin-containing subunit (B subunit; mol wt, 73,000) (Lau et al., 1979, 1980). The native enzyme may be hexameric $(AB)_6$ and is localized exclusively on the inner mitochondrial membrane (Hector et al., 1980).

This enzyme is involved in the oxidation of leucine and catalyzes the carboxylation at the β carbon of 3-methylcrotonyl-CoA to form 3-methylglutaconyl-CoA involving two partial reactions, as shown below:

(x) $\text{E-biotin} + \text{MgATP} + HCO_3^- \rightleftharpoons \text{E-biotin-}CO_2 + \text{MgADP} + \text{Pi}$

(xi) $\text{E-biotin-}CO_2 + \text{3-methylcrotonyl-CoA} \rightleftharpoons \text{E-biotin} + \text{3-methylglutaconyl-CoA}$

Sum: $\text{3-methylcrotonyl-CoA} + \text{MgATP} + HCO_3^- \rightleftharpoons \text{3-methylglutaconyl-CoA} + \text{MgADP} + \text{Pi}$

3-Methylcrotonyl-CoA carboxylase is assayed in crude preparations by a ^{14}C-bicarbonate-fixation procedure (Lau et al., 1980) and in partially purified enzyme preparations by a coupled spectrophotometric assay (Lau et al., 1980). For the ^{14}C-bicarbonate fixation assay, the reaction mixture contains 100 m*M* Tris-HCl, pH 8.0, 100 m*M* KCl, 0.5 m*M* 3-methylcrotonyl-CoA, 2 m*M* ATP, 7.5 mM $MgCl_2$, 12.5 m*M* $NaH^{14}CO_3$ (2.5 μCi), 50 μg bovine serum albumin, and appropriate amount of enzyme in a final volume of 0.1 mL. The reaction can be initiated by adding either ATP or 3-methylcrotonyl-CoA, and incubated for 5–15 min at 37°C. The reaction is terminated by acidification followed by CO_2 gassing, and the determination of acid-stable radioactivity is carried out as described for pyruvate carboxylase. The omission of 3-methylcrotonyl-CoA serves to determine background fixation.

The enzyme is activated by monovalent cations such as K^+, and has the following K_m values for its substrates: 3-methylcrotonyl-CoA, 75 μ*M*; ATP, 82 μ*M*; HCO_3^-, 1.8 m*M* (Lau et al., 1980). (2Z)-3-Ethylcrotonyl-CoA and transcrotonyl-CoA also serve as substrates with K_m values of 22 and 225 μ*M*, respectively. (2E)-3-Ethylcrotonyl-CoA and (2Z)-geranyl-CoA act as competitive inhibitors of the enzyme with K_i values of 45 and 116 μ*M*, respectively (Lau et al., 1980).

Deficiency of 3-methylcrotonyl-CoA carboxylase in at least two children has been reported, and significant clinical and biochemical findings have recently been reviewed (Tanaka and Rosenberg, 1983).

2.5. Mechanism of Action of Biotin-Dependent Carboxylation

The role of biotin as the prosthetic group in all biotin-dependent carboxylases from both eukarytic and prokarytic cells has been studied extensively and reviewed by several investigators (Lane et al., 1974; Achuta Murthy and Mistry, 1977; Wood and Barden, 1977; Keech and Attwood, 1985). All biotin enzymes contain the prosthetic group convalently linked to a lysine residue (ϵ-*N*-biotinyl-*L*-lysine) (biocytin) of the biotin-containing subunit of a multisubunit enzyme. The apoenzyme is synthesized first. A synthetase then adds biotin to the apoenzyme forming the holoenzyme. The synthetase-catalyzed reaction occurs in two steps, as shown below, with the formation of a biotinyl-5'-AMP intermediate followed by transfer of the biotinyl moiety to the apoenzyme to form the holoenzyme (Achuta Murthy and Mistry, 1977; Wallace, 1985). The synthetase is present in both the cytosolic and mitochondrial matrix compartments (Achuta Murthy and Mistry, 1977; Obermayer and Lynen, 1976). Deficiency of biotin holocarboxylase synthetase (also referred to as the biotin-responsive multiple carboxylase deficiencies) has been reported in several patients (Bartlett et al., 1985; Sweetman et al., 1985; Barrit, 1985).

(xii) D-biotin + ATP $\rightleftharpoons$ D-biotinyl-5'-AMP + PPi

(xiii) D-biotinyl-5'-AMP + Apoenzyme $\rightarrow$ biotinyl-enzyme + AMP
(holoenzyme)

Amino acid sequence analysis of biotin-containing peptides generated by tryptic cleavage of pyruvate carboxylases isolated from liver mitochondria from sheep, chicken, and turkey revealed a complete duplication of a 19-residue biotinyl peptide derived from the two avian pyruvate carboxylases, and only three amino acid residues (namely residues 2, 17, and 19) differed in sheep liver pyruvate carboxylase (Rylatt et al., 1977; Wood and Barden, 1977; Wallace and Easterbrook-Smith, 1985). Evidence of considerable homology in the region of biotin-binding site between these three mammalian enzymes and pyruvate carboxylases from *Escherichia coli* and *Propionibacterium shermanii* is also presented (Wood and Barden, 1977; Wallace and Easterbrook-Smith, 1985). The amino acid sequence adjacent to the biocytin of chicken liver pyruvate carboxylase is: Gly-Ala-Pro-Leu-Val-Leu-Ser-Ala-Met-Lys(biotin)-Met-Glu-Thr-Val-Val-Thr-Ala-Pro-Arg.

The reactions catalyzed by all biotin-dependent carboxylases utilize bicarbonate as the "carboxyl" donor and involves a carboxylated enzyme intermediate as ε-*N*-(l'-*N*-carboxy-biotinyl)-L-lysyl enzyme [partial reaction (i) in Fig. 2]. l'-*N*-carboxybiotin serves as the carboxyl donor in reactions catalyzed by all biotin enzymes (Guchhait et al., 1974; Wood and Barden, 1977). The formation of the carboxy-biotin enzyme (first partial reaction, Fig. 2) appears to be a single-step process; however, a possible role for 0-phosphobiotin in the carboxylation reaction is not ruled out (Wood and Barden, 1977). The carboxyl transfer occurs according to the partial reaction (ii) in Fig. 2. The overall carboxylation reaction proceeds via a sequential reaction pathway (Keech and Attwood, 1985). The biotinyl moiety attached to the lysyl residue of the enzyme is free to shuttle between the two substrate-binding sites of the two partial reactions.

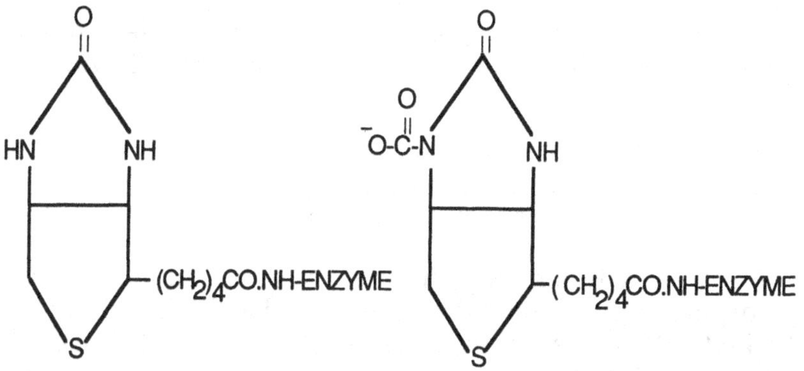

$$\text{BIOTIN-ENZYME} + \text{ATP} + HCO_3^- \rightleftharpoons CO_2\text{-BIOTIN-ENZYME} + \text{ADP} + \text{Pi}$$

$$CO_2\text{-BIOTIN-ENZYME} + \text{ACCEPTOR} \rightleftharpoons \text{BIOTIN-ENZYME} + {}^-\text{OOC-ACCEPTOR}$$

Fig. 2. Mechanism of action of biotin-containing enzymes.

3. Biotin-Independent CO_2 Fixation

Three enzymes, namely phosphoenolpyruvate carboxykinase, NADP-malate dehydrogenase (malic enzyme), and NADP-isocitrate dehydrogenase, are discussed briefly in this section. It is recognized that under physiological conditions these three enzymes function largely in the direction of decarboxylation. However, since these three enzymes can catalyze reversible reactions, they can fix CO_2 under appropriate experimental conditions.

3.1. Phosphoenolpyruvate Carboxykinase

Utter and Kurahashi (1954) were first to describe an enzyme activity in chicken liver mitochondria catalyzing the carboxylation of phosphoenolpyruvate to oxalacetate in the presence of Mn^{2+} and IDP. This enzyme, which catalyzes a reversible reaction as shown below, was called phosphoenolpyruvate carboxykinase (phosphoenolpyruvate carboxykinase [GTP], EC 4.1.1.32).

$$\text{(xiv)}\quad \underset{\text{(or ITP)}}{\text{oxalacetate} + \text{GTP}} \overset{Mn^{2+}}{\rightleftharpoons} \underset{\text{(or IDP)}}{\text{phosphoenolpyruvate} + \text{GDP}} + HCO_3^-$$

In crude extracts the enzyme is assayed conveniently by measuring the fixation of ^{14}C-bicarbonate using the following coupled enzyme system:

(xv) phosphoenolpyruvate + $H^{14}CO_3$ + IDP → ^{14}C-oxalacetate + ITP

(xvi) ^{14}C-oxalacetate + NADH + H^+ → ^{14}C-Malate + NAD^+
NAD-malate dehydrogenase

NAD^+-Malate dehydrogenase and NADH are added in excess of phosphoenolpyruvate carboxykinase so that the overall reaction is limited only by the rate of the phosphoenolpyruvate carboxykinase reaction (Chang and Lane, 1966; Ballard and Hanson, 1969). The reaction mixture (in a final volume of 1 mL) contains 100 m*M* imidazole-HCl buffer, pH 6.6, 2 m*M* $MnCl_2$, 1 m*M* dithiothreitol, 1.25 m*M* sodium IDP, 50 m*M* $KH^{14}CO_3$ (2 μCi), 2.5 m*M* NADH, 1.5 m*M* phosphoenolpyruvate (5 m*M* phosphoenolpyruvate for the brain enzyme, Cheng and Cheng, 1972), and 2 U of NAD-malate dehydrogenase (Chang and Lane, 1966; Ballard and Hanson, 1969). Omission of IDP serves as blank. After pre-equilibration at 37°C in a waterbath under a hood, the reaction is initiated by adding the enzyme preparation in varying amounts. The reaction is stopped by adding 0.5 mL of 10% trichloroacetate after 5–15 min of incubation. Acidification and subsequent gassing with CO_2 are carried out in the hood to remove released $^{14}CO_2$ from the working area, as described above for pyruvate carboxylase, and acid-stable radioactivity in the solution is determined.

The more rapid spectrophotometric assays in the direction of carboxylation (Chang and Lane, 1966) and decarboxylation (Ballard, 1970; Jomain-Baum et al., 1976) can be employed with purified preparations of phosphoenolpyruvate carboxykinase. For the spectrophotometric carboxylation assay, the reaction mixture and conditions are similar to those described above for the $H^{14}CO_3^-$-fixation assay, except unlabeled bicarbonate replaces ^{14}C-bicarbonate and less NADH is used (0.15 μmol) (Chang and Lane, 1966). The initial velocity of NADH oxidation is followed for a period of 2 min at 340 nm after initiating the reaction with phosphoenolpyruvate. The carboxylation reaction follows zero-order kinetics with up to 55 nmol/min of the enzyme activity.

Phosphoenolpyruvate carboxykinase is one of the key enzymes in the gluconeogenic pathway and is present at high levels in rat liver and kidney cortex (2.5 and 1.25 U/g of tissue, respectively). It, however, is also present in rat tissues that do not synthesize glucose to any appreciable extent [mU/g of tissue: 40, adipose tissue; 31, lung; 39, heart; 5, skeletal muscle; 6, brain) (Reshef et al., 1969; Hanson and Garber, 1972)]. The enzyme activity in liver, kidney, adipose tissue, and lung increased three- to five-fold in rats fasted for 24 h (Hanson and Garber, 1972). The intracellular distribution of phosphoenolpyruvate carboxykinase varies widely among species (Hanson and Garber, 1972). Hepatic cytosolic phosphoenolpyruvate carboxykinase is very adaptive to dietary and hormonal manipulations in most species, except the lactating dairy cow (Hanson and Garber, 1972). In contrast, the mitochondrial activity remains largely unaltered by these manipulations. Immunological studies demonstrated that the cytosolic and mitochondrial phosphoenolpyruvate carboxykinase are two distinct proteins in rat tissues (Ballard and Hanson, 1969). Recently the expression of the genes responsible for encoding the cytosolic and mitochondrial forms of phosphoenolpyruvate carboxykinases in rat and chicken has been investigated (Hod et al., 1986), and the gene responsible for encoding the cytosolic enzyme has been identified and partially characterized (Yoo-Warren et al., 1983; Hod et al., 1984).

Rat brain phosphoenolpyruvate carboxykinase is localized in the mitochondria (Cheng and Cheng, 1972; Wilbur and Patel, 1974). This enzyme from rat brain was purified approximately 6000-fold, and kinetic properties of the brain enzyme were found to be largely similar to those of the liver enzyme (Cheng and Cheng, 1972; Utter and Kolenbrander, 1972). The higher K_m for phosphoenolpyruvate (5 m*M*) of the brain enzyme compared with the K_m value of 0.3 m*M* for the liver enzyme (Chang et al., 1966) in the direction of CO_2 fixation led Cheng and Cheng (1972) to suggest that the function of the enzyme in the mitochondria was more for the decarboxylation of oxalacetate with formation of phosphoenolpyruvate.

3.2. NADP-Malate Dehydrogenase (Malic Enzyme)

NADP-Malate dehydrogenase [L-malate:NADP oxidoreductase (decarboxylating), EC 1.1.1.40] catalyzes the following reaction:

$$\text{(xvii)} \quad \text{malate} + NADP^+ \rightleftharpoons \text{pyruvate} + CO_2 + NADPH + H^+$$

The intracellular localization of this enzyme in nonruminant mammals is tissue specific: (1) almost exclusive localization in the cytosol of liver, adipose tissue, and adrenal medulla and (2) presence in both the cytosolic and mitochondrial compartments in heart, kidney, and brain (Brdiczka and Pette, 1971; Salganicoff and Koeppe, 1968; Frenkel, 1972). The enzyme (decarboxylation reaction) is assayed spectrophotometrically by observing the absorbance change at 340 nm (Ochoa, 1955). The reaction mixture contains (final concentration): 100 m*M* N-morpholinopropane sulfonate (pH 7.5) (or suitable buffer such as Tris-HCl, pH 7.5), 0.125 m*M* $NADP^+$, 10 m*M* $MgCl_2$, and suitable amount of enzyme sample. The reaction is started by the addition of L-malate, adjusted to pH 7.5, to give a final concentration of 5 m*M*, and the absorbance change is monitored for 3–5 min at 37°C. High absorbance changes without added malate may possibly represent malate present in crude cellular extracts (Frenkel, 1972). High blanks (minus malate) can be minimized by gel filtration of tissue extracts over a Sephadex G-50 column (0.8 × 5.0 cm) prior to assaying the enzyme activity.

The reductive carboxylation of pyruvate by the enzyme can be determined by a spectrophotometric assay for a partially purified preparation (Frenkel, 1972) or a ^{14}C-bicarbonate fixation assay for a crude extract (Ballard and Hanson, 1967a; Patel, 1974). In the radioactive assay, the concentrations of reactants in a final volume of 1 mL are: 40 m*M* Tris-HCl, pH 7.4, 20 m*M* $KH^{14}CO_3$ (2 μCi), 12 m*M* pyruvate, 0.2 m*M* NADPH, 1 m*M* $MnCl_2$, and suitable amount of enzyme preparation (Patel, 1974). The reaction is started by adding enzyme, and the reaction is allowed to proceed for 5–15 min at 37°C. The omission of enzyme or pyruvate can serve as assay blank. The reaction is stopped by acidification in a hood, followed by CO_2 gassing and counting of the radioactivity as described for pyruvate carboxylase.

Frenkel (1972) partially purified the cytosolic and mitochondrial forms of NADP-malate dehydrogenase from bovine brain. Although the two forms have similar molecular weights (approximately 200,000), they differ in their electrophoretic and chromatographic mobilities and in their kinetic properties. The mitochondrial form shows cooperativity at low concentrations of L-malate. Further, the rates of reductive carboxylation of pyruvate to malate are markedly different for the cytosolic (1/5) and mitochondrial (1/100) enzymes compared with their respective decarboxylation rates (Frankel, 1972). Immunological differences between the two forms from rat heart have been observed (Isohashi et al., 1971). Based on the specific activity of mitochondrial NADP-malate dehydrogenase in brain, inbred strains of mice fall into two classes, high- and low-activity strains (approximately 70 and 30 mU/mg of protein, respectively) (Bernstine et al., 1979). The rate of enzyme synthesis accounts for the inherited level of the brain mitochondrial enzyme and is regulated by a single genetic locus designated as Mdr-1 (Bernstine et al., 1979). The specific activity of mitochondrial NADP-malate dehydrogenase in brains of newborn mice was low and the adult levels of activity were reached in 6–8 wk (Bernstine et al., 1979). In the brain of newborn rats, the activities of cytosolic and mitochondrial NADP-malate dehydrogenases are very low (approximately 0.1 and 0.2 U/g of brain at 37°C, respectively) and reach their respective adult levels (0.42 and 1.3 U/g of brain at 37°C, respectively) at the time of weaning (M. S. Patel, unpublished data). Phenylpyruvate, but not phenylalanine, inhibits both forms of the enzyme in rat brain, causing 50% inhibition of activity at the concentration of 0.3–0.4 m*M* (M. S. Patel, unpublished data).

3.3. NADP-Isocitrate Dehydrogenase

NADP-Isocitrate dehydrogenase (threo-D_s-isocitrate:$NADP^+$ oxidoreductase [decarboxylating]; EC 1.1.1.42) catalyzes the oxidative decarboxylation of D-isocitrate to α-ketoglutarate as shown below:

$$\text{(xviii) isocitrate} + NADP^+ \xrightleftharpoons{Mn^{2+}} \alpha\text{-ketoglutarate} + CO_2 + NADPH + H^+$$

Enzymatic activity in the forward (decarboxylating) direction is measured by monitoring the increase in absorbance of NADPH at 340 nm (Plaut, 1962; Watanabe et al., 1974). The reaction mixture contains: 33 m*M* Tris-acetate buffer, pH 7.2, 1.33 m*M* $MnCl_2$, 0.1 m*M* $NADP^+$, 1.67 m*M* D,L-isocitrate (pH 7.2), 0.33 m*M* EDTA, and enzyme preparation in a final volume of 1 mL. The reaction is started by the addition of D,L-isocitrate. In the reverse (carboxylating) direction, the rate of oxidation of NADPH in the presence of α-ketoglutarate and $NaHCO_3$ is measured spectrophotometrically (Patel, 1974). The reaction mixture contains: 33 m*M* Tris-HCl, pH 7.4; 0.33 m*M* EDTA, 1.33 $MnCl_2$, 0.13 m*M* NADPH, 33 m*M* $NaHCO_3$, 11 m*M* α-ketoglutarate (pH 7.4), and tissue extract in a final volume of 3 mL. The reaction is started by the addition of α-ketoglutarate. Omission of α-ketoglutarate serves as blank.

NADP-Isocitrate dehydrogenase in mammalian tissues exists as two isozymes, one in the cytosol and the other in the mitochondria (Lowenstein and Smith, 1962; Murthy and Rappoport, 1963; Henderson, 1965; Rafalowska and Ksiezak, 1976). The two isozymes differ in electrophoretic mobility as well as immunological characteristics (Henderson, 1965). Two allelic forms of the cytosolic isozyme have been observed in inbred strains of mice (Henderson, 1965). In rat brain, the specific activity of the mitochondrial form is about three times higher than that of the cytosolic form (Rafalowska and Ksiezak, 1976). Subsequently, it was observed that NADP-isocitrate dehydrogenase is localized in both outer and inner mitochondrial membranes and in the mitochondrial matrix in rat brain. Electrophoretic studies showed that the isozyme associated with the outer membrane behaved similarly to that of the cytosolic form (Gromek and Pastuszko, 1977). Postdecapitative anoxia caused about 40% reduction in the specific activity of the outer membrane isozyme with no significant change in the inner mitochondrial membrane isozyme (Gromek and Pastuszko, 1977). The kinetic constants for $NADP^+$ of NADP-isocitrate dehydrogenase in the synaptic and nonsynaptic mitochondria from adult rat brain were similar (Lai and Clark, 1978). During the postnatal period, the specific activity of the cytosolic form in the brain decreases, whereas the activity of the mitochondrial form is maintained (Watanabe et al., 1974; Loverde and Lehrer, 1973). X-irradiation of the newborn rat brain caused a marked reduction in the activity of the cytosolic form of the enzyme (De Vellis et al., 1967). The involvement of the cytosolic form of NADP-isocitrate dehydrogenase in the α-ketoglutarate shunt has been demonstrated, but its contribution in providing the acetyl group for the synthesis of fatty acids in the brain is relatively small (D'Adamo and D'Adamo, 1968).

4. Significance of CO_2-Fixing Enzymes in Brain

The presence of several CO_2-fixing enzymes including both biotin-dependent and biotin-independent enzymes in the mammalian brain has been discussed in this chapter. Among the four biotin-dependent CO_2-fixing enzymes, pyruvate carboxylase is primarily responsible for the fixation of CO_2 in the brain. For instance, the CO_2 fixed in acetyl-CoA by acetyl-CoA carboxylase forming malonyl-CoA is released as CO_2 when the 2-carbon acetyl-moiety is incorporated into the growing chain of the fatty acid on fatty acid synthase. The net result is no net fixation of CO_2 by acetyl-CoA carboxylase. The magnitude of fixation of HCO_3^- by methylcrotonyl-CoA carboxylase in the catabolism of leucine and by propionyl-CoA carboxylase in the catabolism of isoleucine, valine, methionine, and threonine in the brain is difficult to assess, but appears to be relatively low. Similarly, the contributions of biotin-independent CO_2-fixing enzymes, namely NADP-malate dehydrogenase (both cytosolic and mitochondrial forms), NADP-isocitrate dehydrogenase (both cytosolic and mitochondrial forms), and mitochondrial phosphoenolpyruvate carboxykinase to CO_2-fixation in the brain in vivo is apparently very small (or negligible) since the direction of the decarboxylation reaction catalyzed by these enzymes is favored under the physiological condition (Table 4; *also see* Patel, 1974). The cytosolic NADP-malate dehydrogenase and, possibly, cytosolic NADP-isocitrate dehydrogenase are involved in the generation of NADPH for the reductive biosynthesis of long-chain fatty acids and cholesterol (Patel, 1974). The significance of the mitochondrial forms of these two enzymes is less well understood. Similarly, the importance of mitochondrial phosphoenolpyruvate carboxykinase in cerebral metabolism remains unclear since the brain is a nongluconeogenic tissue.

Pyruvate carboxylase, an anaplerotic enzyme, plays an important role by replenishing the C_4 compound, oxalacetate, required in many biosynthetic processes. Pyruvate carboxylase provides oxalacetate for the translocation of the acetyl moeity as citrate from mitochondria to the cytosol for lipogenesis in the developing brain as well as for the synthesis of acetylcholine (Patel and Tilghman, 1973; Patel and Owen, 1976). Oxalacetate is also used for the synthesis of aspartate, glutamate, and glutamine *de novo* in the brain. In the brain, pyruvate derived from glucose is mainly metabolized via the pyruvate dehydrogenase complex and pyruvate carboxylase in the mitochondria. It is estimated that approximately 10% of the total pyruvate metabolized in the brain is utilized via the CO_2-fixation pathway (Cheng et al., 1967) primarily via pyruvate carboxylase. The carboxylation of pyruvate via pyruvate carboxylase in brain mitochondria is regulated by the intramitochondrial concentrations of pyruvate and oxalacetate and the ATP : ADP ratio (Patel, 1972; Patel and Tilghman, 1973).

The existence of metabolic compartmentation, consisting of at least two metabolic pools, a "small" (or "synthetic") glial compartment, and a "large" (or "energetic") neuronal compartment, in the brain, has been well established (Berl and Clarke, 1969; Balazs et

Table 4
Activities of Several CO_2-Fixing Enzymes in the Cytosol and the Mitochondria of Adult Rat Brain[a]

	Enzyme activity, mU/g of brain	
CO_2-Fixing enzymes	Cytosol	Mitochondria
Pyruvate carboxylase		
Carboxylation	<5	720 ± 90
Phosphoenolpyruvate carboxykinase		
Carboxylation	<3	54 ± 6
NADP-Malate dehydrogenase		
Carboxylation	93 ± 11	ND[b]
Decarboxylation	639 ± 51	992 ± 94
NADP-Isocitrate dehydrogenase		
Carboxylation	125 ± 23	ND[b]
Decarboxylation	1177 ± 88	2190 ± 110

[a]Data taken from Patel (1974).
[b]ND, not determined.

al., 1972). Recent evidence shows that pyruvate carboxylase is present predominately, if not exclusively, in astrocytes, and little, if any, enzyme activity is localized in neurons (Yu et al., 1983; Shank et al., 1985). Anaplerotic activity of glial pyruvate carboxylase provides sources of glutamine and α-ketoglutarate that serve as precursors for the replenishment of neurotransmitter pools within the presynaptic terminals. Thus, pyruvate carboxylation occurs largely in astrocytes and provides glutamine and α-ketoglutarate for the release and eventual uptake of these compounds by neurons (Yu et al., 1983; Shank et al., 1985). In this regard, pyruvate carboxylase may play its anaplerotic function via metabolic shuttles between astrocytes and neurons.

Acknowledgment

The support from the U.S. Public Health Service grants DK 20478 and HD 11089 is acknowledged. I thank Dr. Douglas Kerr, Lap Ho, Donna Carothers, and Dr. Thomas Thekkumkara for their careful reading of the manuscript with valuable suggestions and assistance in preparing the art work.

References

Achuta Murthy P. N. and Mistry S. P. (1977) Biotin. *Prog. Food Nutr. Sci.* **2,** 405–455.

Aeberhard E., Grippo J., and Menkes J. H. (1969) Fatty acid synthesis in the developing brain. *Pediatr. Res.* **3,** 590–596.

Alberts A. W. and Vagelos, P. R. (1972) Acyl-CoA Carboxylases, in *The Enzymes* (Boyer P. D., ed.) vol 6 (3rd Ed.) Academic, New York.

Arinze J. C. and Mistry S. P. (1971) Activities of some biotin enzymes and certain aspects of gluconeogenesis during biotin deficiency. *Comp. Biochem. Physiol.* **38B,** 285–294.

Atkin B. M., Buist N. R. M., Utter M. F., Leiter A. B., and Banker B. Q. (1979) Pyruvate caboxylase deficiency and lactic acidosis in a retarded child without Leigh's disease. *Pediatr. Res.* **13,** 109–116.

Attwood P. V. and Keech D. B. (1984) Pyruvate carboxylase. *Curr. Top. Cell. Regul.* **23,** 1–55.

Balazs R., Patel A. J., and Richter D. (1972) Metabolic Compartments in the Brain: Their Properties and Relation to Morphological Structures, in *Metabolic Compartmentation in the Brain* (Balazs R. and Cremer J. E., eds.) Halsted Press, John Wiley, New York.

Ballard F. J. (1970) Kinetic studies with cytosol and mitochondrial phosphoenolpyruvate carboxykinase. *Biochem. J.* **120,** 809–814.

Ballard F. J. and Hanson R. W. (1967a) Changes in lipid synthesis in rat liver during development. *Biochem. J.* **102,** 952–958.

Ballard F. J. and Hanson R. W. (1967b) The citrate cleavage pathway and lipogenesis in rat adipose tissue: Replenishment of oxaloacetate. *J. Lipid Res.* **8,** 73–79.

Ballard F. J. and Hanson R. W. (1969) Purification of phosphoenolpyruvate carboxykinase from the cytosol fraction of rat liver and the immunological demonstration of differences between this enzyme and the mitochondrial phosphoenolpyruvate carboxykinase. *J. Biol. Chem.* **244,** 5626–5630.

Ballard F. J., Hanson R. W., and Reshef L. (1970) Immunological studies with soluble and mitochondrial pyruvate carboxylase activities from rat tissues. *Biochem. J.* **119,** 735–742.

Barritt G. J. (1985) Diseased States in Man and Other Vertebrates, in *Pyruvate Carboxylase* (Keech D. B. and Wallace J. C., eds.) CRC Series in Enzyme Biology, CRC Press, Boca Raton, Florida.

Bartlett K., Ghneim H. K., Stirk J-H., Wastell H. J., Sherratt H. S. A., and Leonard J. V. (1985) Enzyme studies in combined carboxyase deficiency. *Ann. NY Acad. Sci.* **447,** 235–251.

Berl S. and Clarke D. D. (1969) Metabolic Compartmentation of Glutamate in the CNS, in *Handbook of Neurochemistry* (Latjtha A., ed.) vol. 1, Plenum, New York.

Berl S., Takagaki G., Clarke D. D., and Waelsch H. (1962a) Metabolic compartments *in vivo*. Ammonia and glutamic acid metabolism in brain and liver. *J. Biol. Chem.* **237,** 2562–2569.

Berl S., Takagaki G., Clarke D. D., and Waelsch H. (1962b) Carbon dioxide fixation in the brain. *J. Biol. Chem.* **237,** 2570–2573.

Bernstine E. G., Koh C., and Lovelace C. C. (1979) Regulation of mitochondrial malic enzyme synthesis in mouse brain. *Proc. Natl. Acad. Sci. USA* **76,** 6539–6541.

Blass J. P. (1983) Inborn Errors of Pyruvate Metabolism, in *The Metabolic Basis of Inherited Disease* Fifth Ed. (Stanbury J. B., Wyngaarden J. B., Fredrickson D. S., Goldstein J. L., and Brown M. S., eds.) McGraw-Hill, New York.

Brady R. D. (1960) Biosynthesis of fatty acids. II. Studies with enzyme obtained from brain. *J. Biol. Chem.* **235,** 3099–3103.

Brdiczka D. and Pette D. (1971) Intra- and extramitochondrial isozymes of (NADP) malate dehydrogenase. *Eur. J. Biochem.* **19,** 546–551.

Brownsey R. W. and Denton R. M. (1982) Evidence that insulin activates fat-cell acetyl-CoA carboxylase by increased phosphorylation at a specific site. *Biochem. J.* **202,** 77–86.

Chang H.-C. and Lane M. D. (1966) The enzymatic carboxylation of phosphoenolpyruvate. II. Purification and properties of liver mitochondrial phosphoenolpyruvate carboxykinase. *J. Biol. Chem.* **241,** 2413–2420.

Chang H.-C., Maruyama H., Miller R. S., and Lane M. D. (1966) The enzymatic carboxylation of phosphoenolpyruvate. III. Investigation of the kinetics and mechanism of the mitochondrial phosphoenolpyruvate carboxykinase-catalyzed reaction. *J. Biol. Chem.* **241,** 2421–2430.

Cheng S.-C. and Cheng R. H. C. (1972) A mitochondrial phosphoenolpyruvate carboxykinase from rat. brain. *Arch. Biochem. Biophys.* **151,** 501–511.

Cheng S.-C., Nakamura R., and Waelsch H. (1967) Relative contribution of carbon dioxide fixation and acetyl-CoA pathways in two nervous tissues. *Nature* **216,** 928–929.

Chiang G. S and Mistry S. P. (1974) Activities of pyruvate carboxylase and propionyl-CoA carboxylase in rat tissues during biotin deficiency and restoration of the activities after biotin administration. *Proc. Soc. Exp. Biol. Med.* **146,** 21–24.

Cote L., Cheng S.-C., and Waelsch H. (1966) CO_2 fixation in the nervous system. I. CO_2 fixation in the sciatic nerve of the bullfrog. *J. Neurochem.* **13,** 271–279.

Crane R. K. and Ball E. G. (1951) Relationship of $C^{14}O_2$ fixation to carbohydrate metabolism in retina. *J. Biol. Chem.* **189,** 269–276.

D'Adamo A. F., Jr. and D'Adamo A. P. (1968) Acetyl transport mechanisms in the nervous system. The oxoglutarate shunt and fatty acid synthesis in the developing rat brain. *J. Neurochem.* **15,** 315–323.

De Vellis J., Schjeide O. A., and Clemente C. D. (1967) Protein synthesis and enzymatic patterns in the developing brain following head X-irradiation of newborn rats. *J. Neurochem.* **14,** 449–511.

Dhopeshwarkar G. A., Maier R., and Mead J. F. (1969) Incorporation of [1-^{14}C]-acetate into the fatty acids of the developing rat brain. *Biochim. Biophys. Acta* **187,** 6–12.

Felicioli R. A., Gabrielli F., and Rossi C. A. (1966) Intracellular distribution of pyruvate carboxylase in mammalian brain cortex. *Experientia* **22,** 728–729.

Flavin M. and Ochoa S. (1957) Metabolism of propionic acid in animal tissues. I. Enzymatic conversion of propionate to succinate. *J. Biol. Chem.* **229,** 965–979.

Frenkel R. (1972) Isolation and some properties of a cytosol and a mitochondrial malic enzyme from bovine brain. *Arch. Biochem. Biophys.* **152,** 136–143.

Gilbert E. F., Arya S., and Chun R. (1983) Leigh's necrotizing encephalopathy with pyruvate carboxylase deficiency. *Arch. Pathol. Lab. Med.* **107,** 162–166.

Greenspan M. and Lowenstein J. M. (1967) Effect of magnesium ion and adenosine triphosphate on the activity of acetyl coenzyme A carboxylase. *Arch. Biochem. Biophys.* **118,** 260–263.

Gromek A. and Pastuszko A. (1977) The localization of mitochondrial NADP-dependent isocitrate dehydrogenase in normal and hypoxic conditions. *J. Neurochem.* **28,** 429–433.

Gross I. and Warshaw J. B. (1974) Fatty acid biosynthesis in developing brain. Acetyl-CoA carboxylase activity. *Biol. Neonate* **25,** 365–375.

Grover W. D., Auerbach V. H., and Patel M. S. (1972) Biochemical studies and therapy in subacute necrotizing encephalomyelopathy (Leigh's syndrome). *J. Pediatr.* **81,** 39–44.

Guchhait R. B., Polakis S. E., Hollis D., Fenselau C., and Lane M. D. (1974) Acetyl coenzyme A carboxylase system in *Escherichia coli.* Site of carboxylation of biotin and enzymatic reactivity of 1'-*N*-(ureido)-carboxybiotin derivatives. *J. Biol. Chem.* **249,** 6646–6656.

Halenz D. R., Feng J-Y., Hegre C. S., and Lane M. D. (1962) Some enzymic properties of mitochondrial propionyl carboxylase. *J. Biol. Chem.* **237**, 2140–2147.

Hanson R. W. and Garber A. J. (1972) P-enolpyruvate carboxykinase. I. Its role in gluconeogenesis. Comments in Biochemistry. *Am. J. Clin. Nutr.* **25**, 1010–1021.

Hardie D. G., Carling D., Ferrai S., Guy P. S., and Aitken A. (1986) Characterization of the phosphorylation of rat mammary ATP-citrate lyase and acetyl-CoA carboxylase by Ca^{2+} and calmodulin-dependent multiprotin kinase and Ca^{2+} and phospholipid-dependent protein kinase. *Eur. J. Biochem.* **157**, 553–561.

Hector M. L., Cochran B. C., Logue E. A., and Fall R. R. (1980) Subcellular localization of 3-methylcrotonyl-coenzyme A carboxylase in bovine kidney. *Arch. Biochem. Biophys.* **199**, 28–36.

Henderson N. S. (1965) Isozymes of isocitrate dehydrogenase: Subunit structure and intracellular location. *J. Exp. Zool.* **158**, 263–274.

Hod Y., Yoo-Warren H., and Hanson R. W. (1984) The gene encoding the cytosolic form of phosphoenolpyruvate carboxykinase (GTP) from the chicken. *J. Biol. Chem.* **259**, 15609–15614.

Hod Y., Cook J. S., Weldon S. L., Short J. M., Wynshaw-Boris A., and Hanson R. W. (1986) Differential expression of the genes for the mitochondrial and cytosolic forms of P-enolpyruvate carboxykinase. *Ann. NY Acad. Sci.* **478**, 31–45.

Holland R. and Hardie D. G. (1985) Both insulin and epidermal growth factor stimulate fatty acid synthesis and increase phosphorylation of acetyl-CoA carboxylase and ATP-citrate lyase in isolated hepatocytes. *FEBS Lett.* **181**, 308–312.

Hommes F. A., Polman H. A., and Reerink J. D. (1968) Leigh's encephalomyelopathy: An inborn error of gluconeogenesis. *Arch. Dis. Child.* **43**, 423–426.

Isohashi F., Shibayama K., Maruyama E., Aoki Y., and Wada F. (1971) Immunological studies on malate dehydrogenase (decarboxylating) (NADP). *Biochim. Biophys. Acta* **250**, 14–24.

Jomain-Baum M., Schramm V. L., and Hanson R. W. (1976) Mechanism of 3-mercaptopicolinic acid inhibition of hepatic phosphoenolpyruvate carboxykinase (GTP). *J. Biol. Chem.* **251**, 37–44.

Kalousek F., Darigo M. D., and Rosenberg L. E. (1980) Isolation and characterization of propionyl-CoA carboxylase from normal human liver: Evidence for a protomeric tetramer of nonidentical subunits. *J. Biol. Chem.* **255**, 60–65.

Kaziro Y., Ochoa S., Warner R. C., and Chen J.-Y. (1961) Metabolism of propionic acid in animal tissues. VIII. Crystalline propionyl carboxylase. *J. Biol. Chem.* **236**, 1917–1923.

Keech D. B. and Attwood P. V. (1985) The Reaction Mechanism, in *Pyruvate Carboxylase* (Keech D. B. and Wallace J. C., eds.) CRC Series in Enzyme Biology, CRC Press, Boca Raton, Florida.

Keech D. B. and Utter M. F. (1963) Pyruvate carboxylase. II. Properties. *J. Biol. Chem.* **238,** 2609–2614.

Keech D. B. and Wallace J. C. (eds.) (1985) *Pyruvate carboxylase* pp. 1–262. CRC Series in Enzyme Biology, CRC Press, Boca Raton, Florida.

Kelley R. E., Jr. and Joel C. D. (1973) The activity of acetyl-coenzyme A carboxylase in rat brain. *Biochem. Soc. Trans.* **1,** 467–469.

Kim K-H. (1983) Regulation of acetyl-CoA carboxylase. *Curr. Top. Cell. Regul.* **22,** 143–176.

Knappe J., Schlegel H. G., and Lynen F. (1961) Zur Biochemischen funktion des biotins: I, Die beteiligung der β-methylcrotonyl carboxylase un der bildung von β-hydroxy-β-methylglutaryl CoA aus β-hydroxyisovaleryl CoA. *Biochem. Z.* **335,** 101–122.

Lai J. C. K. and Clark J. B. (1978) Isocitrate dehydrogenase and malate dehydrogenase in synaptic and non-synaptic rat brain mitochondria: A comparison of their kinetic constants. *Biochem. Soc. Trans.* **6,** 993–995.

Lane M. D., Moss J., and Polakis S. E. (1974) Acetyl coenzyme A carboxylase. *Curr. Top. Cell Regul.* **8,** 139–195.

Lau E. P., Cochran B. C., and Fall R. R. (1980) Isolation of 3-methylcrotonylcoenzyme A carboxylase from bovien kidney. *Arch. Biochem. Biophys.* **205,** 352–359.

Lau E. P., Cochran B. C., Munson L., and Fall R. R. (1979) Bovine kidney 3-methylcrotonyl-CoA and propionyl-CoA carboxylases: Each enzyme contains non-identical subunits. *Proc. Natl. Acad. Sci. USA* **76,** 214–218.

Lent B. A., Lee K.-H., and Kim K.-H (1978) Regulation of rat liver acetyl-CoA carboxylase. Stimulation of phosphorylation and subsequent inactivation of liver acetyl-CoA carboxylase by cyclic 3':5'-monophosphate and effect on the structure of the enzyme. *J. Biol. Chem.* **253,** 8149–8156.

Ling A.-M. and Keech D. B. (1966) Pyruvate carboxylase from sheep kidney. I. Purification and some properties of the enzyme. *Enzymologia* **30,** 367–380.

Loverde A. W. and Lehrer G. M. (1973) Subcellular distribution of isocitrate dehydrogenases in neonatal and adult mouse brain. *J. Neurochem.* **20,** 441–448.

Lowenstein J. M., and Smith S. R. (1962) Intra- and extra-mitochondrial isocitrate dehydrogenase. *Biochim. Biophys. Acta* **56,** 385–387.

Mahan D. E., Mushahwar I. K., and Koeppe R. E. (1975) Purification and properties of rat brain pyruvate carboxylase. *Biochem. J.* **145,** 25–35.

Majno G. and Karnovsky M. L. (1958) A biochemical and morphological studies of myelination and demyelination. 1. Lipids biosynthesis *in vitro* by normal nervous tissue. *J. Exp. Med.* **107,** 475–496.

Matsuhashi M., Matsuhashi S., and Lynen F. (1964) Zur biosynthese der fettsauren. V. Die acetyl-CoA carboxylase aus rattenleber und ihre aktivierung durch citronensaure. *Biochem. Z.* **340,** 263–289.

McClure W. R., Lardy H. A., and Kneifel H. P. (1971a) Rat liver pyruvate carboxylase. I. Preparation, properties, and cation specificity. *J. Biol. Chem.* **246,** 3569–3578.

McClure W. R., Lardy H. A., Wagner M., and Cleland W. W. (1971b) Rat liver pyruvate carboxylase. II. Kinetic studies of the forward reaction. *J. Biol. Chem.* **246,** 3579–3583.

Mildvan A. S., Scrutton M. C., and Utter M. F. (1966) Pyruvate carboxylase. VII. A possible role for tightly bound manganese. *J. Biol. Chem.* **241,** 3488–3498.

Miller A. L., and Levy H. R. (1969) Rat mammary acetyl coenzyme A carboxylase. I. Isolation and characterization. *J. Biol. Chem.* **244,** 2334–2342.

Moldave K., Winzler R. J., and Pearson H. E. (1953) The incorporation *in vitro* of C^{14} into amino acids of control and virus-infected mouse brain. *J. Biol. Chem.* **200,** 357–365.

Moss J., Yamagishi M., Kleinschmidt A. K., and Lane M. D. (1972) Acetyl coenzyme A carboxylase. Purification and properties of the bovine adipose tissue enzyme. *Biochemistry* **11,** 3779–3786.

Murphy J. V., Isohashi F., Weinberg M. B., and Utter M. F. (1981) Pyruvate carboxylase deficiency: An alleged biochemical cause of Leigh's disease. *Pediatrics* **68,** 401–404.

Murthy M. R. V. and Rappoport D. A. (1963) Biochemistry of the developing rat brain. II. Neonatal mitochondrial oxidations. *Biochim. Biophys. Acta* **74,** 51–59.

Naruse H., Cheng S.-C., and Waelsch H. (1966a) CO_2 fixation in the nervous tissue. IV. CO_2 fixation and citrate metabolism in lobster nerve. *Exp. Brain Res.* **1,** 284–290.

Naruse H., Cheng S.-C., and Waelsch H. (1966b) CO_2 fixation in the nervous tissue. V. CO_2 fixation and citrate metabolism in rabbit nerve. *Exp. Brain Res.* **1,** 291–298.

Numa S. and Yamashita S. (1974) Regulation of lipogenesis in animal tissues. *Curr. Top. Cell. Regul.* **8,** 197–246.

Obermayer M. and Lynen F., (1976) Structure of biotin enzymes. *Trends Biochem. Sci.* **1,** 169–171.

Ochoa S. (1955) Malic Enzyme, in *Methods in Enzymology* (Colowick S. P. and Kaplan N. O., eds.) Academic, New York.

Otsuki S., Geiger A., and Gombos G. (1963) The metabolic pattern of the brain in brain perfusion experiments *in vivo*. I. The quantitative significance of CO_2 assimilation in the metabolism of the brain. *J. Neurochem.* **10,** 397–404.

Patel M. S. (1972) The effect of phenylpyruvate on pyruvate metabolism in rat brain. *Biochem. J.* **128,** 677–684.

Patel M. S. (1974) The relative significance of CO_2-fixing enzymes in the metabolism of rat brain. *J. Neurochem.* **22,** 717–724.

Patel M. S. (1977) Age-dependent changes in the oxidative metabolism in rat brain. *J. Gerontol.* **32,** 643–646.

Patel M. S. and Owen O. E. (1976) Lipogenesis from ketone bodies in rat brain. Evidence for conversion of acetoacetate to acetyl coenzyme A in the cytosol. *Biochem. J.* **156,** 603–607.

Patel M. S. and Tilghman S. M. (1973) Regulation of pyruvate metabolism via pyruvate caraboxylase in rat brain mitochondria. *Biochem. J.* **132,** 185–192.

Patel M. S. and Tonkonow B. L. (1974) Development of lipogenesis in rat brain cortex: The differential incorporation of glucose and acetate into brain lipids *in vitro*. *J. Neurochem.* **23,** 309–313.

Plaut G. W. E. (1962) Isocitric Dehydrogenase (TPN-Linked) from Pig Heart (Revised Procedure), in *Methods in Enzymology* (Colowick S. P. and Kaplan N. O., eds.) vol. V, Academic, New York.

Rafalowska U. and Ksiezak H. (1976) Subcellular localization of enzymes oxidizing citrate in the rat brain. *J. Neurochem.* **27,** 813–815.

Reshef L., Hanson R. W., and Ballard F. J. (1969) Glyceride-glycerol synthesis from pyruvate. Adaptive changes in phosphoenolpyruvate carboxykinase and pyruvate carboxylase in adipose tissue and liver. *J. Biol. Chem.* **244,** 1994–2001.

Rosenberg L. E. (1983) Disorders of Propionate and Methylmalonate Metabolism, in *The Metabolic Basis of Inherited Disease* (Stanbury J. B., Wyngaarden J. B., Fredrickson D. S., Goldstein J. L., and Brown M. S., eds.) Fifth Ed., McGraw-Hill, New York.

Rylatt D. B., Keech D. B., and Wallace J. C. (1977) Pyruvate carboxylases: Isolation of the biotin-containing tryptic peptide and the determination of its primary sequence. *Arch. Biochem. Biophys.* **183,** 113–122.

Salganicoff L. and Koeppe R. E. (1968) Subcellular distribution of pyruvate carboxylase, diphosphopyridine nucleotide and triphosphopyridine nucleotide isocitrate dehydrogenases, and malate enzyme in rat brain. *J. Biol. Chem.* **243,** 3416–3420.

Scrutton M. C. and Mildvan A. S. (1970) Pyruvate carboxylase: Nuclear magnetic resonance studies of the enzyme-manganese-oxaloacetate and enzyme-manganese-pyruvate bridge complexes. *Arch. Biochem. Biophys.* **140,** 131–151.

Scrutton M. C. and Utter M. F. (1965) Pyruvate carboxylase. III. Some physical and chemical properties of the highly purified enzyme. *J. Biol. Chem.* **240,** 1–9.

Scrutton M. C. and Young M. R. (1972) Pyruvate Carboxylase, in *The Enzymes* (Boyer P. D., ed.) Third Edn., Academic, New York.

Scrutton M. C., Olmsted M. R., and Utter M. F. (1969) Pyruvate Carboxylase from Chicken Liver, in *Methods in Enzymology* (Lowenstein J. M., ed.) Academic, New York.

Scrutton M. C., Utter M. F., and Mildvan A. S. (1966) Pyruvate carboxylase. VI. The presence of tightly bound manganese. *J. Biol. Chem.* **241,** 3480–3487.

Seufert D., Herleman E.-M., Allbrecht E., and Seubert W. (1971) On the mechanism of gluconeogenesis and its regulation. VII. Purification and properties of pyruvate carboxylase from rat liver. *Hoppe Seylers Z. Physiol. Chem.* **352,** 459–478.

Shank R. P., Bennett G. S., Freytag S. O., and Campbell G. LeM. (1985) Pyruvate carboxylase: An astrocyte-specific enzyme implicated in the replenishment of amino acid neurotransmitter pools. *Brain Res.* **329,** 364–367.

Sweetman L., Burri B. J., and Nyhan W. L. (1985) Biotin holocarboxylase synthetase deficiency. *Ann. NY Acad. Sci.* **447,** 288–296.

Tanaka K. and Rosenberg L. B. (1983) Disorders of Branched Chain Amino Acid and Organic Acid Metabolism, in *The Metabolic Basis of Inherited Disease* (Stanbury J. B., Wyngaarden J. B., Fredrickson D. S., Goldstein J. L., and Brown M. S., eds.) Fifth Edn., McGraw-Hill, New York.

Utter M. F. and Keech D. B. (1963) Pyruvate carboxylase. I. Nature of the reaction. *J. Biol. Chem.* **238,** 2603–2608.

Utter M. F. and Kolenbrander H. M. (1972) Formation of Oxalacetate by CO_2 Fixation on Phosphoenolpyruvate, in *The Enzymes* (Boyer P. D., ed.) Third Edn. Academic, New York.

Utter M. F. and Kurahashi K. (1954) Purification of oxalacetic carboxylase from chicken liver. *J. Biol. Chem.* **207,** 787–802.

Utter M. F. and Scrutton M. C. (1969) Pyruvate carboxylase. *Curr. Top. Cell. Regul.* **1,** 253–296.

Utter M. F., Keech D. B., and Scrutton M. C. (1964) A possible role for acetyl CoA in the control of gluconeogensis. *Adv. Enzyme Regul.* **2,** 49–68.

Vagelos P. R., Alberts A. W., and Martin D. B. (1963) Studies on the mechanism of activation of acetyl coenzyme A carboxylase by citrate. *J. Biol. Chem.* **238,** 533–540.

Valentine R. C., Wrigley N. G., Scrutton M. C., Irias J. J., and Utter M. F. (1966) Pyruvate carboxylase. VIII. The subunit structure as examined by electron microscopy. *Biochemistry* **5,** 3111–3116.

Volpe J. J. and Vagelos P. R. (1976) Mechanisms and regulation of biosynthesis of saturated fatty acids. *Physiol. Rev.* **56,** 339–417.

Waelsch H., Berl S., Rossi C. A., Clarke D. D., and Purpura D. P. (1964) Quantitative aspects of CO_2 fixation in mammalian brain *in vivo. J. Neurochem.* **11,** 717–728.

Waelsch H., Cheng S.-C., Cote L. J., and Naruse H. (1965) CO_2 fixation in the nervous system. *Proc. Natl. Acad. Sci. USA* **54,** 1249–1253.

Waite M. and Wakil S. J. (1962) Studies on the mechanism of fatty acid synthesis. XII. Acetyl coenzyme A carboxyase. *J. Biol. Chem.* **237,** 2750–2757.

Wallace J. C. (1985) Distribution and Biological Functions of Pyruvate Carboxylase in Nature, in *Pyruvate Carboxylase* (Keech D. B. and Wallace J. C., eds.) CRC Series in Enzyme Biology, CRC Press, Boca Raton, Florida.

Wallace J. C. and Easterbrook-Smith S. B. (1985) The Structure of Pyruvate Carboxylase, in *Pyruvate Carboxylase* (Keech D. B. and Wallace J. C., eds.) CRC Series in Enzyme Biology, CRC Press, Boca Raton, Florida.

Watanabe T., Goto H., and Ogasawara N. (1974) Specific development of isocitrate dehydrogenases in rat brain. *Biochim. Biophys. Acta* **358,** 340–346.

Wilbur D. O. and Patel M. S. (1974) Development of mitochondrial pyruvate metabolism in rat brain. *J. Neurochem.* **22,** 709–715.

Witters L. A., Tipper J. P., and Bacon G. W. (1983) Stimulation of site-specific phosphorylation of acetyl coenzyme A carboxylase by insulin and epinephrine. *J. Biol. Chem.* **258,** 5643–5648.

Wood H. G. and Barden R. E. (1977) Biotin enzymes. *Ann. Rev. Biochem.* **46,** 385–413.

Yeh L. A. and Kim K.-H. (1980) Regulation of acetyl-CoA carboxylase: Properties of CoA activation of acetyl-CoA carboxylase. *Proc. Natl. Acad. Sci. USA* **77,** 3351–3355.

Yoo-Warren H., Monahan J., Short J., Short H., Bruzel A., Wynshaw-Boris A., Meisner H. M., Samols D., and Hanson R. W. (1983) Isolation and characterization of the gene coding for cytosolic phosphoenolpyruvate carboxykinase (GTP) from the rat. *Proc. Natl. Acad. Sci. USA* **80,** 3656–3660.

Yu A. C. H., Drejer J., Hertz L., and Schousboe A. (1983) Pyruvate carboxylase activity in primary cultures of astrocytes and neurons. *J. Neurochem.* **41,** 1484–1487.

High-Energy Phosphates in Brain from Dissected Freeze-Dried Regions to Single Cells

Suzanne Dworsky, Marian Namovic,
and David W. McCandless

1. Introduction

The successful measurement of cerebral high-energy phosphates (ATP, ADP, AMP, and phosphocreatine) by the enzymatic techniques to be described in this chapter is dependent on: (1) adequate tissue fixation, (2) preparation, dissection, and weighing of samples to be analyzed, and (3) fluorometric enzymatic analysis of the sample for high-energy phosphates. The fact that the brain is organized in a highly complex fashion necessitates the measurement of metabolites in very small pieces of tissue, down to and including single cells. Therefore, the methodology for dissection, weighing, and analysis of tissue varies to some extent with the size of the sample.

Historically, the original development of micromethods was the result of work in the Carlsberg Laboratories in Copenhagen by Kaj Linderstrom-Lang and Heinz Holter. An early pupil of Linderstrom-Lang and Holter was Oliver Lowry, who has continued to develop the methodologies. Notable was the invention of the quartz-fiber fishpole balance, which enabled each piece of tissue to be weighed with great accuracy, and the development of the oil-well technique and enzymatic cycling.

The present chapter is limited to the measurement of high-energy phosphates using these micromethods; however, most metabolites of intermediary metabolism can be measured. All of these techniques take advantage of the fact that TPNH—reduced triphosphopyridine nucleotide (DPNH—reduced diphosphopyridine nucleotide) fluoresces, whereas TPN—triphosphopyridine nucleotide (DPN—diphosphopyridine nucleotide) does not. Any metabolite that can be linked to a TPN/TPNH or DPN/DPNH reaction can theoretically be measured. The classic work by Lowry and Passonneau details analysis of many metabolites (Lowry and Passonneau, 1972).

2. Adequate Tissue Fixation

The finest analytical techniques are wasted on tissue that has been suboptimally fixed. Therefore, the first issue the investigator must face is the question of the lability of the metabolite(s) to be measured. From the standpoint of the present chapter, fixation is a primary concern. Several methods have been developed to provide a rapid inactivation of tissue enzymes prior to metabolite analysis, and the reader is referred to a thorough discussion in this volume of the advantages and disadvantages of each. For this chapter, a brief description of the most widely used methods will be provided. These methods include: rapid submersion of the animal in liquid N_2, funnel freezing, freeze blowing, and microwave irradiation. The rationale behind these methods is to effectively inactivate tissue enzymes as close to instantaneously as possible. The problem is that brain is well encased in bone and has a rapid metabolic rate. These two factors rule out the use of altered pH to inactivate enzymes, a method that is used in other tissues such as liver.

2.1. Rapid Submersion in Liquid N_2

Briefly stated, in this method a Dewar flask (2–4-L capacity) is filled with liquid N_2, and the experimental animal is quickly plunged into the coolant and stirred until bubbling ceases. Liquid N_2 freezes at about –210°C and boils at –196°C. Therefore, when a warm object is placed in liquid N_2, the conversion of the liquid to a gas tends to form an insulating layer around the animal, which slows somewhat the cooling process. Stirring breaks this layer and enhances cooling. Extensive studies have been performed timing the rate of the cooling front and comparing labile metabolite values with other sacrifice methods. The concensus is that for *small* animals (<30 g) and/or superficial structures (cortex), submersion in liquid N_2 is adequate. For deeper structures such as brainstem or even hippocampus, values of metabolites are usually reported lower than those obtained using, for example, microwave irradiation. Advantages of the submersion method include low cost and ease, as well as the fact that no restraining devices are used. One disadvantage, in common with all freezing methods, is that subsequent dissection of the frozen tissue may be difficult.

2.2. Funnel Freezing

In this method, the usual procedure is to pour liquid N_2 onto the exposed brain through a funnel, which serves to direct the N_2. The pouring of liquid N_2 is continued until the brain is frozen (Welsh, 1980). At that time, the head is usually removed into a Dewar flask of N_2 to ensure cooling of adjacent tissue. This method has the advantage of being able to be used on animals that are maintained on a respirator. This ensures oxygen delivery to the brain during the freezing episode. The funnel freezing method can also be used on large animals. Metabolite values derived from brains of animals sacrificed in this manner are high or higher than any other sacrifice technique. This method is cost effective and easy to use. There are no disadvantages to this method provided the experimental animal is anesthetized and artifically ventilated.

2.3. Freeze Blowing

In the freeze blowing technique, two hollow metal probes are driven by compressed air into the cranial vault of the animal, and a jet of air through one probe blows the brain out of the cranium through the other probe (Veech et al., 1973). The rapidly extruded brain is impelled upon a disc of steel or aluminum prechilled in liquid N_2. The usual result is that the entire supratentarial portion of the brain is frozen as a thin wafer in less than 1 s. Using this device has resulted in values for labile metabolites equal to or surpassing those obtained by other methods. The advantage of this method is the very rapid inactivation of metabolic activity. Disadvantages include the inability to do regional determinations and the cost of the freeze-blowing device.

2.4. Microwave Irradiation

In this technique, a focused microwave exposure, usually at 2450 or 915 MHz, and with a power output of from 6.5 to 25 KW, is delivered to the animal's head (Medina et al., 1980). This results in an increase in brain temperature from 37 to over 90°C within 1 s. Problems with this technique center on the even distribution of the irradiation through the brain. However, when properly set up, the microwave device is probably the best sacrifice technique in terms of high values for labile metabolites and the relative ease at which regional dissection is possible. Early fears that the heating might result in the movement of metabolites from one microregion to another have been dispelled. If the head is allowed to come to room temperature (30+ min) before removing the brain, sections can be prepared in a cryostat without loss of structural integrity. Disadvantages include high price and the necessity for restraining the animal before sacrifice and the inability to measure enzymes in the samples.

To summarize, if the experimental design is such that small animals (mice) can be used and only cortical layers examined, rapid submersion with stirring is the method of choice for tissue enzyme inactivation. If cost is not a factor, then a high-power microwave oven is clearly the favored choice. Its versatility in terms of animal size, ease of dissection, and ability to analyze all levels of brain make the microwave device unique. A more thorough review of sacrifice methods appears elsewhere in this volume.

3. Preparation, Dissection, and Weighing of Samples for Analysis

Once the experimental animal has been sacrificed, the brain can be stored at –70°C for several months without detectable loss of high-energy phosphates. The analytical techniques described in this chapter are intended for samples weighing from 4 ng (single cells) to 1 μg. Samples this size are best prepared by first removing the brain from the cranial vault, then sectioning it at 20 μm. The sections are then freeze dried and free hand dissected into appropriate regions. The microsamples are individually weighed before placing them in an oil well rack for the initial steps of the analytical procedures.

3.1. Removal of Brain from Cranial Vault

The removal of the brain from the cranial vault can be a tricky procedure. If the animal has been sacrificed using a microwave device, then after a suitable cooling period (<30 min) the brain can be removed, divided into regions, and stored in scintillation vials until use. This represents a major advantage of the microwave irradiation sacrifice technique.

For brain samples stored at –70°C in the cranial vault, removing the brain is best accomplished by using the Wedeen-type cryostat. This apparatus is a cold chamber into which the hands can be placed. The brains can be removed at a controlled temperature. At temperatures below about –30°C, the brain is quite brittle, and chips tend to fly off when the bone is removed. At temperatures around –5 to –10°C, the brain can be readily removed; however, some autolytic loss of metabolites may occur, even if the brain is exposed to the warmer temperature for as little as 30 min. Each laboratory should develop its own technique in terms of temperature and validate it by comparing values obtained from tissue warmed up to –10°C to values from tissue maintained at –70°C. Many investigators choose to use slabs of dry ice as the working surface. A small chisel and hammer can be used effectively after some practice. Scalpel blades tend to break frequently. By proceeding slowly, and with practice, the entire brain, or regions thereof, can be removed with no undesired loss of tissue or metabolites.

3.2. Sectioning the Brain

It is important to prevent the formation of ice crystals on the samples during storage and during the handling prior to sectioning. Ordinarily, the samples are sectioned immediately after removal from the cranial vault. If the dissection is done on slabs of dry ice, the pieces of brain to be sectioned should be mounted and placed in the cryostat rather than left on the dry ice, since CO_2 could be absorbed by the tissue, which could affect the assay. The samples for sectioning are mounted on cryostat metal discs using brain paste. Brain paste has the advantage of being chemically inert. It is prepared by mashing unused brain tissue with about one third volume of water. The paste is kept just above freezing prior to use, and can be stored frozen at –20°C. Enough paste is placed on the cryostat disc so that the lower 25% of the tissue sample will be submerged. The disc and sample is then picked up with forceps and submersed into a small beaker containing hexane and dry ice. The dry ice should be added to the hexane slowly to prevent boiling over. The sample/disc can now be transferred to the cryostat. Sectioning at 20 μm is best accomplished at a temperature of about –18°C. The sample/disc should be allowed to equilibrate with the temperature of the cryostat before attempting to section.

The 20-μm thick sections will be transferred to aluminum racks (Fig. 1) prior to freeze drying; therefore the racks as well as the ace glass holders (Fig. 2) should be placed into the cryostat about 15–20 min before sectioning. If they are placed in the cryostat too early, ice crystals will form. Each investigator must determine the best way to achieve good, flat, chatter-free sections from the cryostat. A sharp blade and good even antiroll plate are essential. Sections are picked up from the blade using a small brush or Teflon probe with a sharp end. A hair point (Fig. 3) can also be used for moving small sections. Ordinarily, three to six sections can be placed in each hole in the aluminum holders. When full, the top cover-slide is placed on and tightened. Care must be exercised not to let the hand warm up the holder or the sections will melt. Following this, the aluminum holders are placed in the ace glass holders and transferred rapidly to a freezer at –35 to –40°C with a port. The vacuum pump is connected to the ace glass holder through the port using vacuum tubing. Vacuum dissection should proceed as soon as possible, and a minimum of 18 h at –35°C is needed for thorough freeze drying. Following the freeze drying, the ace glass holders can be closed while the pump is on, and the sections stored indefinitely at –35 to –40°C under vacuum. When ready for analysis, the ace glass holders are removed from the freezer and the vacuum pump attached and pumping initiated. The ace glass holders should be opened to the vacuum pump as soon as possible and allowed to come to room temperature (30–45 min) while the pump is running. Opening the ace glass holder before it and its contents have reached room temperature allows warm humid air to rush in, and the samples will absorb the moisture in the air and be ruined. Remember that the metal section holders may be colder than the ace glass holder during the first 30 min or so after removal from the freezer. An ace glass holder with four aluminum section holders takes 15 min longer to reach room temperature than one with only one aluminum holder.

3.3. Dissection of the Samples

It is imperative that the ace glass holders be opened only when they and their contents have reached room temperature. The humidity can be a problem when it is over 50–60%. Therefore, it is best to open the holders in the room where dissection and weighing of the samples is to take place in order to minimize the chance that the samples might pick up H_2O from the room air.

The sections are removed from the aluminum holders and placed on a viewing box (Fig. 4). The box has a translucent top and

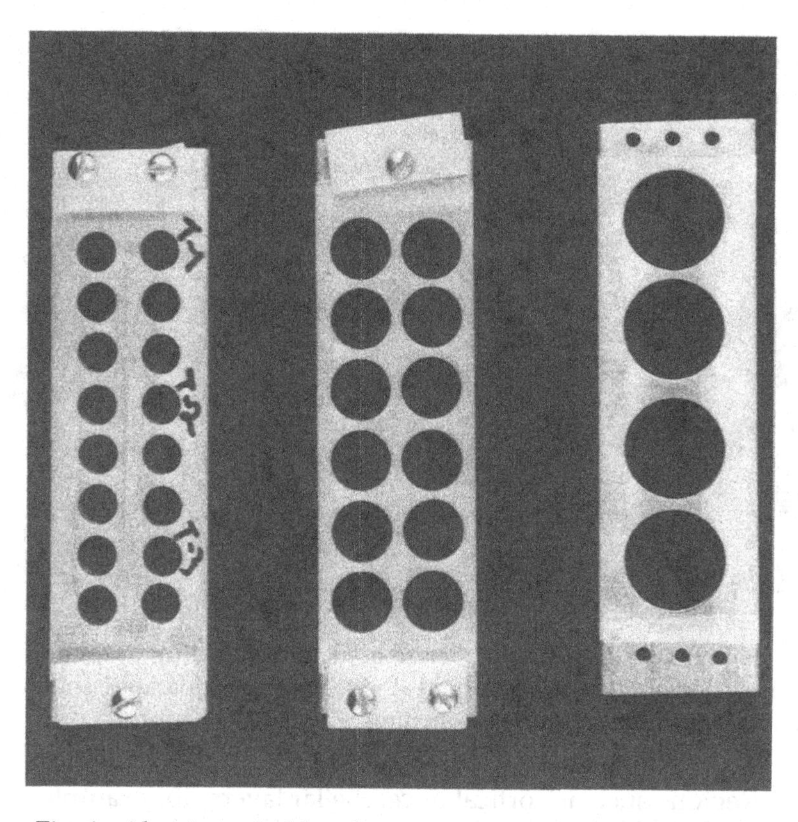

Fig. 1. Aluminum holders for sections. The size is about $3\frac{1}{2} \times 1$ in. The plastic end retainers tighten down and serve to hold 3×1 in. microscope slides in place. Holders with large holes are used for whole brain sections.

is illuminated from beneath using a microscope light. The sections are best viewed through a zoom stereoscopic dissecting microscope. Specially constructed microscalpels (Fig. 3) are used to cut the tissue into desired samples. These scalpels are made using dowel rod for the handle and a stiff flexible bristle to which is attached a razor blade fragment. Small razor blade fragments suitable for dissecting purposes can be made by snapping off pieces from a larger blade using hemostats. Be sure to wear protective goggles when making these fragments. They are attached to the flexible bristle using epoxy. The bristle is in turn attached to a small piece of copper wire, which in turn is epoxyed to the wooden dowel (*see* Fig. 3). The copper wire can be bent to provide the best cutting angle.

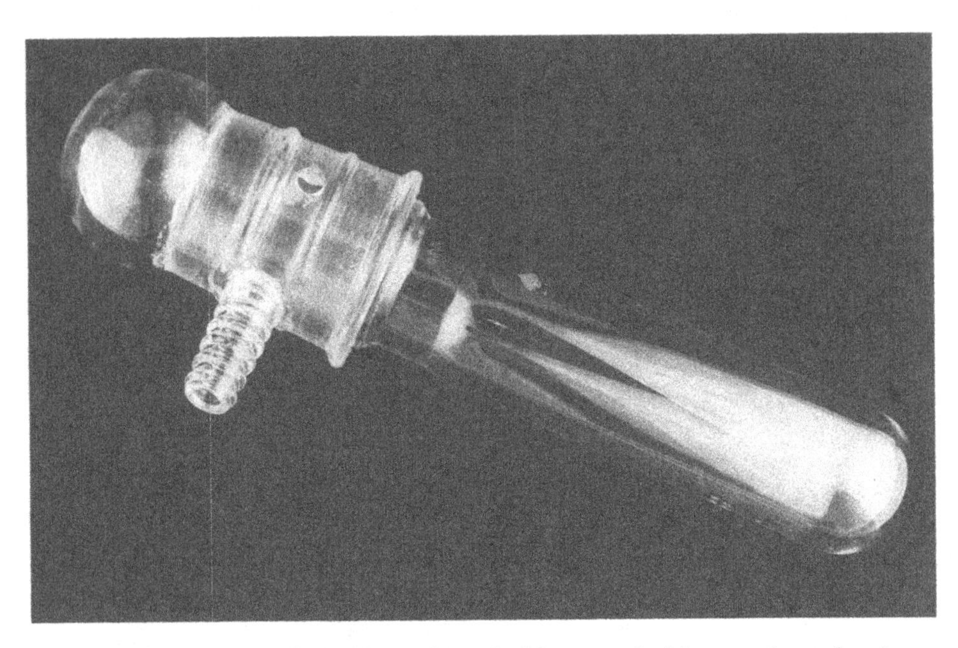

Fig. 2. Ace glass holder. These holders can hold up to four aluminum section holders. When stored at –40°C, a piece of parafilm should be placed over the vacuum tube opening.

Regions such as cortical or cerebellar layers, for example, can be visualized in the freeze dried sections. Several good atlases are available to aid in localization of areas of interest. Alternate sections can be picked up and placed on microscope slides at the time of sectioning and stained with H&E or Nissl.

The area in which dissection and weighing of the tissue takes place must be dehumidified as best as possible. Humidity of greater than 50% can cause changes in tissue metabolites. The temperature recommended is 20°C or as cool as comfort allows. If the room temperature is too hot, greater than 25°C in some cases, the tissue can become sticky and difficult to maneuver. Another important environmental factor are air drafts that can easily blow the very light, dry tissue samples. It may be necessary to enclose the dissecting and weighing area behind curtains or glass doors to prevent air drafts.

Static electricity can be a problem if not controlled. Brushing your hand over the microscope stage, touching your hair, or wearing wool-type sweaters can cause increased problems with static electricity. Radium or polonium are sources that can help to dis-

Fig. 3. Wooden pipet holder containing a microscoped, hair point, and three constriction pipets. The end constriction pipet is the oil well type with a 45°C bend and fine tip. These pipets are all calibrated using gravimetric or radioisotope methods.

sipate charge when hung near the dissecting stage. *(Note radiation safety.)*

Hand tremor is an important limiting factor when doing free-hand dissection. This tremor can be minimized by bracing both hands together against the dissecting surface. The tip of the blade is used as a pivot point. Once the tip is placed just in front of the area to be cut, it can be moved slightly in a lateral motion and then a downward motion to make the desired cut. Using this procedure, samples 15–20 μm in diameter can be easily free-hand dissected. The successfully dissected sample can be picked up and transferred using a hair point. A large sample will adhere to a stiff and sharp hair point when pressed into the sample. For smaller samples it is only necessary to merely touch the hair point to the tissue or to try and pick the sample up from underneath. It is once again important that static charge be controlled. The desired tissue samples are lifted on to the sample carrier (Fig. 4) in the appropriate order.

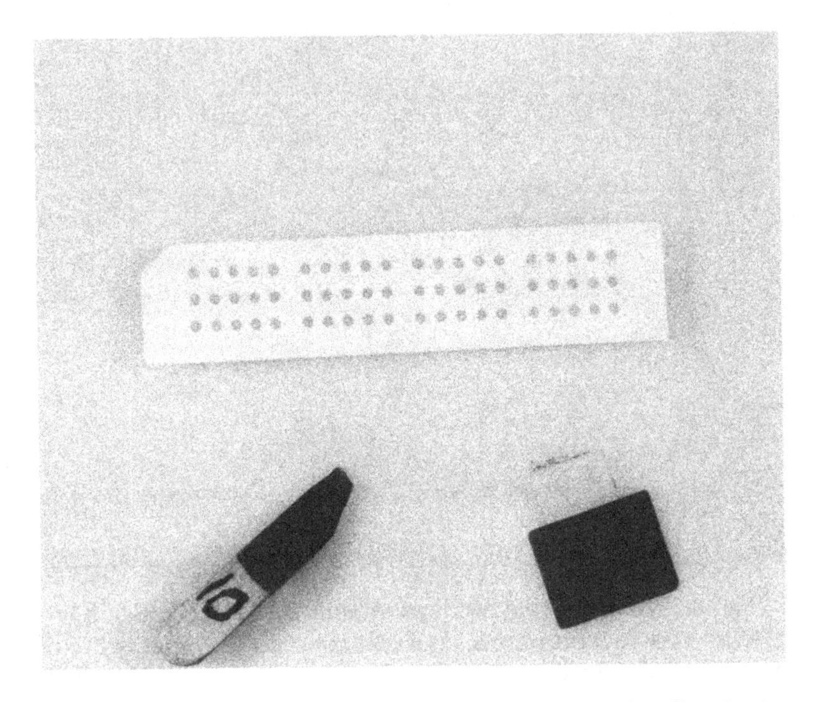

Fig. 4. Tissue dissection viewing box showing an oil well rack, tissue carrier, and block paper with hole to improve visualization. The tissue dissection viewing box can be illuminated from beneath using a mirror and microscope illuminator. The tissue carrier has a small piece of glass (microscope slide) epoxied to the end at a 45° angle. The black paper is placed under the oil well rack to improve visualization.

3.4. Weighing Microsamples

Dissected microsamples are individually weighed using quartz-fiber fishpole balances (Fig. 5). The range of each balance is somewhat limited, so three to four balances are needed to span a range of from 2 to 1000 ng. The balances are made from quartz fibers that are commercially available. Fibers small enough for weighing single cells can be drawn from thicker fibers. For a thorough explanation of quartz fibers, *see* Lowry and Passonneau (1972). The fibers are mounted in leurlok syringes in which the lock has been removed (Fig. 5). This allows the plunger to be placed in backward. The fiber is epoxyed onto the end of the plunger.

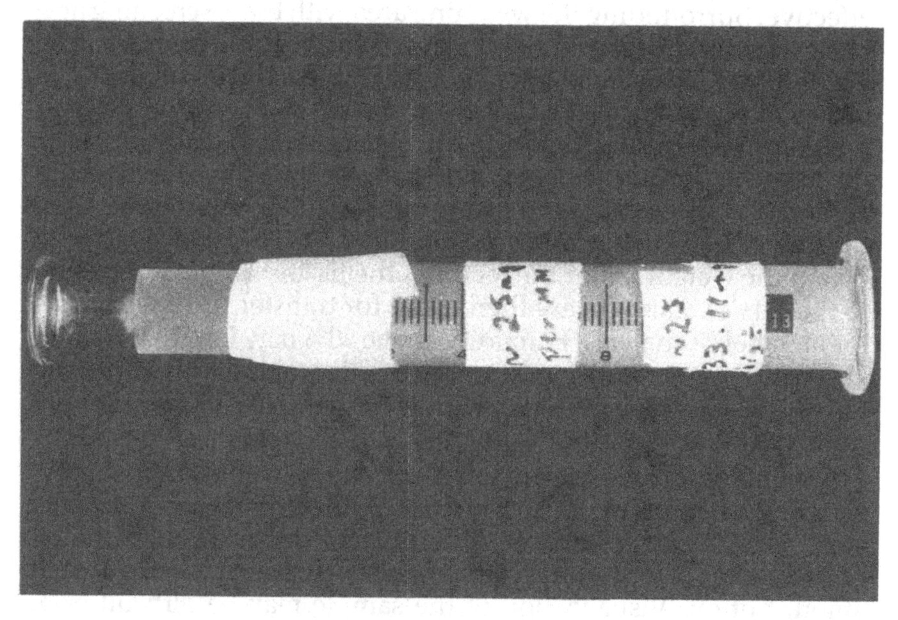

Fig. 5. Quartz-fiber fishpole balance. The "lock" end of the syringe has been removed and the plunger placed in backward. A copper wire is epoxied to the end of the plunger, and the quartz fiber in turn is epoxied to the end of the copper wire. The plunger/quartz fiber can be turned for the best droop angle. A 1 × 1 in. piece of glass to which are attached rubber bands serve to cover the end of the balance when readings are taken. The balance case can be held by clamps to a stand. The rubber bands of the cover glass are in turn attached to a second stand behind the balance stand.

A coverslip is made from a microscope slide to seal the end when weighing samples. The fiber is viewed through a stereoscope that can be orientated horizontally. A reticule in one eyepiece is used to measure the droop of the fiber when a sample is placed on it. The balances are calibrated using *p*-nitrophenol, quinine, or cut-up pieces of tissue weighed on a larger balance. The reader is again referred to Lowry and Passonneau (1972) for a more thorough review of quartz-fiber fishpole balance construction and calibration.

Problems encountered in weighing include static electricity, hand tremor, and sticking tissue. The static electricity problem can usually be controlled by placing a small strip of radium in the balance case. Commerically available "static monitors" may also be effective. Surrounding the weighing area with brass screening and being sure the area is grounded may also help. Hand tremor can be averted by bracing the hands as they approach the balance. The pieces of tissue are picked up from the tissue carrier with a hair point and gently moved next to the quartz fiber. Usually touching the tissue to the quartz fiber is sufficient for transfer. The coverslip is placed on the end of the balance case, and a reading from the reticule is taken. This value is multiplied by the balance factor to obtain the weight. After the reading, the tissue sample is returned to the carrier. Samples are then ready for transfer to oil-well racks for metabolite analysis. Humidity above 50% may lead to the slight absorption of moisture by the freeze dried samples. This artificially raises the weight as well as making the samples sticky and very difficult to transfer.

Samples for analysis are transferred to oil well racks for analysis using a hair point. It is best to use a different hair point for this step since the samples should be pushed down through the oil onto the aliquot for the first step, and the hair point picks up a small amount of oil. Visualization of the sample may be difficult once contact with the oil occurs. This problem increases as the sample size decreases. A small 1-in. square piece of black paper with a small hole in it placed under the oil well rack aids in visualizing small samples (Fig. 4). It is critical to see the sample make contact with the aliquot. The sample may appear to "explode" when contact with the surface of the aliquot occurs. In the oil well technique, the initial steps are performed under oil in order to prevent evaporation of the small volumes used. The standards and blanks can be added before the tissue is added if the weighing and loading procedure is short. Static electricity can be a real problem when loading samples. In this case, the sample shoots off to the side as soon as it comes into contact with the oil and frequently cannot be found. Placing radium strips on the oil well before loading may help this problem. Use of a static master may also be beneficial. When placing the initial aliquot into the oil well rack holes, care must be taken to ensure that the aliquot is totally submersed, or evaporation will occur through the "window." Light blowing through a constriction pipet (Fig. 3) may serve to force the aliquot beneath the surface. Similarly, subsequent additions must make contact with the existing aliquot and remain submersed. Incubations above room temperature are achieved by covering the oil well with aluminum foil, placing it in a small pan, and floating it in a waterbath at the appropriate temperature.

4. Fluorometric Enzymatic Analysis of Samples for High-Energy Phosphates

Several different methods exist for the measurement of metabolites. These include radioimmunoassays, binding assays, radioactive derivative formation assays, and recently developed highly sensitive luciferin-luciferase assays. With the increase in commercially available specific enzymes, there has been an increase in the number of metabolites that can be easily measured. These assays are based on the appearance or disappearance of reduced pyridine nucleotides. Reduced forms of pyridine nucleotides absorb light at 340 nm, whereas the oxidized pyridine nucleotides do not. Some of the absorbed light is reemitted as fluorescence that can be measured at 460 nm.

Certain "tips" regarding the conduct of enzymatic fluorescence assays seem warranted. "Stock" solutions of most components of each reagent can be made and stored indefinitely at –40°C. Generally, these can be 10× the concentration required for the assay and diluted appropriately prior to use. Remember to mix well anything that has been frozen and thawed. Glassware should be scrupulously cleaned and rinsed well, since contamination can be a real problem. The use of disposable glassware whenever possible is advised. Fluorometer tubes can be 10 × 75 mm disposable tubes and should be handled so as to avoid finger prints and scratching. Plastic racks can serve to avoid scratching of the tubes (Fig. 6). At the end of the cycling step, the rack of tubes is placed in boiling water, which should completely cover the fluid in the tubes and should be boiling vigorously. During this step and the incubation period, cover the rack with aluminum foil. After boiling, the tubes are wiped and transferred to a dry rack.

4.1. ATP and PCr

$$\text{ATP} + \text{glucose} \xrightarrow{\text{hexokinase}} \text{ADP} + \text{glucose-6-P}$$

$$\text{glucose-6-P} + \text{TPN} \xrightarrow{\text{glucose-6-P-dehydrogenase}} \text{6-P-gluconolactone} + \text{TPNH} + \text{H}$$

$$\text{phosphocreatine} + \text{ADP} \xrightarrow{\text{creatine kinase}} \text{creatine} + \text{ATP}$$

Both ATP and PCr must be standardized in the 10 m*M* range. This is best accomplished spectrophotometrically. The first step is

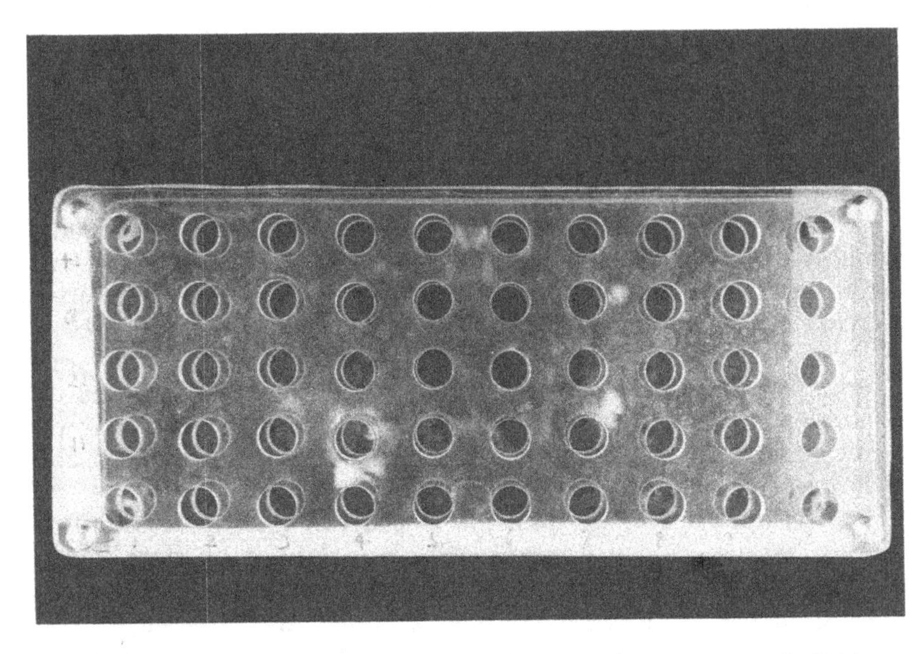

Fig. 6. Plexiglass 10 × 75 mm test tube rack. This type rack eliminates tube scratches associated with metal racks. Note numbering on the rack.

to prepare a 10× ATP reagent that can be used later for fluorometric analysis. The 10× reagent can be stored for several months at –40°C. It can be made in 100-mL volumes with the following constituents: Tris-HCl, pH 7.5, 500 m*M*; $MgCl_2$, 10 m*M*; dithiothreitol, 5 m*M*; glucose, 1 m*M*; and BSA, 02%. For ATP standardization purposes, to 1 mL of 10× ATP reagent and 9 mL of H_2O, add TPN, 500 μ*M*, and glucose, 1 m*M*. Hexokinase, 2 μg/mL (0.28 U/mL), and glucose-6-P dehydrogenase, 0.15 μg/mL (0.07 U/mL) can be added next. It is convenient to make a solution of these two enzymes and add them together. Following a reading, a 10 m*M* standard is added (1% of total volume). The reaction can be followed for 10 min, at which time it should be complete. Phosphocreatine is standardized in a similar fashion except that 500 μ*M* ADP is added. If four spectrophotometer cuvets are used, fill one with H_2O, the other three with reagent. Next, add the HK/G-6-P dehydrogenase enzyme, mix and read. Then add phosphocreatine standard to two cuvets (10 m*M*, 5 μL in a 500 μL total volume), read again. Next add creatine kinase (50 μg/mL, 0.9 U/mL) and follow the reaction for about 10 min. Calculate the concentrations according to the formula:

$$\frac{\text{OD}}{6270} \times \frac{\text{total volume}}{\text{volume of standard}} = \text{m}M$$

The ADP should be added to the reaction with the creatine kinase. Commercially available sources of ADP frequently contain traces of ATP, which can be removed (see end of chapter).

4.1.1. Cycling 1–10 × 10^{-12} mol. Tissue ATP and PCr (500 ng Range, Table 1)

Reagents

ATP

Tris-HCl buffer, pH 7.5, 50 m*M* (10 m*M* base, 40 m*M* Tris-HCl)
$MgCl_2$, 1 m*M*
Dithiothreitol, 0.5 m*M*
Glucose, 0.1 m*M*
Bovine plasma albumin, 0.02%
Hexokinase, 2 µg/mL
Glucose-6-P dehydrogenase, 0.4 µg/mL (0.15 U/mL)

P-Creatine

Tris-HCl buffer, pH 8.6, 200 m*M* (150 m*M* base, 50 m*M* Tris-HCl)
ADP (free of ATP), 80 µ*M*
Bovine plasma albumin, 0.02%
Hexokinase, 4 µg/mL
Glucose-6-P-dehydrogenase, 0.8 µg/mL (0.35 U/mL)
Creatine kinase, 100 µg/mL

TPN Cycling reagent

Tris-acetate buffer, pH 8.0, 100 m*M* (50 m*M* base, 50 m*M* Tris-acetate)
α-ketoglutarate, 5 m*M*
Glucose-6-phosphate, 1 m*M*
Ammonium acetate, 10 m*M*
5'-ADP, 100 µ*M*
Glutamate dehydrogenase, 4 µg/mL
Glucose-6-P-dehydrogenase, 0.25 µg/mL

6-P-Gluconate reagent

Tris-HCl, pH 8.0, 40 m*M* (20 m*M* base, 20 m*M* Tris-HCl)
EDTA, 100 µ*M*

Table 1
1–10 × 10^{-12} mol, 100–1000 μg samples[a]

Metabolite	Step 1	Step 2	Step 3	Step 4	Step 5
ATP	1 μL of 0.1*N* NaOH, 30 min at 60°C	5 μL: Tris-HCl, pH 7.5 50 m*M*; $MgCl_2$, 1 m*M*; DTT, 0.5 m*M*: glucose, 0.1 m*M*; BSA, 0.02%; HK, 2 μg/mL; G6PDH, 0.4 μg/mL, 20 min at RT	3 μL of 0.2*N* NaOH, 30 min at 60°C	Entire volume to 200 μL TPN cycling reagent, 60 min at 38°C, 5 min at 100°C	1 mL of 6-P-gluconate reagent, 30 min at RT
PCr	Same as for ATP Step 1	3 μL from ATP step 2, 1 μL 0.125*N* HCl, 10 min at RT 4μL: Tris-HCl, pH 8.6, 200 m*M*; ADP, 80μM; G6PDH, 0.8 μg/mL; HK, 4 μg/mL; G6PDH, 0.8 μg/mL; CK, 100 μg/mL 30 min at RT	3 μL of 0.2*N* NaOH, 30 min at 60°C	Entire volume to 200 μL TPN cycling reagent, 60 min at 38°C, 5 min at 100°C	1 mL of 6-P-gluconate reagent 30 min at RT

AMP	1 μL of 0.02*N* HCl, 20 min at 80°C	4 μL: Imidazol-acetate, pH 7.1, 100 m*M*; BSA, 0.1%; NaSCN, 20 m*M* β-mercaptoethanol 5 m*M*; Magnesium acetate, 4 m*M*; glucose 1,6-P_2, 1 μM; glycogen, 2 m*M*; G6PDH, 4 μg/mL; P-glucomutase, 4 μg/mL; Phosphorylase a, 2 μg/mL; TPN, 75 μ*M*; Ap5A, 10 μ*M*. 2.5 μL 1 m*M* Na_2HPO4 to each well. 60 min at RT	3 μL of 0.04*N* NaOH, 30 min at 80°C	4 μL of aliquot 100 μL TPN cycling reagent, 60 min at 38°C 5 min at 100°C	1 mL of 6-P-gluconate reagent, 30 min at RT

[a]TPN cycling reagent, Tris-acetate buffer, pH 8.0, 100 m*M*; α-ketoglutarate, 5 m*M*; Glucose-6-phosphate, 1 m*M*; ammonium acetate, 10 m*M*; 5' ADP, 100 μ*M*; glutamate dehydrogenase, 4 μg/mL; glucose-6-P-dehydrogenase, 0.25 μg/mL. 6-P-Gluconate reagent: Tris-HCl, pH 8.0, 40 m*M*; EDTA, 100 μ*M*; ammonium acetate, 30 m*M*; $MgCl_2$, 5 m*M*; TPN^+, 50 μ*M*; 6-P-gluconate dehydrogenase, 1 μg/mL. Abbreviations used: Ap5A, P^1,P^5-Di(adenosine-5'-)pentaphosphate; BSA, bovine plasma albumin; Ck, creatine kinase; DTT, dithiothreitol; G6PDH, glucose-6-P dehydrogenase; HK, hexokinase; RT, room temperature.

Ammonium acetate, 30 m*M*
$MgCl_2$, 5 m*M*
TPN^{2+}, 50 μ*M*
6-P-gluconate dehydrogenase, 1 μg/mL (0.006 U/mL)

ATP Dilution
5-20 μL, 1 m*M* ATP
1 mL, 0.1*N* NaOH

P-Creatine dilution
5-20 μL, 1 m*M* P-Creatine
1 mL, 0.1*N* NaOH

4.1.1.1. COMMENTS. The bovine plasma albumin is added to the various reagents to protect the enzymes that are present in such small volumes. The various buffers used for the ATP and P-creatine reagents are designed so that upon addition to the samples, the resulting pH will be close to 8. The reaction times can be checked directly in the fluorometer using all reagents in 200-fold larger volumes.

The TPN cycling reagent without the glutamate dehydrogenase and the glucose-6-P-dehydrogenase can be stored at –50°C with minimal loss of activity. At higher temperature there is a loss of activity of the α-ketoglutarate. Therefore, if the reagent is frozen and thawed several times, a loss of activity or unstable measurements may be noted because of the deterioration of the α-ketoglutarate. The 6-P-gluconate reagent can also be prepared in a 10-fold concentrated stock reagent, without the TPN^+, and 6-P-gluconate dehydrogenase, and stored indefinitely at –50°C.

Reaction vessels
Oil well racks
Fluorometer tubes, 10 × 75 mm

Procedure
Add 1 μL of 0.1*N* NaOH to oil wells for samples and blanks. Add 1 μL of ATP dilution or P-creatine dilution to wells for standards. Wrap the oil well rack in aluminum foil, place the rack in a small covered pan, then incubate in a hot water bath 30 min at 60°C.
Add 5 μL of ATP reagent to all wells. Incubate 20 min at room temperature. Then transfer a 3-μL aliquot from each oil well into a second set of oil wells for P-creatine analysis.
To the second set of wells for P-creatine, add 1 μL of 0.125*M* HCl. Incubate 10 min at room temperature.

Then add 4 μL of P-creatine reagent to the second set of wells. Incubate 30 min at room temperature.

To both sets of wells add 3 μL 0.2*N* NaOH. Wrap rack as in the first step of this procedure. Incubate 30 min at 60°C.

Transfer the total volume from the oil wells into fluorometer tubes (10 × 75 mm) on ice containing 200 μL of TPN^+ cycling reagent. Cover the tops of the tubes with aluminum foil. Incubate in a warm water bath for 1 h at 38°C. Follow this incubation with 5 min at 100°C (boiling water bath).

Add 1 mL of 6-P-gluconate reagent to all tubes and take fluorometer readings when reaction has stopped, approximately 30 min at room temperature.

4.1.2. Cycling 1–5 × 10^{-14} mol. Single Cells ATP and PCr (4 ng Range, Table 2)

Reagents

ATP Reagent

Tris-HCl buffer, pH 7.5, 60 m*M* (10 m*M* base, 50 m*M* Tris-HCl)
$MgCl_2$, 1 m*M*
Dithiothreitol, 0.5 m*M*
Glucose, 100 μ*M*
TPN^+, 20 μ*M*
Bovine plasma albumin, 0.03%
Hexokinase, 2 μg/mL
Glucose-6-P-dehydrogenase, 0.4 μg/mL (0.15 U/mL)

P-Creatine reagent

Tris-HCl buffer, pH 8.6, 200 m*M* (150 m*M* base, 50 m*M* Tris-HCl)
ADP (free of ATP), 70 μ*M*
Bovine plasma albumin, 0.06%
Hexokinase, 5 μg/mL
Glucose-6-P-dehydrogenase, 0.5 μg/mL
Creatine kinase, 250 μg/mL

TPN Cycling reagent

Tris-acetate buffer, pH 8.0, 100 m*M* (50 m*M* base, 50 m*M* Tris acetate)
α-ketoglutarate, 5 m*M*
Glucose-6-P, 1 m*M*
Ammonium acetate, 10 m*M*
5'-ADP, 100 μ*M*

Table 2
$1–5 \times 10^{-14}$ mol, Single Cells[a]

Metabolite	Step 1	Step 2	Step 3	Step 4	Step 5
ATP	20 nL of 0.1*N* NaOH, 20 min at 80°C	80 nL; Tris-HCl, pH 7.5, 60 m*M*; MgCl, 1 m*M*; DTT, 0.5 m*M*; glucose, 100 μ*M*; TPN^+, 20 μM; BSA, 0.03%; HK, 2 μg/mL; G6PDH, 0.4 μg/mL, 15 min at RT	200 nL 0.1*N* NaOH, 30 min at 80°C	5 μL TPN cycling reagent, 2 h at 38°C, 1 μL 1.0*N* NaOH, 10 min at 90°C	4.5 μL aliquot into 1 ml 6-P-30 min at R.T.
PCr	Same as for ATP Step 1	45 nL aliquot from ATP, Step 2. 7 nL 0.5*N* HCl, 10 min at RT, 30 nL: Tris HCl, pH 8.6, 200 m*M* ADP, 70 μ*M*; BSA, 0.06%; Hk, 5 μg/mL; G6PDH, 0.5 μg/mL; CK, 250 μg/mL. 45 min at RT	220 nL 0.1*N* NaOH, 30 min at 80°C	5 μL TPN cycling reagent 2 h at 38°C, 1 μL 1.0*N* NaOH 10 min at 90°C	4.5 μL aliquot into 1 mL of 6-P-gluconate reagent, 30 min at RT

AMP	10 of nL 0.1*N* NaOH, 30 min at 38°C	40 nL; Imidazol-acetate, pH 7.0, 100 m*M*; BSA, 0.1%; NaSCN, 20 m*M*; β-mercaptoethanol, 5 m*M*; magnesium acetate, 4 m*M*; glucose-1-6-P_2, 1 m*M*; glycogen, 2 m*M*; P-glucomutase, 4 μg/mL; G6PDH, 2 μg/mL; Phosphorylase a, 2 μg/mL TPN^+, 75 μ*M*, 25 nL 1 m*M* Na_2HPO_4 to each well. 60 min at RT	3 μL 0.02*N* NaOH, 30 min at 80°C	2 μL aliquot into 50 μL TPN cycling reagent, 60 min at 38°C, 5 min at 100°C	1 mL of 6-P-gluconate reagent, 30 min at RT

[a]TPN cycling reagent, Tris-Acetate buffer, pH 8.0, 100 m*M*; α-ketoglutarate, 5 m*M*; glucose-6-P, 1 m*M*; ammonium acetate, 10 m*M*; 5'-ADP, 100 μ*M*; glutamic dehydrogenase, 400 μg/mL; glucose-6-P dehydrogenase, 30 μg/mL. 6-P-Gluconate reagent, Tris-HCl, pH 8.1, 40 m*M*; EDTA, 100 μ*M*; ammonium acetate, 30 m*M*; $MgCl_2$, 5 m*M*; TPN^+, 30 μ*M*; 6-P-gluconate dehydrogenase, 0.5 μg/mL. Abbreviations used: BSA, bovine plasma albumin; Ck, creatine kinase; DTT, dithiothreitol; G6PDH, glucose-6-P dehydrogenase; Hk, hexokinase.

Glutamic dehydrogenase, 400 μg/mL
Glucose-6-P-dehydrogenase, 30 μg/mL (12 U/mL)

6-P-Gluconate reagent
Tris-HCl, pH 8.1, 40 m*M* (20 m*M* base, 20 m*M* Tris-HCl)
EDTA, 100 μ*M*
Ammonium acetate, 30 m*M*
MgCl, 5 m*M*
TPN^{2+}, 30 μ*M*
6-P-gluconate dehydrogenase, 0.5 μg/mL

ATP Dilution
5–20 μL, 1 m*M* ATP
1 mL, 0.1*N* NaOH

P-Creatine dilution
5–20 μL, 1 m*M* P-creatine
1 mL, 0.1*N* NaOH

Reaction vessels
- Oil well racks
- Fluorometer tubes 10 × 75 mm

Procedure
- Add 20 nL of 0.1*N* NaOH to oil wells for samples and blanks.
- Add 20 nL of ATP dilution or P-creatine dilution to wells for standards. Wrap oil well rack in aluminum foil, place the rack in a small covered pan, then incubate in a hot water bath 20 min at 80°C.
- Add 80 nL of ATP reagent to all wells. Incubate 15 min at room temperature.
- Transfer a 45-nL aliquot from each oil well into a second set of oil wells for P-creatine analysis. Add 7 mL of 0.5*N* HCl to a second set of wells. Incubate 10 min at room temperature.
- Add 30 nL of PCr reagent to the second set of wells. Incubate 45 min at room temperature.
- Add 220 nL, 0.1*N* NaOH to both sets of wells. Wrap rack as in the first step of this procedure. Incubate 30 min at 80°C.
- To all wells add 5 μL of TPN cycling reagent. Wrap rack as in the first step of this procedure. Incubate 2 h at 38°C.
- Add 1 μL of 1.0*N* NaOH to all wells. Wrap rack as in the first step. Incubate 10 min at 90°C.
- Transfer 4.5 μL aliquots into 1 mL of 6-P-gluconate reagent in fluorometer tubes. Take fluorometer reading when reaction has stopped, approximately 30 min at room temperature.

4.2. Fluorometer Direct Assay, 0.1– 8 × 10^{-9} mol. ADP

ADP Reagent

Imidazole-HCl buffer, pH 7.0, 50 m*M* (30 m*M* base, 20 m*M* imidazole-HCl)
$MgCl_2$, 2 m*M*
KCl, 75 m*M*
ATP, 5 μ*M*
DPNH, 0.2–10 μ*M*
P-pyruvate, 20 μ*M*
Pyruvate kinase, 5 μg/mL
Lactic dehydrogenase, 2 μg/mL (0.4 U/mL)

Reaction vessel

Fluorometer tubes, 10 × 75 mm

Procedure

Add 1 mL of ADP reagent to fluorometer tubes for samples, blanks, and standards. Add 1 μL of a suitable ADP dilution to the tubes being used for the standards. Take fluorometer readings in 3–6 min.

4.2.1. Comments

The fluorometer direct assay method given here is used for measuring concentrations of ADP in larger pieces of tissue. For the measurement of ADP concentration in the 1–10 × 10^{-12} and 1–5 × 10^{-14} mol range using a cycling method no problems are foreseen, although it has not been done first-hand by the authors. The final steps of the cycle would use a DPN cycling reagent and an indicator reagent.

4.3. Cycling 1–10 × 10^{-12} mol. Tissues, AMP, Barbehenn et al. (1985) (Table 1)

Reagents

AMP

Imidazole acetate buffer, pH 7.1, 100 m*M* (35 m*M* base, 65 m*M* acid)
Bovine plasma albumin, 0.1%
NaSCN, 20 m*M*
β-mercaptoethanol, 5 m*M*
Magnesium acetate, 4 m*M*
Glucose 1,6-P_2, 1 μ*M*

Glycogen, 2 m*M*
Glucose-6-P-dehydrogenase, 4 μg/mL
P-glucomutase, 4 μg/mL
Phosphorylase a, 2 μg/mL
TPN^+, 75 μm
P', P^5-di(adenosine-5'-)pentaphosphate, 10 μ*M*

TPN Cycling reagent
Tris-acetate buffer, pH 8.0, 100 m*M* (50 m*M* base, 50 m*M* Tris-acetate)
α-Ketoglutarate, 5 m*M*
Glucose-6-phosphate, 1 m*M*
Ammonium acetate, 10 m*M*
5'-ADP, 100 μ*M*
Glutamate dehydrogenase, 4 μg/mL
Glucose-6-P-dehydrogenase, 0.25 μg/mL

6-P-Gluconate reagent
Tris-HCl, pH 8.0, 40 m*M* (20 m*M* base, 20 m*M* Tris-HCl) EDTA, 100 μ*M*
Ammonium acetate, 30 m*M*
$MgCl_2$, 5 m*M*
TPN^+, 50 μ*M*
6-P-gluconate dehydrogenase, 1 μg/mL

AMP Dilution
5–20 μL, 1 m*M* AMP
1 mL, 0.02 *N* HCl

4.3.1. Comments

The bovine plasma albumin is added to the AMP reagent to protect the enzymes that are present in such small volumes.

The AMP reagent without the P-glucomutase, glucose-6-P-dehydrogenase, phosphorylase a, and the TPN^+ can be stored at –50°C with minimal loss of activity. The TPN cycling reagent without the glutamate dehydrogenase and the glucose-6-P-dehydrogenase can also be stored at –50°C with minimal loss of activity. The 6-P-gluconate reagent can also be prepared in a 10-fold concentrated stock reagent without the TPN^+ and 6-P-gluconate dehydrogenase, and stored indefinitely at –50°C.

Reaction vessels
Oil well racks
Fluorometer tubes, 10 × 75 m*M*

Procedure

Add 1 μL of 0.02*N* HCl to oil wells for samples and blanks.

Add 1 μL of AMP dilution to oil wells for standards. Wrap oil well rack in aluminum foil, place the rack in a small covered pan, then incubate in a hot water bath 20 min at 80°C.

Add 4 μL of AMP reagent to all wells. Then add 2.5 μL of 1 m*M* Na_2HPO_4 to each well. Incubate 60 min at room temperature.

Add 3 μL of 0.04*N* NaOH to all wells. Wrap rack as in the first step of this procedure. Incubate 30 min at 80°C.

Transfer 4 μL aliquots into fluorometer tubes (10 × 75 mm), on ice, containing 100 μL TPN cycling reagent. Cover the tops of the tubes with aluminum foil. Incubate in a warm water bath for 1 h at 38°C. Follow this incubation with 5 min at 100°C (boiling water bath).

Add 1 mL of 6-P-gluconate reagent to all tubes, and take fluorometer readings when the reaction has stopped, approximately 30 min at room temperature.

4.3.2. Cycling 1–5 × 10^{-14} mol. Single Cells, AMP, Barbehenn et al. (1976) (Table 2)

Reagents

AMP Reagent

Imidazole-acetate buffer, pH 7.0, 100 m*M* (35 m*M* base, 65 m*M* acid)
Bovine plasma albumin, 0.1%
NaSCN, 20 m*M*
β-mercaptoethanol, 5 m*M*
Magnesium acetate, 4 m*M*
Glucose-1-6-P_2, 1 m*M*
Glycogen, 2 m*M*
P-glucomutase, 4 μg/mL
Glucose-6-P-dehydrogenase, 2 μg/mL
Phosphorylase a, 2 μg/mL
TPN^+, 75 μ*M*

TPN Cycling reagant

Tris-acetate buffer, pH 8.0, 100 m*M* (50 m*M* base, 50 m*M* Tris-acetate)
α-ketoglutarate, 5 m*M*
Glucose-6-P, 1 m*M*
Ammonium acetate, 10 m*M*

5'-ADP, 100 μ*M*
Glutamic dehydrogenase, 400 μg/mL
Glucose-6-P-dehydrogenase, 30 μg/mL

6-P-Gluconate reagent
Tris-HCl, pH 8.1, 40 m*M* (20 m*M* base, 20 m*M* Tris-HCl)
EDTA, 100 μ*M*
Ammonium acetate, 30 m*M*
$MgCl_2$, 5 m*M*
TPN^+, 30 μ*M*
6-P-gluconate dehydrogenase, 0.5 μg/mL

AMP Dilution
5–20 μL, 1 m*M* AMP
1 mL, 0.1*N* NaOH

Reaction vessels
Oil well racks
Fluorometer tubes, 10 × 75 mm

Procedure
Add 10 nL of 0.1*N* NaOH to oil wells for samples and blanks. Add 10 nL of AMP dilution to oil wells for standards. Wrap oil well rack in aluminum foil, place the rack in a small covered pan, then incubate in a warm water bath 30 min at 38°C.
Add 40 nL of AMP reagent to all wells. Then add 25 nL of 1 m*M* Na_2HPO_4 to each well. Incubate 60 min at room temperature.
Add 3 μL of 0.02*N* NaOH to all wells. Wrap rack as in the first step in this procedure. Incubate 30 min at 80°C.
Transfer 2 μL-aliquots into fluorometer tubes (10 × 75 mm) on ice containing 50 μL of TPN cycling reagent. Cover the tops of the tubes with aluminum foil. Incubate in a warm water bath for 1 h at 38°C. Follow this incubation with 5 min at 100°C (boiling water bath).
Add 1 mL of 6-P-gluconate reagent to all tubes and take fluorometer readings when the reaction has stopped, approximately 30 min at room temperature.

4.3.2. Oil for Oil Well Racks

60% USP Light mineral oil
40% Hexadecane

Mix in a separatory funnel with 2 vol of 1.0*N* NaOH. When oil and water phase separate, wash oil with water and 0.02*N* HCl until washings are neutral. Centrifuge at 10,000 rpm until oil phase is separated from water phase. Dry oil in a vacuum desiccator.

4.3.3. ATP-Free ADP

200 μL of .5*M* Tris-HCl, pH 8.1
10 μL of 100 m*M* Glucose
10 μL of 100 m*M* $MgCl_2$
50 nL of Hexokinase stock, 100 mg/10 mL
0.5 μL of Glucose-6-P-dehydrogenase stock, 5 mg/mL
3 μL of 100 m*M* TPN^+
0.7 mL of H_2O

Mix the above. Add 10 μL of 100 m*M* ADP. Incubate 30 min at room temperature.
Add 20 μL of 5*N* HCl. Incubate 15 min at room temperature.
Add 20 μL of 5*N* NaOH. Incubate 10 min at room temperature.

Acknowledgment

The authors wish to express their appreciation to Ms. Cheryll Johnson for her expert secretarial assistance.

References

Barbehenn E. K., Noelker D. M., Chader G. J., and Passonneau J. V. (1985) Adenine nucleotide and P-creatine levels in layers of frog retina as a function of dark and light adaption. *Exp. Eye Res.* **40,** 675–686.

Barbehenn E. K., Law M. M., Brown J. G., and Lowry O. H. (1976) Measurement of 5'-adenylic acid by stimulation of phosphorylase a. *Anal. Biochem.* **70,** 554–562.

Lowry O. H. and Passonneau J. V. (1972) *A Flexible System of Enzymatic Analysis* Academic, New York.

Medina M. A., Deam A. P., and Stavinoha W. B. (198) Inactivation of Brain Tissue by Microwave Irridation, in *Cerebral Energy Metabolism and Neural Function* (Passonneau J. V., Hawkins R. A., Lust W. D., and Welsh F. A., eds.) Williams and Wilkins, Baltimore.

Veech R. L., Harris R. L., VeLoso D., and Veech E. H. (1973) Freeze-blowing: A new technique for the study of brain *in vivo*. *J. Neurochem.* **20,** 183–187.

Welsh F. A. (1980) *In situ* Freezing of Cat Brain, in *Cerebral Energy Metabolism and Neural Function* (Passonneau J. V., Hawkins R. A., Lust W. D., and Welsh F. A., eds.) Williams and Wilkins, Baltimore.

Index